Models for Interdisciplinary Mobile Learning:

Delivering Information to Students

Andrew Kitchenham
University of Northern British Columbia, Canada

Information Science
REFERENCE

Senior Editorial Director: Kristin Klinger
Director of Book Publications: Julia Mosemann
Editorial Director: Lindsay Johnston
Acquisitions Editor: Erika Carter
Development Editor: Michael Killian
Production Coordinator: Jamie Snavely
Typesetters: Keith Glazewski, Natalie Pronio & Milan Vracarich, Jr.
Cover Design: Nick Newcomer

Published in the United States of America by
Information Science Reference (an imprint of IGI Global)
701 E. Chocolate Avenue
Hershey PA 17033
Tel: 717-533-8845
Fax: 717-533-8661
E-mail: cust@igi-global.com
Web site: http://www.igi-global.com/reference

Library of Congress Cataloging-in-Publication Data

Models for interdisciplinary mobile learning: delivering information to students / Andrew Kitchenham, editor.
p. cm.
Includes bibliographical references and index.
Summary: "This book presents a global perspective on mobile learning and augments that perspective with examples and applications from around the world written by scholars who are leaders in their countries and in the world"--Provided by publisher.
ISBN 978-1-60960-511-7 (hardcover) -- ISBN 978-1-60960-512-4 (ebook) 1. Mobile communication systems in education--Cross-cultural studies. I. Kitchenham, Andrew, 1963-
LB1044.84.M64 2011
371.33--dc22
2010033437

British Cataloguing in Publication Data
A Cataloguing in Publication record for this book is available from the British Library.

Grant Potter, *University of Northern British Columbia, Canada*
Ghislaine Povey, *University of Wolverhampton, UK*
Kalyani Premkumar, *University of Saskatchewan, Canada*
Ruzita Ramli, *Open University Malaysia, Malaysia*
Yanjie Song, *The University of Hong Kong, China*
Danielle Sloane, *The University of Queensland, Australia*
Dawn Stevens, *University of Northern British Columbia, Canada*
Christopher Tassava, *Carleton College, USA*
Lynn Woodhouse, *The University of Queensland, Australia*

Table of Contents

Section 3
Extending Mobile Learning

Detailed Table of Contents

reliable pedagogies were formed, how subsequent research testing distilled factors noted from this method into both unique and generic pedagogical principles, and how the principles formed can be used in any context or discipline to produce effective and enjoyable learning. The authors include analysis of a worked example using this approach (in this instance from nursing) in order to illustrate how each stage of the model may be performed, and to make clear how the process may be replicated and incorporated into many different settings.

This chapter discusses the appropriate uses of mobile and social media for post-secondary students.The growth of social media and mobile communication provides educators with an opportunity to transmit course-related information to students in new ways. But are students willing to accept course information through those channels, typically seen as "fun" and "social?" The study examines the reasons that students use different types of personal media and how appropriate certain types of communication channels are for academic information. Results show that students prefer to get their academic information through "official" channels, such as email and course management systems. However, they are willing to accept certain types of information through social channels (mobile devices, social networking) as long as they do not have to share personal information.

This chapter presents a sample course design model for effectiveness and efficiency using mobile learning. The demand for mobile, electronic learning environments has increased, but so, too, has the demand for performance-based outcomes. Within this context, efficiency and effectiveness have become the gold standards for quality. The design of electronic learning environments, both blended and fully mobile, requires unique considerations, particularly in regards to self-regulated learning, cognitive load, and learner characteristics. Repeated development of an effective and efficient electronic learning environment can be facilitated through the use of a standardized, flexible course design model. A sample course design model that promotes efficiency and effectiveness, while catering to the unique considerations for mobile learning in an electronic learning environment are presented along with suggestions for future conversations and research.

This chapter discusses pervasive learning at the workplace, namely, work-based learning. Our proposition is based on two complementary learning strategies, situation-based learning and activity-based learning, to fulfill seamless learning across contexts and worked-based learning requirements. In situation-based learning, relevant activities and/or resources are recommended to the user. In activity-based learning, the user has to search and to select his/her activities and the corresponding resources. We propose a pervasive learning environment where learners may follow different learning strategies. They may switch from one strategy to the other one according to their needs and/or the context change. These facilities are possible thanks to a set of models and adaptation processes developed for the P-LearNet project (Pervasive Learning Network). To illustrate this proposal, an example (or a use case) from this project is used.

This chapter presents three m-learning case studies. The first is a school district in Northern Canada that uses one-to-one computing from intermediate to high school. The second case study is an autoethnographic approach to using m-learning in post-secondary education. The last deals with a United States school district that is planning to implement m-learning technologies across three schools in an attempt to address the needs of the Net Generation. The chapter concludes with a synopsis of the findings across the three case studies.

This chapter reports on an in-depth, one-year empirical research into examining five undergraduate student mobile device uses in context. Data collection methods include: student reflective e-journals, student artifacts, observations, interviews, field notes, and memos. Three complementary streams were involved in the data analysis. Seven interacting factors in context that could either facilitate or inhibit mobile device use were identified and discussed.

This chapter describes the medical context and characteristics of medical students, residents and medical professionals and implications for m-learning. Some technologies used and examples of usage, benefits, outcome and barriers at the undergraduate, postgraduate and continuing medical education are explored. Mobile learning (m-learning) is particularly important in medical education and the major users of mobile devices are in the field of medicine. The contexts and environment in which learning occurs necessitates m-learning. Medical students are placed in hospital/clinical settings very early in training and require access to course information and to record and reflect on their experiences while on the move. The work of postgraduates and physicians involves a high degree of mobility between distributed sites and instant communications within work environments. Distributed sites where physicians work and in which students are placed are often in remote and rural areas. The technological advances can be capitalized to promote and facilitate situated learning.

This chapter describes how the Open University Malaysia conceptualized, planned, and created a mobile learning environment using SMS to enhance its current blended learning model in general, and in particular, one of its courses with over 1,000 students enrolled. The chapter also describes the categories used for formulating the SMS content, use of Twitter and Facebook to support the SMS sent, and it discusses the feedback received on the initiative, as well as the issues and challenges.Malaysia has a population of about 28 million people but there are, incredibly, more than 30 million mobile phone subscriptions. Sixth in the world in terms of SMS (Short Message Service) volume, Malaysians appear to be addictive SMS texters. With over 98 percent of its students having mobile phones and 82 percent of the students ready for learning through mobile phones, Open University Malaysia initiated a project that first experimented with podcasts, and SMS texts later.

The chapter aims to investigate the use of podcasting as a means of facilitating learner-generated content in hospitality management at post-graduate level. The research aimed to instigate critical skills development at this postgraduate level, which has been used in delivery at University of Wolverhampton, UK for two years, and addresses a gap in knowledge. Theoretical perspectives of this technique are explored in the chapter and used to analyse the student experiences of generating a 'mockcast' for a new

The chapter draws on a case study method using factor and regression analysis to interpret the questionnaire responses about the uses of wireless handheld devices in higher education. The principal findings included that behavior and attitude contribute strongly to the perceived performance of using such devices in the chosen context, and that facilitating conditions have a more complex and mediated relationship with behavior and attitude on the one hand and perceived performance on the other. The authors elaborate the implications of those findings for increasing alignment across several different interfaces related to blended and mobile learning in the early 21st century.

This chapter will discuss the appropriation of students' own mobile devices to support the development of clinical competency for speech pathology students in a standardised patient clinic. The need to adequately prepare students for the workplace competencies of a health professional in the 21st century demands exploration of alternative learning opportunities. Two such examples are the appropriation of mobile technologies and the use of standardised patients to support clinical learning. The chapter includes descriptions of a project that focussed on the role of mobile technologies in supporting learning across different contexts. The results indicated that the use of mobile technologies in a clinical practice setting can make a positive contribution to clinical competency development. Issues for future integration of mobile technologies in clinical practice are raised.

This chapter will profile a case study exploration involving the possible use of library classification and subject headings as an element of contextual identification – evidence that a particular material's subject matter and content (as a whole or a portion of it) may satisfy an informational need. One of the challenges of this task is to develop a materials evaluation process using library classification that is both user-friendly and technologically savvy. In this case, mobile technology has been selected as the possible mode of information delivery.

Foreword

Mobile learning is a well-established research area that has an impressive track record of innovation in both technology and pedagogy. One of the first major mobile learning projects, Wireless Coyote, opened up new ideas of collaborative, situated, mobile learning with what was then cutting edge technology of 'duct tape, velcro, microprocessors and radios' (Grant, 1993). Similar themes of exploratory, situated learning were developed further in the Ambient Wood project a decade later, embedding technologies including RFID tags into the domain of mobile learning (Rogers et al., 2004), and this type of environmentally focused mobile learning continues to evolve (e.g,. Rogers & Connelly, 2010; Spikol et al, 2009). Other concepts that have gained traction in the mobile learning community have included mobile educational games (e.g., Dugstad Wake & Baggetun, 2009; Schwabe & Göth, 2005), technologies for augmented learning in context (e.g., Facer et al., 2004; Ogata, 2008), and contextual interactive learning in museums, galleries and historic places (e.g., Vavoula et al., 2009). We have also seen major European mobile learning projects such as MOBILearn and M-Learning (Kukulska-Hulme et al., 2009) and the HP Mediascapes platform that enables anyone to rapidly develop geotagged multimedia mobile learning tools (Stenton et al., 2007).

In parallel with these significant and ambitious projects, which have often pushed the boundaries of technology, we have also seen a groundswell of more modest but also more widespread mobile learning initiatives, often leveraging commonly available and cheaper technologies such as SMS messaging (e.g., Petrova, 2010; Scornavacca et al, 2009) and podcasting (e.g., Bell et al, 2007; Wilson et al., 2009). Small mobile applications have also been developed in many areas that can be easily deployed on a wide range of standard mobile devices, for example, Java Midlets for mathematics learning (Weizman, 2005).

The development of mobile learning over the last 20 years, embracing both the present and the future, can be seen as a synergistic range of initiatives that have involved commercial organizations, researchers, and educators in universities and schools, and independent developers, giving us both inspirational projects and practical, everyday tools. This rich heritage raises the question: what are the major issues that need to be addressed in contemporary mobile learning? In that context, the focus of this volume is how mobile learning can deliver information to students, and the concerns of its authors outline some essential themes and questions that the mobile learning community must continually address, such as: how can mobile learning be encouraged? What tools are most appropriate for its delivery? What is best practice in mobile pedagogical design? And how does mobile learning work in practice?

The first of these major themes, which emerges from several chapters in this book, is questioning how learners may be encouraged to use certain types of mobile learning. Some authors have focused on identifying factors that can facilitate or inhibit mobile device use, since it is essential to have an understanding of the learners' perspective. The importance of student perceptions cannot be underestimated.

Learner behaviour, attitude, and perceived performance have a complex relationship with the learning conditions that we provide. One such issue discussed is that students prefer to get their academic information through "official" channels, such as email and course management systems. There is, potentially, resistance towards receiving information through channels that are perceived to be outside the usual realm of educators, including mobile devices and social networking. However, it does appear that they are willing to accept certain types of information through social channels, as long as they do not have to share personal information.

When we start to look at the tools most appropriate for mobile learning delivery, it is clear that there are many choices, and, as indicated at the beginning of this foreword, technologies will often be chosen for their availability and economy, as well as their utility. One of the chapters in this book uses an extensive literature review to indicate that a range of mobile device types (e.g., Smartphone, iPod, PDA etc.) are in widespread use for mobile learning across the world, emphasizing the value of approaches that are not linked to a single platform. Two commonly applied technologies that work across many mobile applications, which are also economical and reliable, are SMS messaging and podcasting, and some discussion of these approaches is included here. Of course, just because a technology is simple and widespread does not mean that its use alone is helpful. Brabazon (2007) has expressed some reservations about the headlong rush towards online resources, without critical thinking about their application: "Education is not a hobby to be slotted into a lifestyle. Without care in the construction of curriculum, the fun and flexibility of sonic mobility will crush the discipline required for motivated learning" (p. 30). Clearly, it is essential to continue the debate about how such technologies should be employed. One important aspect of this debate covered here is the role of learning objects and learner-generated content in podcasting implementations. This is not to say that these common and basic technologies are the only concern of this volume, as more ambitious approaches, such as augmented reality, are also explored.

From conditions of technology, and the awareness that applying a technology alone does not address learning issues, we move naturally on to considering what are best practices in mobile pedagogical design. A number of the chapters in this book address various issues of design, including the importance of overall course design models, and the application of pedagogical principles to specific learning situations. There is also discussion of the differences between situation based learning and activity based learning (push and pull), the value of both perspectives, and the design of pervasive learning environments than can support both.

Finally, a number of the authors address how mobile learning works in practice, including various case studies, covering topics as disparate as library classification, second language learning, hospitality, and medicine, including speech pathology. The medical context seems to be a useful domain for case studies, due to the prevalence of mobile devices in that profession, the physical mobility of medical staff who often work across multiple sites, and their need for frequent communication. The scope for situated learning is also very strong and a compelling example of where mobile learning can be usefully applied.

In summary, this volume addresses a range of important contemporary issues in mobile learning research, and provides the research community with valuable additional resources in the growing canon of mobile learning literature.

David Parsons, Massey University, New Zealand
Editor in Chief, International Journal of Mobile and Blended Learning

David Parsons *is the founding editor-in-chief of the International Journal of Mobile and Blended Learning and holds an academic post at Massey University, Auckland, New Zealand. His work on mobile learning has been published in a range of journals, including the International Journal of Mobile Learning and Organisation and IEEE Transactions on Learning Technologies, and he has presented at many major conferences including mLearn, IADIS Mobile Learning and the IEEE International Conference on Advanced Learning Technologies. He acted as Chair for the Conference on Mobile Learning Technologies and Applications (MoLTA) in 2007. He was co-editor (with Hokyoung Ryu) of 'Innovative Mobile Learning: Techniques and Technologies' (Information Science Reference, 2009) and is the author of a number of texts on software development covering Java, C++, and Web-based applications. He is a member of the International Association for Mobile Learning and a professional member of the British Computer Society.*

REFERENCES

Bell, T., Cockburn, A., Wingkvist, A., & Green, R. (2007). Podcasts as a supplement in tertiary education: An experiment with two computer science courses . In Parsons, D., & Ryu, H. (Eds.), *Mobile Learning Technologies and Applications (MoLTA) 2007* (pp. 70–77). Massey University.

Brabazon, T. (2007). Mobile learning: The iPodification of universities. *Nebula*, *4*, 119–130.

Dugstad Wake, J., & Baggetun, R. (2009). Premierløytnant Bielke: A mobile game for teaching and learning history. *International Journal of Mobile and Blended Learning*, *1*(4), 12–28. doi:10.4018/jmbl.2009090802

Facer, K., Stanton, D., Kirk, D., Joiner, R., Reid, J., & Hull, R. (2004). Savannah: Mobile gaming and learning? *Journal of Computer Assisted Learning*, *20*, 399–409. doi:10.1111/j.1365-2729.2004.00105.x

Grant, W. (1993). Wireless Coyote: A computer-supported field trip. *Communications of the ACM*, *36*(5), 57–59. doi:10.1145/155049.155062

Kittl, C., Edegger, F., & Petrovic, O. (2009). Learning by pervasive gaming: An empirical study . In Ryu, H., & Parsons, D. (Eds.), *Innovative mobile learning: Techniques and technologies* (pp. 60–82). Hershey, PA: Information Science Reference.

Kukulska-Hulme, A., Sharples, M., Milrad, M., Arnedillo-Sánchez, I., & Vavoula, G. (2009). Innovation in mobile learning: A European perspective. *International Journal of Mobile and Blended Learning*, *1*(1), 13–35. doi:10.4018/jmbl.2009010102

Ogata, H. (2008). Computer supported ubiquitous learning: Augmenting learning experiences in the real world. *Proceedings of the Fifth IEEE International Conference on Wireless, Mobile and Ubiquitous Technologies in Education.*

Petrova, K. (2010). An implementation of an m-learning scenario using short text messaging: An analysis and evaluation. *International Journal of Mobile Learning and Organisation*, *4*(1), 83–97. doi:10.1504/IJMLO.2010.029955

Rogers, Y., & Connelly, K. (2010). Enhancing learning: A study of how mobile devices can facilitate sensemaking. *Personal and Ubiquitous Computing*, *14*(2), 111–124. doi:10.1007/s00779-009-0250-7

Rogers, Y., Price, S., Fitzpatrick, G., Fleck, R., Harris, E., & Smith, H. … Weal, M. (2004). Ambient Wood: Designing new forms of digital augmentation for learning outdoors. *Interaction Design and Children 2004* (pp. 3-10). University of Maryland.

Schwabe, G., & Göth, C. (2005). Mobile learning with a mobile game: Design and motivational effects. *Journal of Computer Assisted Learning*, *21*(3), 204–216. doi:10.1111/j.1365-2729.2005.00128.x

Scornavacca, E., Huff, S., & Marshall, S. (2009). Understanding the value of interactive SMS for large classes . In Ryu, H., & Parsons, D. (Eds.), *Innovative mobile learning: Techniques and technologies* (pp. 48–59). Hershey, PA: Information Science Reference.

Spikol, D., Kurti, A., & Milrad, M. (2009). Collaboration in context as a framework for designing innovative mobile learning activities . In Ryu, H., & Parsons, D. (Eds.), *Innovative mobile learning: Techniques and technologies* (pp. 60–82). Hershey, PA: Information Science Reference.

Stenton, S. P., Hull, R., Goddi, P. M., Reid, J. E., Clayton, B. J., & Wee, S. (2007). Mediascapes: Context-aware multimedia experiences. *IEEE MultiMedia*, *14*(3), 98–105. doi:10.1109/MMUL.2007.52

Vavoula, G., Sharples, M., Rudman, P., Meek, J., & Lonsdale, P. (2009). Myartspace: Design and evaluation of support for learning with multimedia phones between classrooms and museums. *Computers & Education*, *53*, 286–299. doi:10.1016/j.compedu.2009.02.007

Weizman, A. (2005). *Math4mobile*. Retrieved July 16, 2010, from http://www.math4mobile.com/

Wilson, D., Andrews, B., & Dale, C. (2009). Choreopod: Dance and the iPod towards blended learning. *International Journal of Mobile and Blended Learning*, *1*(1), 49–60. doi:10.4018/jmbl.2009010104

Preface

INTRODUCTION

Mobile learning, or m-learning, is one of the fastest-growing fields of research as this book demonstrates with its myriad definitions and applications from around the world and its treatment in recent books from varied disciplines (Babcock, 2010; Brooks-Young, 2010; Collins & Halverson, 2009; Montgomery, 2007; Palfrey & Gasser, 2008; Prensky, 2010; Tapscott, 1998, 2009; Tomei, 2009). It can been seen in technology-driven mobile learning, miniature but portable e-learning, connected classroom learning, mobile training and performance support, large-scale implementation, inclusive and assistive technologies, informal and situated learning, and remote, rural, and development mobile learning (see Kukulska-Hulme & Traxler, 2007 for a more developed discussion). Mobile learning is not, as the name implies, a combination of mobile and learning but rather is the mobilization of e-learning and has grown out of the literature on e-learning and blended learning. For the purposes of this chapter, I have adopted Traxler's (2009) definition of the concept as "the provision of education and training on PDAs/palmtops/handhelds, smartphones and mobile phones" (p. 2). In other words, it is learning that can take place in any environment using technologies that fit in the palm of the hand or can be easily carried from one place to another.

M-learning addresses the needs of the Net Generation of learners who we have in our classrooms, regardless of the level of education. To be sure, there are baby boomers and older students; however, the skills they need are typical of what is expected by the Net Geners around the world. According to Brooks-Young (2010), 21st century skills "comprise both content knowledge and applied skills that today's students need to master to thrive in a continually evolving workplace and society" (p. 6). In particular, she argued that there are three seminal documents that discuss these 21st century skills: *enGauge 21st Century Skills for 21st Century Students*, the National Educational Technology Standards for Students (NETS*S), and the *Framework for 21st Century Skills*.

The first document presents four broad areas of applied skills which students must acquire: (1) digital age literacy, (2) inventive thinking, (3) effective communication, and (4) high productivity. The recommended skills in the second document are similar in nature: (1) creativity and innovation, (2) communication and collaboration, (3) research and information fluency, (4) critical thinking, problem solving, and decision making, (5) digital citizenship, and (6) technology operations and concepts. Lastly, the third document specifies four content area knowledges: (1) core subjects and 21st-century themes, (2) learning and innovation skills, (3) information, media, and technology skills, and (4) life and career skills. As can be seen from these documents, the teacher must still include content knowledge, but the application of skills, based in m-learning technologies, has become paramount. Tapscott (2009) and Prensky (2010) have made similar arguments.

UNIQUENESS OF THE BOOK

Kukulska-Hulme and Traxler (2005), pioneers in mobile learning, published an informative volume for educators and trainers that changed how we saw m-learning beyond those initial stages in its infancy. When Ryu and Parsons (2009) produced their book "about providing a comprehensive survey of mobile learning research and projects that both academics and practitioners may utlize in their work (p. xiv), it presented one of the first books to bring the theoretical and practical implications of m-learning to the forefront. Similarly, Ally's (2009) work brought together a group of academics as they discussed their perspectives on m-learning across the world. Vavoula, Pachler, and Kukulska-Hulme's (2009) volume outlined the frameworks, tools, and research designs for m-learning which have been adopted and adapted to present new models and a deeper understanding of the phenomenon. Most recently, Macdonald and Creanor (2010) have created a handbook for students as they learn about and interact with online and mobile technologies.

This present book is unique in that it is one of the few that presents a global perspective on mobile learning and augments that perspective with examples and applications from around the world, written by scholars who are leaders in their countries and in the world. It highlights examples from the school system, from undergraduate classes, and from graduate classes. In fact, this volume takes much of was offered by our predecessors and adds on to the perspectives, application, theories, and philosophies so that it has become cutting-edge in its presentation of recent research on m-learning. Most notably, we are witnessing applications in developing countries as they embrace mobile systems in place of the less-reliable Internet connections and witness the explosion of knowledge as avenues for knowledge acquisition open up. In developed countries, we see that same explosion, but we also see a finetuning of the m-learning framework. Lastly, this book is unique in that it demonstrates how blended learning has developed into mobile learning opportunities. In fact, much of that argument is continued in another editted book of mine, *Blended Learning across Disciplines: Models for Implementation*.

TARGET AUDIENCE

The prospective audiences for this volume will be academics and practitioners in the areas of distance learning, e-commerce/e-government, healthcare, business, education, engineering, and science, to name but a few. This volume contains chapters from leading experts in the field, which will be immensely helpful for all stakeholders, and will aid them in all aspects of teaching and learning.

The potential uses for this publication are vast. The volume could be used as a prescribed text in graduate schools across the world since there is a great deal of information on the latest trends in mobile learning. The book can used as a bookshelf book for academics, since much of the current research on mobile learning is encapsulated in these pages from myriad respected scholars. The book can be a frequently-used library reference book, since it contains trends, recent research, and seminal studies on mobile learning in an easy-reading style. The volume is pertinent to higher education administrators as both a source for change and for faculty discussion. Lastly, this book is perfect for anyone who is interested in reading about the next stages of mobile learning as we begin to experiment and discuss with our colleagues from around the world. Once again, having chapters from leading experts in the field will be helpful and will aid readers in all aspects of teaching and learning in the mobile age.

The potential benefits for the reader of this publication are that he or she will have cutting-edge research on mobile learning, written by key academics in the respective areas of expertise (see the next section and the Tables of Contents for chapter headings and abstracts). Additionally, the benefit of this edited volume to enhance the available literature is that it brings together the writers from other books and journals into one volume. It also leads to opportunities for new and experienced researchers to meet at a common venue, based on what is written in the chapters.

THE STRUCTURE OF THE BOOK

The book begins with an informative Foreword by David Parsons who outlines the history of the terminology and explains its evolution. The book itself is divided into three natural sections. The first, *Theorizing About Mobile Learning*, includes five key chapters dealing with theoretical and philosophical arguments for mobile learning. The second section, *Practicing Mobile Learning*, contains six chapters that exemplify mobile learning in various contexts. The last, *Extending Mobile Learning*, includes four chapters that demonstrate how mobile learning can be applied in innovative ways.

Theorizing About Mobile Learning

The book opens with a chapter that contextualizes the book and subsequent chapters. In "*An Analysis of Mobile Learning in Education, Business, and Medicine*," Dawn Stevens and I examine m-learning within the fields of Education, Business, and Medicine. Specifically, three types of mobile devices were examined within the three subcategories of m-learning: the mobile phone or smartphone, the iPod, and the PDA. A mixed-method design was used to review 40 m-learning articles and to synthesize the literature to explore m-learning projects around the world. The literature revealed that m-learning was used in many parts of the world, but mostly in North America, within all three fields. There were also numerous projects in Europe, Asia, the United Kingdom, and in Oceania. Mobile phones, smartphones, iPods, and PDAs were used in all three fields.

The second chapter, "*Producing Generic Principles and Pedagogies for Mobile Learning: A Rigorous Five Part Model*," Davina Calbraith and Reg Dennick present their own applied research on learning objects. The chapter outlines how the five-part model was designed and developed to create a robust research model that can be used by others interested in m-learning. The step-by-step presentation of the model will assist new and seasoned m-learning researchers, as it has applications across myriad disciplines. They present an example from nursing in which they deconstruct the model to show how a real-life example would work at each stage.

In the third chapter, "*But They Want Us in 'Their' World? Evaluating the Types of Academic Information Students Want Through Mobile and Social Media*," Tim Brown and Amanda Groff present an argument for the appropriate uses of mobile and social media for post-secondary students. Purporting that the growth of social media and mobile communication provide post-secondary educators opportunities to present course-related information in a manner that would appeal to the Net Generation, they pose the question: But are students willing to accept course information through those channels, typically seen as "fun" and "social?" The chapter expands on the reasons that students use different types of personal media and outlines the appropriate communication channels for transferring academic information. Brown and Groff report that the students much prefer to receive information through email and content management systems (CMSs) as they are much more official channels of communication. They authors do argue, however, that the students are open to receiving specific forms of information through social networking or mobile devices if they are not asked to share any personal information in those media.

In the next chapter, "*Standardized, Flexible Design of Electronic Learning Environments to Enhance Learning Efficiency and Effectiveness,*" Jennifer Banas presents a solid argument for creating course design models that are effective and efficient using m-learning. In explaining the design of electronic learning environments (ELEs), she outlines the importance of self-regulated learning, cognitive load, and learner characteristics. She presents a sample course design model for an eight-week span and demonstrates how using the standardized format with inherent flexibility, a course can be designed so that it is both effective and efficient. She concludes the chapter with suggestions for future conversations and research regarding m-learning.

In the last chapter of this section, "*Situation-Based and Activity-Based Learning Strategies for Pervasive Learning Systems at the Workplace*," Amel Bouzeghoub, Serge Garlatti, Kien Ngoc Do, and Cuong Pham-Nguyen discuss pervasive learning in the workplace, or work-based learning. Basing their argument on situation-based and activity-based learning, they posit that choice and relevancy are paramount in work-based learning so that the individual chooses from resources (situation-based) and also seeks out and selects activities and the corresponding resources (activity-based) which are described as push and pull strategies (Cheverst, Mitchell, & Davies, 1998). Proposing a pervasive learning environment model in which the learner can choose specific strategies as their needs and contexts dictate, the authors present a use model based on the P-LearNet project (Pervasive Learning Network).

Practising Mobile Learning

In the opening chapter of this section, "*Mobile Learning in Action: Three Case Studies with the Net Generation*," I outline three m-learning case studies that are designed for the Net Generation. The first case is a school district in Northern Canada that has embraced m-learning vis-à-vis 1:1 computing with Grade 4 to 10 students. This case is followed by an autoethnographic approach to using m-learning in post-secondary education with the technologies of a netbook, Iphone, and a portable printer. The last outlines the plans of a United States school district that is proposing to implement m-learning technologies across three schools in an attempt to address the needs of the Net Generation. The chapter concludes with a synopsis of the findings across the three case studies.

In Chapter 7, "*Investigating Undergraduate Student Mobile Device Use in Context*," Yanjie Song presents her data on an in-depth, one-year empirical study on five undergraduate student's mobile device uses. She used reflective e-journals, artifacts from each student, observations of mobile devices in use, interviews with the students, her own field notes, and memos. She concluded that there were seven interacting factors that could either promote or inhibit mobile device use: (1) goals, (2) tasks, (3) learning resources, (4) time and place, (5) social factors, (6) the mobile device, and (7) individual interpretation. She concludes the chapter with the assertion that this study was the first to demonstrate the interactivity among these seven factors in relation to m-learning.

In Chapter 8, "*Mobile Learning in Medicine*," Kalyani Premkumar describes the use of m-learning in the field of medicine. She begins with an overview of the medical context in Canada and with a description of the medical student, resident, and medical professional characteristics that sets the context for the chapter. She explores the uses, benefits, outcomes, and barriers of m-learning for undergraduate, graduate, and continuing education participants. She argues that medical practitioners are particularly good candidates for m-learning, since they are often placed in contexts and environments that necessitate access to m-learning technologies. Specifically, medical students are in hospital and clinical settings that not only require access to course information, but also systems for recording information and reflecting on their cases; postgraduates and physicians require similar access to interact between distributed sites and offices—especially in rural and remote areas.

In the next chapter, "*Unleashing the Potential of Mobile Learning through SMS Text for Open and Distance Learners*," Zoraini Wati Abas, Tina Lim, and Ruzita Ramli outline how the Open University Malaysia began with a concept of using SMS text, planned a solid model for delivery, and created the learning environment to augment its current blended learning model in general. In particular, they describe how it works one of its courses with over 1,000 students enrolled. The authors describe the categories used for formulating the SMS content and their use of Twitter and Facebook to support the SMS sent, discuss the feedback received on the initiative, and outline the issues and challenges.

In Chapter 11, "*Promoting Learner-Generated Content and Podcasting in Postgraduate Hospitality Education*," Crispin Dale and Ghislaine Povey investigate the use of podcasting as a means of facilitating learner-generated content in hospitality management at the post-graduate level. They describe how the students created content as "learning objects," using podcasts as the main medium in which the information was shared with their peers and tutors at the University of Wolverhampton. Further, the authors explore the theoretical underpinning of the technique that are explored and are used to analyze the students' experiences of generating a "mockcast" for a new gastronomic concept in a post-graduate hospitality management course.

In the concluding chapter for this section, "*Closing in on Vocabulary Acquisition: The Use of Mobile Technologies in a Foreign Language Classroom*," Carly Born, Andrea Nixon, and Christopher Tassava explore the use of mobile technologies in the second language classroom with a concentration on vocabulary building. The authors conducted a study with 39 students enrolled in an introductory-level French class in which some students were given Ipod Touches (n = 10) to practise vocabulary through the use of flashcards, and some student were not given Ipods (n = 29). Comparing the results of the two groups, the authors conclude that the mobile technologies improved vocabulary acquisition. They conclude with pedagogical and logistical implications and recommendations for future research.

Extending Mobile Learning

In Chapter 12, "*Augmented Reality and Mobile Technologies*," Grant Potter expands on the potential of augmented reality and mobile technologies. He argues that the blend of both the virtual and the real allow AR application interfaces on mobile devices to display information that is dependent on users' time and location. To exemplify the applications of AR, he writes about its use in business, tourism, and education. He concludes the chapter with the prediction that augmented reality will become a major focus on m-learning research in the next few years.

Raj Gururajan, Abdul Hafeez-Baig, Patrick Danaher, and Linda De George-Walker in "*Student Perceptions and Uses of Wireless Handheld Devices: Implications for Implementing Blended and Mobile Learning in an Australian University*," draw on a case study to discuss the uses of wireless handheld devices in post-secondary education. Using factor analysis and regression analysis to explain the results of a questionnaire, the authors conclude that behaviour and attitude are strong contributors to the perceived performance of using mobile devices in the specific contexts, and that facilitating conditions have a more complex and mediated relationship with behavior and attitude on the one hand and perceived performance on the other. They finish the chapter with a thorough discussion of the implications of their research as it pertains to both mobile and blended learning.

In Chapter 14, "*Using Students' Own Mobile Technologies to Support Clinical Competency Development in Speech Pathology*," Trish Andrews, Bronwyn Davidson, Anne Hill, Danielle Sloane, and Lynn Woodhouse discuss students' uses of mobile technologies in the field of speech pathology. The chapter focuses on the role of m-learning technologies in supporting learning across varied contexts and provides a description of a specific project conducted by the authors. They conclude that mobile technologies have a clear and positive impact on the clinical competency development of the students.

In the concluding chapter, "*The New Age 'Information Dowser'* and mobile learning opportunities: The use of library classification and subject headings in K-20 education – today and tomorrow," Tom Adamich profiles a case study of using m-learning in post-secondary libraries. He purports that m-learning technologies could be a possible mode of information delivery that will address the need for a library classification system that is both user friendly and technologically savvy to address the needs of the Net Generation of learners.

CONCLUSION

This book represents months of hard work from a group of dedicated scholars who are passionate about mobile learning. It is a truly collaborative and international effort on the part of 32 academics from seven countries and four continents. When I was asked by IGI Global to edit a book dealing with international perspectives on e-learning, I was deeply honoured and rose to the challenge of soliciting chapters from colleagues across the world. In total, there were over 50 submissions from which 28 were chosen. The book chapters were submitted to a double-blind review and the successful authors wrote their final chapters. As it turned out, the quality and quantity of the book chapters were so outstanding that we decided to make the original book into two excellent books. This one, *Models for Interdisciplinary Mobile Learning: Delivering Information to Students,* represents the mobile learning scholars but much of their content deals with the arguments outlined in the second book, *Blended Learning across Disciplines: Models for Implementation.* Although the decision to include a chapter was certainly not arbitrary, many times the decision was difficult. I believe that the end product will provide an extremely valuable resource to those students, researchers, and scholars interested in the topic of blended learning.

In the end, this book has become an excellent resource for any person interested in mobile learning: the definitions, the concept, examples from around the world, and applications from secondary school to graduate school. It will be a valuable addition to any person's library.

Andrew Kitchenham
University of Northern British Columbia, Canada

REFERENCES

Ally, M. (Ed.). (2009). *Mobile learning: Transforming the delivery of education & training*. Athabasca, AB: Athabasca University Press.

Babcock, C. (2010). *Management strategies for the cloud revolution*. Toronto, Canada: McGraw-Hill.

Brooks-Young, S. (2010). *Teaching with the tools kids really use: Learning with Web and mobile technologies*. Thousand Oaks, CA: Corwin.

Collins, A., & Halverson, R. (2006). *Rethinking education in the age of technology: The digital revolution and schooling in America*. New York, NY: Teachers College.

Kukulska-Hulme, A., & Traxler, J. (2007). Design for mobile and wireless technologies . In Beetham, H., & Sharpe, R. (Eds.), *Rethinking pedagogy for the digital age* (pp. 180–192). London, UK: Routledge.

Macdonald, J., & Creanor, L. (2010). *Learning with online and mobile technologies: A student survival guide*. Burlington, VT: Gower Publishing.

Montgomery, K. C. (2007). *Generation digital: Politics, commerce, and childhood in the age of the Internet*. Cambridge, MA: MIT Press.

Palfrey, J., & Gasser, U. (2008). *Born digital:Understanding the first generation of digital natives*. NewYork, NY: Basic Books.

Prensky, M. (2010). *Teaching digital natives: Partnering for real learning*. Thousand Oaks, CA: Corwin Press.

Ryu, H., & Parsons, D. (Eds.). (2009). *Innovative mobile learning: Techniques and technologies*. Hershey, PA: IGI Global.

Tapscott, D. (1998). *Growing up digital: The rise of the Net Generation*. Toronto, Canada: McGraw-Hill.

Tapscott, D. (2009). *Grown up digital: How the Net Generation is changing your world*. Toronto, Canada: McGraw-Hill.

Tomei, L. (Ed.). (2009). *Information Communication Technologies for enhanced education and learning: Advanced applications and developments*. Hershey, PA: IGI Global.

Traxler, J. (2009). Learning in a mobile age. *International Journal of Mobile and Blended Learning*, *1*(1), 1–12.

Vavoula, G., Pachler, N., & Kukulska-Hulme, A. (Eds.). (2009). *Research mobile learning: Frameworks, tools, and research designs*. New York, NY: Peter Lang.

Acknowledgment

I would like to wholeheartedly thank all who were part of the collection and double-blind review process for the chapters contained in this volume. The Editorial Advisory Board members were always ready to assist when asked, and I truly appreciated the support provided by each member when called upon. This project would not have been so successful without the help extended by all and the incredible abilities of all to meet the required deadlines. I would like to especially thank Dr. Colin Chasteauneuf for his support as a colleague and a listener. I wish to express my gratitude to Dr. David Parsons for authoring the Foreword to this volume.

Many of the authors in this volume served as reviewers for their peers, and I thank them for their commitment to the high-quality scholarship that emerged. You were constructive and comprehensive, and I know that each author appreciated the detailed feedback. This book is yours as much as it mine, and I hope that you have an opportunity to share it with your colleagues at home and further afield.

I would also like to thank Mike Killiam at IGI Global for his patience with my myriad emails and for always having an answer, no matter how mundane the question. You own part of this book as well, and I cannot express my gratitude enough. I also thank IGI Global for the invitation to edit this book and for the chance to edit another as a direct result of this project.

Lastly, I would like to thank my wife, Leah Jackson, for her constant support and understanding for 25 years and my son, Davy, who amazes me constantly with his patience, wit, and sense of humour.

Andrew Kitchenham
University of Northern British Columbia, Canada

Section 1
Theorizing About Mobile Learning

Chapter 1
An Analysis of Mobile Learning in Education, Business, and Medicine

Dawn Stevens
University of Northern British Columbia, Canada

Andrew Kitchenham
University of Northern British Columbia, Canada

ABSTRACT

This chapter examines m-learning within education, business, and medicine. Specifically, three types of mobile devices were examined within the three subcategories of m-learning: the mobile phone or smartphone, the iPod, and the PDA. A mixed method design was used to review 40 m-learning articles and to synthesize the literature to explore m-learning projects around the world. The literature revealed that m-learning was used in many parts of the world, and most in North America, within all three fields. There were also numerous projects in Europe, Asia, the United Kingdom, and in Oceania. Mobile phones, smartphones, iPods, and PDAs were used in all three fields.

INTRODUCTION

Following on the academic heels of blended learning, mobile learning, or m-learning, is the zeitgeist of teaching and learning in the 21st century. From kindergarten to graduate school, learners are using m-learning hardware and software to enhance their own and others' learning processes (Laouris & Eteokleous, 2005). The idea of m-learning was rooted in the early development of hardware

DOI: 10.4018/978-1-60960-511-7.ch001

devices but has been only in the last decade that we have seen the software, the applications, take flight in the hands of the Net Generation and take off in the hands of the Gen Xers and Baby Boomers (Ally, 2009; Horn & Johnson, 2008; Kelly, McCain, & Jukes, 2009; Palfry & Gasser, 2008; Tapscott, 2009).

Consider these typical scenarios. Imagine you are a university student studying a second language. While sitting in a coffee shop over a leisurely cup of coffee, you decide you have some time to study, so you pull out your cell phone. Using your high speed Internet Explorer Mobile software, you download a grammar lesson, review it, and then proceed to test your knowledge with a self-assessment, also downloaded to your cell phone. Or you are a family physician with a very busy practice. While examining a patient you decide to prescribe a new drug. You are not sure of possible drug interactions with the patient's existing medication, so you use your Personal Digital Assistant to browse a bookmarked pharmaceutical website, and check for drug contraindications. You do not need to excuse yourself to consult a large, cumbersome volume of drug listings, and can be on to your next patient that much sooner. You could be a business person waiting in the airport for your flight home. You have an hour before you leave, so out comes your iPod and your earphones. You listen to the latest podcast from your company's president, with the details of the last quarter's performance. The picture being painted here is one of m-learning, otherwise known as mobile learning or handheld learning.

Laouris and Eteokleous (2005) ran a Google search for mobile learning in January 2005 and received 1,240 items. After running the same search five months later, it resulted in 22,700 items, proving that the interest in mobile learning was growing rapidly and exponentially. At the same time, a Google Scholar search resulted in only 231 items. The researchers concluded that mobile learning means different concepts to different people, depending on context. They collated numerous definitions of mobile learning, and found that researchers such as Pinkwart, Hoppe, Milrad, and Perez (2003) and Traxler (2005) generally agreed that the basic premise of mobile learning involved e-learning that used mobile devices and wireless transmission. Our own search on the words "mobile learning" in January 2010 resulted in 99,600,000 items on Google and 1,250,000 items on Google Scholar. This indicates an enormous increase in the interest in mobile learning over the last five years.

A mere two years after asserting that mobile learning was basically a more transportable version of e-learning, Traxler (2007) opined that mobile learning was difficult to define, and his opinion had altered:

Some advocates of mobile learning attempt to define and conceptualise it in terms of devices and technologies; other advocates define and conceptualise it in terms of the mobility of learners and the mobility of learning, and in terms of the learners' experience of learning with mobile devices. (p. 1)

He went on to argue that mobile devices create not only new forms of knowledge and new ways of accessing it, but also new forms of art and performance, commerce and economic activity. He theorized that mobile learning is not really about "mobile" or "learning", but is part of a new mobile conception of society.

Quinn (2000) theorized that m-learning was the intersection of mobile learning and e-learning. It was learning independent of location in time or space. Quinn (2000) predicted that one day mobile devices would have wireless networking, would always be "on" (as opposed to dial-up), would have high resolution colour screens, and would act as tiny yet powerful computers. Of course these devices now exist, not many years later, as a testament to the fact that the technology enabling m-learning to exist is growing exponentially.

For the purposes of this chapter, we define mobile learning as the use of a wireless handheld device; a cell phone, personal digital assistant (PDA), mini-computer, or iPod to engage in some form of meaningful learning. This learning could be a component of formal education, or it could be information gained from a mobile device in a business or medical setting. As we will demonstrate, there is a definite crossover of m-learning use, in which not only is it used in the education sector, but is also used in Business or Medicine at the same time.

E-LEARNING TO M-LEARNING

M-learning is considered to be a subset, or a natural evolution of e-learning, with a new set of terminology to go with it. Common e-learning terms are *multimedia*, *interactive*, and *media rich*. M-learning is characterized by such terms as *spontaneous*, *connected*, and *informal* (Laouris & Eteokleous, 2005). Table 1 demonstrates how the language pertaining to learning is changing with the evolution of blended and m-learning.

Prensky (2005) reported that 1.5 billion people worldwide used cell phones, and that the high-end phones of the day were like powerful computers in the purse or pocket. The UN Millennial Goals Report (2008) estimated that 2.2 billion mobile phones were in use by the end of 2006. While m-learning and its related technologies in North America lag behind that of Asia and Europe, it is a growing field. As more and more of the technologies are put into place, Canada and the U.S. are poised to take advantage of improved cellular wireless service and new models of cell phones and PDAs.

Tapscott (2009) surmised that Canada and the United States can expect to follow the mobile phone usage trends seen outside North America. In many countries access to the internet is cheaper by mobile phone than it is by desktop or laptop computer. In every country in Africa mobile phones outnumber landlines. While in the U.S. just over half of consumers subscribe to wireless data plans, that number rises to 90 percent in many Asian countries. Asians employ their mobile phones for many other purposes than for phone calls and text messaging. They use their phones as train passes, for watching movies, and for buying food at vending machines. Their mobile phones are an integral part of their daily lives. Additonally, he called today's phones "sleek digital Swiss Army Knives that do a lot more than make a phone call" (p. 48). They are small, powerful computers

Table 1. Terminology comparisons between e-learning and m-learning

e-learning	m-learning
Computer	Mobile
Bandwidth	GPRS, G3, Bluetooth
Multimedia	Objects
Interactive	Spontaneous
Hyperlinked	Connected
Collaborative	Networked
Media-rich	Lightweight
Distance learning	Situated learning
More formal	Informal
Simulated situation	Realistic situation
Hyperlearning	Constructivism, situationism, collaborative

that combine voice communication, music player, web browser, texting device, digital camera, and video camera into one portable device. The newest mobile devices, called smartphones, are capable of browsing the internet at high speeds, sending and receiving email, as well as playing video and text messaging. They feature full QWERTY keyboards to enable quick email and texting. Every new generation of the cell phone and the PDA has increased resolution for improved graphics and video (Prensky, 2005).

To accompany the latest mobile hardware, computer technology companies are researching and developing ever-increasing mobile communications software adapted for the small screen. Microsoft offers mobile versions of its ubiquitous Word, PowerPoint, and Excel and has its own Windows Mobile operating system. Apple, Blackberry, and Google have all created their own mobile operating systems and are developing long lists of software applications as the use of smartphones becomes more widespread (Tapscott, 2009).

According to apple.com (2007), 100 million iPods had been sold worldwide as of April 2007. iPods are used with increasing frequency to download and play recorded lectures via podcasts, making lectures accessible to those who have difficulty attending face-to face-classes (Doolittle,, Lusk, Byrd, & Mariano, 2009). In our experiences, they are also useful for students with memory retention or note-taking problems who benefit from listening to a lecture over and over. A virtual visit to the iTunes web store lists dozens of universities offering free lecture downloads to iPod. Included are prestigious institutions such as MIT, Cambridge, and Harvard universities.

The stage is set for using cell phones, PDAs, and smartphones as mobile classrooms. Metcalf and De Marco (2006) noted that mobile devices are beneficial in making small amounts of knowledge available when we are not otherwise occupied. The ability to have a mobile device capable of reactivating knowledge as often as needed provides new support for effective learning.

NET GENERS AND DIGITAL NATIVES

Tapscott (2009) labelled American children born between 1977 and 1997 the *Net Generation* (Net Geners). This demographic is comprised of 81.1 million children and young adults, or 27 percent of the U.S. population. Net Geners have been immersed in computer technology all of their lives and assimilate it, rather than having to accommodate technology, as older generations are forced to do. Due to "always on" technology, Net Geners actually think differently than do older generations. They expect information to be readily at their fingertips.

The Net Generation is not limited to America. Although the percentage of each country's population represented by Net Geners varies from country to country, there are many similarities among them. Tapscott (2009) presented eight norms, or distinctive attitudinal and behavioural characteristics that differentiate Net Geners around the world from their "baby boomer" parents. The eight norms are freedom, customization, scrutiny, integrity, collaboration, entertainment, speed, and innovation. Tapscott theorized that Net Geners are changing our world thanks to digital and mobile technology, and that we must change to accommodate their needs. In particular, education systems must change to engage and stimulate the Net Generation, and companies will have to reconfigure themselves to attract and retain this generation entering the workforce.

Similarly, Prensky (2001, 2010) gave the term *digital natives* to students from kindergarten to college who had spent their entire lives surrounded by and using technology in the form of computers, videogames, cell phones, and many other digital tools and toys. They are accustomed to receiving information quickly, to multi-tasking, and networking. Prensky (2001) theorized that these digital natives think and process information in a fundamentally different way from their predecessors. He encouraged educators to invent digital native methodologies to properly teach their students or the students would have to wait until they grew up to educate themselves.

It is clear that m-learning has a promising future; however, while educational institutions and corporations are called upon to meet the technological needs of the digital natives, issues can arise which may hamper the implementation of m-learning. The conversion of educational materials to be suitable for mobile devices requires careful planning and consideration of human-computer interaction. For example, information should be displayed in limited chunks of text to accommodate small display screens. Websites must be formatted in a such a way as to enable viewing and navigation in a miniaturized fashion. As well, the diverse mobile hardware, software, and internet connection speeds of the distributed users must be recognized when planning content delivery (Amin et al., 2006; Kroeker, 2005).

STUDY OVERVIEW

This section will outline the research design. We begin with an explanation of how we selected the articles for our research. Next, we explain the methodology for sorting the selected articles into the three main groups. Finally, we lay out our methodology for summarizing the articles in two synthesis matrices.

This substantive literature review and analysis will provide a basis for future m-learning researchers to collate previous research conducted in the field. As well, since m-learning is in its infancy, by our collating much of the information to date and describing how it can be used, some readers who were not aware of its potential may be inspired to employ it in their own fields or practices. A literature review is essential to advance knowledge and to facilitate theory development (Webster & Watson, 2002). We argue that we will influence both theory and practice with our work by providing an informative, useful summary of m-learning definitions and uses.

Research Procedures

The purpose of this research project was to synthesize the literature on m-learning. Our goal was to analyze and summarize a cross-section of m-learning literature to investigate what m-learning was, how it was being used, and where m-learning research was being conducted. We employed an exploratory research methodology to analyze qualitative data which, according to Mauch and Park (2003), is used for "investigations into new or relatively unknown territory" (p. 129). We collected and analyzed qualitative data (the m-learning articles we reviewed), categorized the data quantitatively, and identified themes. We then created two synthesis matrices to help summarize the fields, to locate the geographical areas where m-learning is researched most, and to determine the hardware and software used.

To guide us in the research process, we followed Cooper's (1982) five-stage model for conducting integrative literature reviews. Cooper's stages include problem formulation, data collection, data evaluation, analysis and interpretation, and public presentation. Cooper theorized that integrative reviewing contains many decision points, as are presented below. The processes we followed are set forth in the following paragraphs.

Problem Formulation

Although the field of m-learning is growing exponentially, a large body of scholarly research on the subject has not yet been produced. In order to engage students, business people, and healthcare workers, and to keep up with the digital generation, it is important to further study m-learning. Our research purpose was to analyze and summarize a variety of literature on the topic of m-learning. We chose to group the literature into the three fields of Education, Business, and Medicine. The reasons for this choice will be explained in the *Analysis and Interpretation* section of the project.

Data Collection

We began searching for m-learning articles in March 2008 and collected them up until April 2010. Prior to starting the literature search, we decided to concentrate on the use of m-learning in Education, Business, and Medicine, as we were familiar with the concept in these areas, and we believed they would be most relevant to our audience. We also decided to search only for articles that discussed cell phones or smartphones, iPods, or PDAs since we wanted to explore the m-learning possibilities with these most ubiquitous mobile devices. We selected scholarly articles wherever possible, but when it seemed data were lacking, we relied on m-learning articles from such sources as m-learning industry websites and computer software websites for a well-rounded view of m-learning. For example, we found only a small number of scholarly articles on m-learning in Business, so we reviewed a report from a mobile learning company promoting their m-learning software in India (Deltecs InfoTech, 2009). At the conclusion of our literature search we had 178 m-learning articles and book chapters on hand.

According to Suri (2002), "methodological inclusivity is perceived as an important step to enhance the compatibility between the contemporary methods of primary research and research synthesis" (p. 4). We searched predominantly primary research articles; however, in conducting a preliminary literature review we noted that some of the valuable scholarly literature on m-learning consisted of secondary research. For example, Pozzi (2007) provided an in-depth review of m-learning in school contexts, citing some of the researchers whose work we reviewed for the research (e.g., Hoppe, 2003; Naismith, 2009). As Pozzi was discussing valuable research from well-known m-learning researchers, we considered her work to be relevant to our m-learning explorations. Thus, we supplemented the primary research with a small amount of secondary research and with articles and reviews from industry in order to analyze a substantial cross section of the literature. We believe a sizable analysis of articles from a variety of sources allowed for well-informed conclusions about the use of m-learning today.

Data Evaluation

The criteria we used for inclusion or exclusion of articles for the final review were: that the research seemed to be of good, scholarly quality, or in the case of a technical report, that it was from a reputable company; that the report discussed m-learning in one or more of the fields of Education, Business, or Medicine; and that the article discussed mobile phones, smartphones, iPods, or PDAs, devices that are carried most of the time by most people. Some articles reported research on m-learning via laptop computers; however, we did not consider a laptop to be a true m-learning device unless used for the purposes of m-learning. Most people do not carry laptops with them habitually, and therefore they cannot gain knowledge anywhere and anytime (Traxler, 2007).

We discarded articles that were of a very technical nature and beyond the scope of the research. The purpose for our literature sampling was to provide a good cross-section of the scholarly literature from around the world, from a number of researchers, and from a variety of sources: books, journals, conferences, and university reports. This type of purposeful sampling is described by Cresswell (2005) as "a qualitative sampling procedure in which researchers intentionally select individuals and sites to learn or understand the central phenomenon" (p. 596). The "individuals" to whom Cresswell alluded in relation to our project are the m-learning articles we chose to review. Purposeful selection was also an important qualitative research method to Suri (2002), who called for "diversity, complexity, and richness of purposes as central to educational research" (p. 5).

Not surprisingly, we discovered that the scholarly literature from the field of Education was much more substantial than in the other two fields, and therefore we purposefully selected articles on a range of m-learning themes within Education to portray the variety of research projects in this area.

Analysis and Interpretation

In this stage we allocated 40 articles to the three main groups; a total of 25 articles in Education, nine in Business, and six in Medicine. We sorted the literature, both scholarly and industry, into three main categories: Education, Business, and Medicine. We knew that we would be able to find enough scholarly literature within these groups to justify a relevant and interesting research project for potential m-learning stakeholders. We also knew that the interest in this research would include national and international venues such as this book, as the impact of this literature review is greater than meeting the needs of our own institution.

We defined m-learning in Education as acquisition of knowledge with a mobile device in an educational setting. The location could be an elementary school, a high school, or a college or university. It could also be a museum, historical park, or other cultural setting where knowledge is traditionally accessed. As well, this definition included the support of acquisition of knowledge such as in organizing educational information with a mobile device's calendar, word processor, or spreadsheet software.

The definition of m-learning in Business took an unexpected twist. We had originally intended it to encompass strictly m-learning in the workplace. However, as we analyzed the literature, we discovered several researchers whose definitions of m-learning included acquiring formal knowledge, as well as other types of information. For instance, Traxler (2007) posited that:

Mobile devices are creating new forms of commerce and economic activity as well. So mobile learning is not about 'mobile' as previously understood, or about 'learning' as previously understood, but part of a new mobile conception of society (p. 5).

Vavoula, Pachler, and Kukulska-Hulme (2009) argued that learning is mobile between areas of life, and that it may relate to work, to self-improvement, or to leisure. It can happen on work days or on weekends. To this end, we included m-commerce in the field of m-learning in Business. Acquiring information such as bank account balances, checking stock activity, flight information, or the location of the nearest restaurant are all business-related learning activities that can be performed with a mobile device. Our m-learning in Business definition also included using mobile devices to acquire knowledge as a form of workplace training, or as a form of knowledge gained while in the field. It also incorporated the management of knowledge gained with a mobile device.

We defined m-learning in Medicine as the use of a mobile device in a health care setting such as a doctor's office or a hospital to gain medical knowledge, to provide point-of-care service, or to assist with the management of healthcare information.

After the literature reviews were completed, we used the highlighter feature in Microsoft Word to colour-code the commonalities among and between the three main m-learning categories. We chose to use this simple procedure rather than qualitative software such as NVivo as the data were simple in nature and did not require the sophisticated coding procedures available in more robust software programs. Coding is a common method to analyze text for themes (Creswell, 2005; Hramiak, 2005). These themes will be discussed later in the chapter.

Public Presentation

Upon completing the article reviews, we created a synthesis matrix, Numerical Listing of M-Learning Articles (see Appendix A) to help summarize the data and clarify its complexity. Each article was assigned a number from 1 to 40, in no particular order, and its author, title, m-learning discipline, country in which the research was performed, and hardware and software described were entered into a table in numerical order. With a brief scan of the Numerical Listing of M-Learning Articles, the reader can get a sense of who the major m-learning researchers are, where and in which field they performed research, and the types of hardware and software discussed. Data from the appendix were then transferred to Table 2.

Table 2 represents the cross tabulation results of the literature review categorization. The numbers 1 to 40 corresponding to the m-learning articles appear in the table as many times as appropriate. For example, in the appendix, Trifonova's (2003) article was assigned the number 6, and it discussed Educational m-learning research in the United Kingdom, Finland, the United States, and Taiwan, using the mobile phone and the PDA, and the software SMS, MMS, and the Web. The number 6 appears in the table 10 times, to represent all the areas of the world the m-learning occurred, in which field it was used, and which hardware and software types were used.

In summary, 178 articles on m-learning in Education, Business, and Medicine were collected, scanned for relevant qualitative data, and subsequently condensed to 68 candidates for reading and possible review. The 68 articles were evaluated for research quality and content. A total of 40 articles were analyzed and reviewed. They had been purposefully selected to portray a wide scope of m-learning uses. Next, two synthesis matrices were created. The first was a text matrix listing the reviewed articles, with their authors, main field, country where research was performed, types of hardware and the types of software discussed. A shorter, tabular matrix was then created to summarize the data in a more succinct manner. The matrices were created for easy reference and comparison of the m-learning data. The reader does not need to read the entire literature review to get an overview of the literature findings.

BRIEF REVIEW OF THE LITERATURE

Because m-learning is a relatively new field, there is not a large amount of scholarly literature to review. Judging by the m-learning articles found to date, it appears that more research has been performed on m-learning in Education than in Business or Medicine; however, m-learning is burgeoning in the latter two fields and we are beginning to see more literature written on the subject. This section will present a brief overview of the professional literature as a full discussion is beyond the scope of this chapter.

M-Learning in Education

Prensky (2005, 2010) argued that even the simplest voice-only cell phones are more complex and more powerful than the computer that landed a spaceship on the moon in 1969. He added that we need to think of our cell phones as computers, just as our desktop computers and laptops are. As well, he noted that the U.S. and Canada were the only places in the world where PCs outnumber cell phones. Some countries have up to 10 times the number of cell phones than PCs.

He asserted that in education cell phones complement the short-burst, casual, multitasking style of today's digital native learners. He noted that while SMS has been in use in North America for a short period of time, Europeans and Asians have enjoyed this technology for several years. SMS can be used in learning environments to give pop quizzes, spelling or math tests, or to

Table 2. Cross tabulation of m-learning articles

	Education	Business	Medicine
Geographical area of research	4, 6, 7, 28, 32, 33, 34, 36, 40	*8, 10, 13, 27*	2, 9, 14, 15, 16, 29
North America	1, 6, 7, 19, 20, 38	*11, 12, 37*	17
Europe	3, 5, 6, 7, 18	*10, 35, 37*	
Asia	6, 23, 24, 25	*26*	
United Kingdom	18		
Africa	21, 22, 30, 31, 39		
Oceania			
Hardware	1, 3, 4, 6, 7, 18, 19, 28, 30, 31,	10, 11, 27, 37	9, 17
Mobile phone	38, 39, 23	8, 12, 13, 26, 27, 35	2, 14, 29
Smartphone	5, 21, 22	8, 26, 27	2, 14, 15, 29
PDA	3, 4, 5, 6, 20, 24, 32, 38, 39, 40	13	16
iPod	21, 25, 33, 34, 36		
Software	3, 4, 5,6, 7	8, 10, 11, 13	14, 15, 29
Palm OS	6, 7	10	9, 14, 15, 29
Windows Mobile	6, 24, 28, 36, 38	27, 37	15, 16, 17, 29
Web	7, 18, 30, 31, 39	27	17
SMS	3, 4, 39	12	14
MMS	3, 4, 5, 19, 23, 25, 32, 33, 38	8, 10, 26, 37	14, 15, 29, 36
GPS	1, 18, 20, 24	8, 27	17
Specialized software	21, 22, 23, 24, 39, 40	35	16
Email	33, 34	12	15, 29
Video	33, 40		16
Game	21, 24, 39		16
iTunes			
Facebook			
Adobe Reader			
PowerPoint			
RSS Feeds			
Camera			
Calculator	28		
Clock	28		
MS Office	36		

Note. Windows Mobile-enabled PDAs were formerly known as Pocket PCs.

poll students' opinions. Students can use SMS messages in real time to analyze and diagnose a problem. This strategy would be particularly useful to medical students, noted Prensky (2005).

The bright, high resolution screens of most cell phones are appropriate for meaningful amounts of text to be displayed, with users able to adjust the display of text to their own reading speed. Entire novels written on and read on cell phones are popular in Japan. Now that cell phones have memories or memory card slots, educational programs may be downloaded to the phones. Prensky (2005) noted that this would be useful for studying for specialized exams such as the MCAT medical school entrance exam or the GRE for entrance to graduate school.

Prensky (2005) highlighted the value of GPS built into some cell phones. He believed GPS would have clear applications in geography, math, orienteering, archaeology, and architecture. He noted that some colleges use cell phone GPS for their orientation programs, allowing students to find their way around campus. He (2005) concluded that educators must develop ways for students to use the cell phones already in their pockets as exciting, innovative tools. Rather than punish cell phone use in school, they must encourage students' creativity and ingenuity.

In a subsequent work, Prensky (2010) argued that partnering was a concept that would address the students' need for multitasking and using technology and the teacher's responsibility to use his or her skills to present information to the students and to learn how to use technology better by learning from the students. This notion has much promise in Education, to be sure, but also in Business and Medicine.

Thornton and Houser (2005) conducted a three-part study on mobile phone use by university students studying English as a second language in Japan. They theorized that learning a foreign language involves memorization and practice of a large number of vocabulary words and that much exposure to the vocabulary is necessary. Since face-to-face class time is very limited in Japanese universities, the researchers wanted to know if mobile phones could aid in teaching English to the students.

In the first part of the study 333 Japanese university students were polled regarding their use of mobile devices. 100 percent of the students owned mobile phones capable of viewing standard web pages and sending and receiving email. Email was the most utilized mobile phone feature, with an average of almost 200 email messages exchanged each week. The students were asked to rate the desirability of several types of mobile phone-enabled educational functions. The functions students most wanted to use were receiving notifications for class cancellations and room changes, receiving and submitting assignments, and receiving quiz and exam grades.

For the second part of the research, three times per day Thornton and Houser emailed mini vocabulary lessons to 13 students studying English as a second language, and posted the identical lessons on a mobile-compatible website that another 13 students were to study. The lessons consisted of word definitions, using each word in multiple contexts, and they also reviewed previously introduced vocabulary. After two weeks the groups switched media for another two weeks of studying vocabulary. Vocabulary tests were administered before and after the study and the researchers found a marked increase in the test scores after the mobile learning took place. The most gains were in the group who received the emailed lessons at regular intervals, as they were prompted to study more often than the students who were encouraged to view a website once a week.

In part three of the study, the researchers wanted to investigate the usability of multimedia mobile devices to study English idioms. Students used mobile phones and PDAs to study web pages and 15-second videos explaining idiomatic meanings. They then answered questions evaluating various aspects of the hardware, web pages, videos, and educational effectiveness. The students' evaluations were positive, citing such aspects as hardware and software ease of use and that they were good for studying and remembering English idioms.

Thornton and Houser concluded that when educational materials were designed for mobile phones, students evaluated them positively, and test results proved that they were able to learn via mobile devices. They found that rich multimedia captured the students' interest, and that "pushing" study opportunities by mobile email resulted in students learning English vocabulary more effectively.

Lefoe, Olney, Wright, and Herrington (2009) argued that while mobile learning was an important new pedagogy, educators needed to move beyond training to use the mobile technologies. They identified a need for teachers to spend time planning for mobile integration in learning activities, as often pedagogical aspects of m-learning were forgotten due to funding and workload structures. They provided an overview of a staff development project at an Australian university's Faculty of Education.

Faculty members were tasked with personal use of mobile technology for a period of six months before implementing the technology in their teaching. With an action learning approach they used smartphones and iPods to immerse themselves in the technologies adopted by their millennial learner students. The goal of the program was not only to learn how to use the mobile technologies themselves, but also how to employ the technologies as part of their everyday teaching to engage and keep up with their digital-minded students.

After the six-month technology familiarization period was over, the faculty implemented m-learning projects for their students for a further twelve months. During this time they met regularly to share ideas and to collaboratively explore m-learning possibilities. They used the digital technologies to support their learning and reflective activities and to provide photographic and audio reflections for later analysis by the researchers. In summarizing this research project, Lefoe et al. asserted that comprehensive staff development and support were critical in implementing mobile learning activities. It was imperative that teachers use and understand digital technologies before implementing them in their curriculums.

An experiment by Herrington (2009) was aimed at proving that smartphones had the potential to be used as effective teaching tools in higher education. Research was carried out on 14 teachers of higher education courses to evaluate smartphones as data collection tools to capture video, pictures, and audio. This digital data was gathered to create digital narratives, where learners collaboratively created and edited a story and used movie editing software to post a video on a social networking website such as YouTube. Herrington thought that this method of mobile learning would provide a social constructivist alternative to more common methods of knowledge construction.

Before beginning the study the students had the opportunity to attend smartphone workshops to learn how to use common features. Their task was to create a two to three minute digital narrative by writing a storyboard demonstrating a skill they used to teach, capture pictures and videos using a smartphone, download the multimedia into movie editing software on a personal computer, add narration and music, and upload the completed video to a social networking website.

The results of the study indicated that students initially felt overwhelmed upon learning of their task, but they were surprised at how easy the devices and the software were to use. All of the students were successful in carrying out the smartphone assignment. They cited portability and ease of use as important factors in adopting smartphones for mobile learning. They also liked that smartphones allowed for spontaneity in taking pictures and videos for educational purposes. As well, they appreciated the chance to learn how to use the smartphones before beginning their projects.

Most of the students agreed that they would continue to use smartphones to develop similar tasks for their classes. The researchers noted that in this experiment the technology's spontaneous use was the most appreciated feature, whereas mobility of the smartphone technology was not a major factor. They planned to study the affordances of mobility in a future research project.

Moura and Carvalho (2008) asserted that in the current knowledgeable and internet-devoted society, mobility and ubiquitous learning become more relevant. They called new ways of learning a pedagogical paradigm which required changes in the design of educational materials and the way they were made available. They developed the Mobile Generation project to study the use of mobile phones and iPods by 15 secondary students, all of whom owned a mobile phone or MP3 player. The objective of the study was to give the students the opportunity to learn at their own pace, time, and location, and to provide feedback by answering questionnaires.

In the first component of the study the researchers created diversified educational activities to be conducted with cell phones. They received positive feedback from the students in all sections of the questionnaire. The majority of students found that the cell phones aided collaborative work, provided motivation for school activities, allowed quick access to course material, and allowed necessary information to be accessed at anytime and anywhere.

The students downloaded mp3 podcasts of Portuguese language lectures for the experiment's second component. Again, they reported positive outcomes and said the podcasts were a complement to the classroom lectures because they aided in preparation for tests, in memorization of course details, and in stimulation to learn. Moura and Carvalho (2008) concluded that mobile devices were important educational tools, extending the boundaries of the classroom and providing students with more learning options. They predicted that mobile devices and wireless technologies would become routine both inside and outside the classroom.

Lu, Lin, Lin, and Su (2007) theorized that mobile learning would change the traditional one-way learning model where teachers teach and students learn passively. Based on Realistic Mathematics Education, a Dutch theory that math must be connected to reality and be relevant to society, the researchers designed an experiment to test whether K-12 students could learn math more easily by going outside with mobile phones or PDAs and applying mathematic principles to the real world, rather than being stuck in a classroom.

The students played a competitive arcade-style game on their mobile devices which involved moving around the school grounds to collect data about real world items such as trees and buildings. The mobile game was downloaded from the internet. They were encouraged to work collaboratively to solve the math problems. As well, exploration of the school grounds was necessary to solve the problems, resulting in "real time" learning, rather than learning from a textbook. The researchers concluded that the interactive and collaborative activity with the mobile devices supported learning and integrated constructive learning. They asserted that by using mobile devices and situation-based mathematic problem solving activities, the students were better able to grasp mathematical concepts. They suggested that learning with the internet is attractive to students and engages them, and mobile devices help learning outside of the traditional classroom.

M-Learning in Business

Tapscott (2009) cautioned that the tech-savvy Net Geners entering the workforce will not be satisfied with the hierarchical model of most businesses. Companies who want to be successful will have to implement the networked structures and peer collaboration valued by younger generations. He added that the Net Generation was already transforming the workforce and that their new approaches must be welcomed by businesses and governments around the world. They are, in fact, smart mobs (Rheingold, 2002).

Metcalf and De Marco (2006) theorized that with the growing trends of mobility in our society, it was important to gain back some of the time we spend commuting, in airports, and waiting in line. M-learning with mobile phones and PDAs allows you to have connective, online access even while you are on the go in a mobile setting, which is important for a large portion of professionals. The researchers asserted that the goal of m-learning was to develop learning content that integrates with mobile applications and provides learning and performance in a just-in-time, just-in-place dynamic. They decided to explore the ways that just-in-time learning could be enhanced by the use of new technologies such as audio and multimedia, accessing enterprise systems while on the go, reference materials, and small courses for business professionals.

They reported on corporations using m-learning to enable their salespeople and technicians to stay up to date on client information, training, and technical data no matter where they were. Using mobile email, the sales team leader could keep track of employee sales. The technician out on a service call could use his smartphone web browser to look up how to do a complicated repair. It was easy to keep information not only easily accessible, but constantly updated, rather than referring to an outdated manual in hard copy. The researchers concluded that m-learning was here to stay and that the next generation would include augmented reality, mobile collaboration, and decision support tools. They predicted that mobile gaming would become important for getting people interested in using mobile performance support and handheld collaboration. They asserted that users would need to spread awareness of the time savings, efficiency, and greater results and profits offered by m-learning technology.

Although mobile phones were once considered a luxury, Kumar and Zahn (2003) noted that they were quickly taking the place of conventional phones and that wireless networks were freeing people from their desks, allowing them to live and work in more flexible ways. They thought that analysis of mobile communications was important, since they were profoundly effecting business operations. As noted by other researchers in my meta-analysis, Kumar and Zahn added that the U.S. had more personal computers per 100 habitants than any other country, but that the U.S. lagged far behind Europe and Asia in cell phone use. The U.S. seemed to want to catch up, as the researchers reported that both the U.S. and Europe wanted to emulate the I-Mode, a type of mobile phone in widespread use in Japan. Users of the I-Mode could send emails, transfer funds between bank accounts, book plane tickets, play interactive games, and download music.

They described how Britannica wanted to push their encyclopedia products to the mobile phone and PDA market. They faced some obstacles in that the phone screens were small and they didn't want users to become frustrated with having to punch too many of the phone's buttons. The problem was solved by breaking Britannica's content down into paragraphs that could fit the mobile screens. Britannica introduced its first cell phone application in September 2000. The authors also reported on mobile business uses in Japan and Finland. Japanese consumers could purchase goods with a cell phone and be billed on their monthly cell phone bill. Cell phone users in Finland could pay for mobile purchases by sending a text message to a telephone company, receive a code number, and give the number to the merchant. The merchant would then contact the phone company for payment from the customer's account. The researchers concluded with predictions that we would soon see such mobile phone functions as bar code scanning, optical character recognition, and digital cameras. They encouraged businesses to pay attention to mobile commerce to enhance operational efficiency by distributing information to employees remotely, and by offering new ways to interact with their clients. They surmised that mobile communications could lead to improved business operations, and happy employees, customers, and suppliers all over the world.

Kleijnen, Wetzels, and de Ruyter (2004) explored the factors contributing to the adoption of mobile services in the context of wireless finance. They wished to study perceived cost, system quality, and social influence. Believing that marketers should understand consumers' motivations that will lead to the adoption of wireless technology, they wanted to provide a deeper insight into what was needed for consumers to accept the new technology. The proven TAM (Technology Acceptance Model) was their basis for studying whether consumers believed mobile services could be integrated into their daily activities. A questionnaire was designed to ask consumers about their views on mobile services usefulness, ease of use, costs, system quality, attitude, intention to use, computer skills, mobile technology readiness, and social influence. Each of the 105 research subjects owned a mobile phone with WAP technology and actually used the technology.

The researchers found that perceived cost of mobile services played a less important role than was expected. They attributed this to the consumers' expectation of high quality content, making up for cost. The effects of computer skills and mobile technology readiness were critical, as were situational contexts rather than functional aspects of mobile services. They found social influence to be an important factor, as positive word of mouth played a key role in persuading consumers to use mobile financial services.

This or That (ToT) was a research project on a social shopping application for iPhones created by Boardman, Casalegno, McMurray, and Pomeroy (2008). ToT was a way to explore the potential of mobile social shopping. They theorized that shopping is a rich-user experience that has evolved from a needs-based activity to an emotional and rich social experience for the shopper and his or her community. They speculated that for social shoppers the actual purchase was secondary to the tactical goals and the social interaction with others while shopping. The ToT application integrated Facebook with the Apple iPhone to allow the lone shopper to connect with their social network for collaboration and informed decision making on products and services to purchase. This could happen anywhere and at any time. The shopper could take pictures of the product, add a short description, create a survey to gain their friend's opinions, set a survey expiration date, and send a notification to the group whose feedback was considered relevant. Multiple groups could be set up ahead of time or at the spur of the moment.

Once the "experts" received the notification, they could log onto the application to view the photographs, complete the survey, and leave comments. The shopper would receive notice when the survey expiration time had been reached. The ToT application integrated with Facebook completely, using the shopper's already-existing friends lists, profiles, and pictures. Boardman et al. conducted a study of 12 subjects in Barcelona and Madrid to test the ToT prototype. The testers had to own an iPhone with data plan and have an active Facebook account. In addition to the 12 initial testers, another 39 users participated in the study after being invited through the Facebook application. After the two-week study the researchers found the ToT application to be promising for wide-scale distribution. They planned to enhance it by adding SMS capability to reduce the feedback time lapse and by enabling it for mobile platforms other than the Apple iPhone.

According to Wagner and Wilson (2005), mobile devices were becoming more and more affordable for everyone, and no matter age, gender, national identity, or socio-economic status, broad mobile device adoption knew no bounds. New mobile technologies were quickly being adapted to as well, and no sooner did a new technology come along than it was immediately embraced. Wagner and Wilson stressed that the growing adoption of mobile devices made mobile learning logical and that m-learning was for people who need access to information and performance support when out in the field or on the job.

They asserted that mobile learning allowed workers to take advantage of place-independent flexibility because they had the ability to connect with the right content on the right device at the right time. They stated that mobile learning was not e-leaning on a cell phone, as bandwidth and processing power were limited. However, the mobile phone was appropriate for conversations and information exchange, performance support, and real-time collaboration. Rather than the traditional behavioral, hierarchical, lecture-recitation models of training, Wagner and Wilson surmised that effective training for current and future mobile professionals needed to be based on communications and social learning. This could be realized through instant messaging and blogs to create new knowledge. They argued that learning professionals would have to help shape the mobile learning movement to avoid substandard mobile applications.

The researchers were excited that the masses could enjoy ubiquitous connectivity, multigenerational uses and users, and services for the mobile worker, especially useful for anyone who had to spend more than 20 percent of their workday away from the office, and services for the mobile learner, where the professional could remain connected and informed. In closing, Wagner and Wilson reminded the reader that the current models for training were based on a model of "command and control" (p. 43), with an instructor in charge, goals to be met, and criteria to be mastered. They stressed that with the rise of mobile device adoption a foundation of connectedness, communication, collaboration, and competitiveness would come for the mobile professional.

M-Learning in Medicine

Over half of U.S. physicians owned a PDA or smartphone in 2008 (Manhattan Research, 2008). Mobile devices are quickly becoming mainstream in the medical field, and of the physicians who do use PDAs, most consider them essential to their practice. As well, more medical schools are requiring PDAs or smartphones in the classroom. PDAs are used routinely as decision support tools for drug reference databases, drug dosage calculators, for clinical references, and for continuing medical education programs. Manhattan Research reported that some of the top handheld resources used by physicians were the drug reference database Epocrates, as well as clinical information resources MerckManual and 5-Minute Clinical Consult.

Osborne (2008) discussed a pilot text messaging project at Children's Hospital Medical Center in Cincinnati, Ohio. Clinicians at the hospital had noticed that teenagers with serious asthma were often forgetting to take their regular medicine and were subsequently suffering with ill health. They also noticed that while these teens were undergoing examinations, they would frequently send and receive text messages on their cell phones. A program was started where a staff member sent text messages to a group of about twenty patients between ages 12 and 21, reminding them to take their medication. The messages were sent to each patient once or twice per day, depending on how often the individuals needed their controller medication. The teenagers reported that the quick text messages were very useful in helping them to develop good self-care habits. They felt much better and appreciated not being "nagged at" to take their medication regularly. Initially a staff member sent the text messages manually on her own cell phone at pre-determined times; however, Osborne (2008) reported that the hospital administrators were testing a commercial system for automated text messaging. The project was deemed a success and the hospital prepared to launch a larger study aimed at people of all ages with chronic disease.

Scherr, Zweiker, Kollmann, Kastner, Schreier, and Fruhwald (2006) conducted a study on the mobile monitoring of cardiac patients at home. Two categories of patients, those with chronic heart failure and those with arterial hypertension, were at increased risk for hospitalization or even death. The researchers wanted to determine if self-monitoring of patients' health and transmission of the data to their physicians could reduce the risk factors. A telemonitoring system was developed and for a total of 1735 days the critical variables of heart rate, blood pressure, and body weight were measured at home by 14 cardiac patients and 6 hypertension patients.

The system consisted of a mobile phone with WAP technology at the patient's home, a personal computer at the doctor's office, and a computer server. Each participant was equipped with an automatic blood pressure device and a digital weight scale. The patients were asked to monitor their blood pressure, heart rate, and weight every day at the same time. After doing this they connected to the internet with the mobile phone, entered the data, and sent it to the doctor's computer for processing. The entire process took less than five minutes. If a patient's values exceeded a predefined limit, the doctor was automatically notified by a text or email message. The system also allowed the doctor to set automatic SMS reminders to have the patients take their medication, weigh themselves, and measure their blood pressure and heart rate. The patients were asked to fill out a questionnaire at the end of the monitoring period to provide feedback on system usability, acceptability, reliability, and effectiveness. The feedback was positive. Patients felt that the self-monitoring system helped them comply with their treatment program and made them more aware of their blood pressure and body weight. Scherr et al. reported that other studies have shown that telemonitoring of weight and blood pressure have significantly reduced mortality compared with standard care. They proposed that telemonitoring has clinical utility for patients with heart failure or hypertension.

Adatia and Bedard (2003) produced a very informative document to outline handheld software for physicians. They opined that more than a quarter of Canadian physicians used a PDA, and that number was expected to increase to more than half within two years. The ever-growing library of mobile medical software was becoming difficult to sort through, and the researchers aimed to categorize and provide an updated and extensive summary of the most widely used programs. The article focused on software available for Palm PDAs, but it was noted that many of the programs could be operated on other operating systems such as Microsoft Pocket PC. As a preamble to the categorization of programs, Adatia and Bedard clarified that most medical programs for the Palm operating system could be downloaded from the internet; some for free and some for purchase. Many free trial versions were also available. They warned that some popular titles may be abridged versions with important content omitted, reducing the usefulness of the product. They also cautioned that some programs could be too large for the PDA's memory storage capacity and that it might be helpful to purchase hardware with memory expansion capability. As well, they suggested it was important to evaluate the manner in which information was presented, since having to scroll through large volumes of text would be cumbersome.

They summarized seven main categories of medical software for PDAs: General medical reference programs, downloadable journal content, pharmacopoeias, medical calculators, patient-tracking programs, billing and coding software, and handheld word processing and office programs. Seemingly the most important category was pharmacopoeias, which allow physicians to easily look up indications, side effects, and dosages of medications. A check can also be run to look for drug interactions. Of physicians surveyed by Harvard University to study a specific brand of software, 50 percent indicated that the pharmacopoeia program helped them avoid at least one adverse drug event per week. The authors predicted that in addition to the wide range of mobile medical software available, the streamlining of pager, cell phone, Dictaphone and email messaging functions into a single device would come to the world of handheld technology. They announced that the upcoming Palm operating system upgrade would allow improved support for audio recording and playback, meaning that physicians would be able to dictate clinical notes, letters and email messages directly into their handheld devices.

Luo (2004) contributed a similar article regarding PDA use in medicine, however his focus was on portable computing in psychiatry. He provided a brief history of the PDA, noting that the modern, ultra portable PDA appeared and was widely adopted in 1996 with the Palm Pilot. Luo touted the PDA's sharp screens, powerful processors, external memories, and the built-in cameras, MP3 players, cell phone service, and wireless internet capability belonging to some models. He pointed out that most PDAs were based on the Palm operating system or Windows Mobile. Like Adatia and Bedard (2003), Luo also cautioned that processor speed and memory size and type must be considered when choosing a PDA, depending on the software to be used. He separated PDA use for psychiatry into six categories: general use of common features such as the calendar and reminder alarms, document editing, databases and spreadsheets, presentations, email, and medical uses. The medical uses section was further categorized into seven sections, providing a comprehensive overview.

Patient tracking, medical texts, drug reference guides, medical education, prescription writing, research, psychiatry specific applications, and security were discussed, with drug reference guides being singled out as the most popular medical use of the PDA. An advantage of the PDA-based guide was that regular updates could be obtained from the internet. Luo (2004) noted that medical schools and residency training programs were increasingly requiring students to purchase PDAs. The specialized software allowed for rich learning experiences and even for highlighting gaps in curriculum as students tracked all of their activities.

In the psychiatry-specific category Luo (2004) described applications for screening dementia, for diagnosing psychiatric illness, for special reference texts, and for psychotropic medication calculators. He concluded by asserting that PDAs were increasingly able to support physicians managing complex information and that implementation of the devices was increasing every year. He predicted that the PDA would become an essential tool in medicine.

According to Maag (2006), at the same time that increased numbers of students were enrolling in nursing colleges, the number of nursing educators was decreasing. Maag presented her research on m-learning with iPods to highlight its benefits and to encourage educators to adopt this alternative learning technology. She argued that the Net Generation has been raised in a media-rich environment and an information-centric world. They expected educators to provide innovative technological tools that complemented their inherent skills and characteristics. Maag opined that innovative technology was altering students' and healthcare providers' expectations of learning. Additionally, effective learning models and knowledge of the Net Generations's characteristics were necessary to provide effective, reflective learning. She provided an overview of how podcasting works. Two forms of podcasting are possible. Simple podcasts are a digital audio event, or MP3 format such as a conversation, lecture, or interview delivered to content management software, such as iTunes. Enhanced podcasts add multimedia so that PowerPoint slides, video clips, and images can be added to the iPod. To publish a podcast the digital file is posted to a website in an RSS feed. The subscriber downloads an RSS reader to subscribe to that website and subsequently receives automatic downloads of updated materials.

Specifically, she conducted a research project to discover nursing students' opinions of using iPods as educational tools. During two academic semesters she recorded and uploaded traditional face-to-face nursing lectures to a website and RSS feed. Students were able to download the lectures to their iPods to listen to as many times as desired. She also provided constructive feedback to her students via a five-minute MP3 audio file. Upon being surveyed about their experiences with the iPod m-learning, students indicated that the podcasts assisted them in retaining information, that they had opportunities to learn while performing other activities, and that they were useful for reviewing material before exams. Overall, the results were positive. Maag also noted that the availability of lecture podcasts had no significant effect on class attendance, which was contrary to the expectation of critics. In conclusion, Maag (2006) summarized her theory that open-source broadcast technologies support the busy lifestyles of learners and allowed for the reinforcement of learning materials. She called for ongoing evaluation of lecture podcasts to guide plans for the development of distance education nursing programs.

SUMMARY

Mobile, digital technology use is growing exponentially around the globe. Researchers have begun to study the m-learning phenomenon. International conferences and academic journals are now devoted entirely to m-learning. Research indicates that due to constant immersion in technology, the brains of youth born around 1977 and later are wired differently than those of the older generations (Prensky, 2001, 2010; Tapscott, 2009). They know nothing other than to have "always on" access to whatever information they require. Further research such as this meta-analysis of m-learning is required to learn how to keep up with and grow with the technology revolution.

M-learning is an exciting new use of technology gaining increasing attention. Used by educational institutions, corporations, and the medical field, mobile handheld devices are invaluable tools for accessing up to date information any time and any place. Through reviewing the literature on m-learning, as well as the diverse hardware and software applications we have demonstrated how information can be disseminated in a modern, technically-sophisticated manner. We believe that this analysis will assist those readers unfamiliar with m-learning, or with some interest but lacking information, to see its potential and possibly incorporate it into their field of work or study to make their lives easier.

REFERENCES

Adatia, F., & Bedard, P. (2003). Palm reading: 2. Handheld software for physicians. *Canadian Medical Association Journal*, *168*(6), 727–734.

Ally, M. (Ed.). (2009). *Mobile learning: Transforming the delivery of education & training*. Athabasca, Canada: Athabasca University.

Amin, A. H. M., Mahmud, A. K., Abidin, A. I. Z., Rahman, M. A., Iskandar, B. S., & Ridzuan, P. D. (2006). M-learning management tool development in campus-wide environment. *Issues in Informing Science and Information Technology*, *3*, 423–434.

Apple. (2007). *100 million iPods sold*. April 9, 2007. Retrieved January 3, 2010, from http://www.apple.com/pr/library/2007/ 04/09ipod.html

Boardman, D., Casalegno, F., McMurray, B., & Pomeroy, S. (2008). Rethinking the mobile social shopping experience. Retrieved February 4, 2010, from http://mobile.mit.edu/research/community/or

Brittain, S., Glowacki, P., Van Ittersum, J., & Johnson, L. (2006). Podcasting lectures: Formative evaluation strategies helped identify a solution to a learning dilemma. *Educause Quarterly*, *3*. Retrieved May 28, 2008, from http://connect.educause.edu/Library/ EDUCAUSE+Quarterly/ Podcasting Lectures/39987

Christensen, C. M., Horn, M. B., & Johnson, C. W. (2008). *Disrupting class: How disruptive innovation will change the way the world learns*. New York, NY: McGraw Hill.

Cooper, H. (1982). *Scientific guidelines for conducting integrative literature reviews*. Paper presented at the annual meeting of the American Educational Research Association, New York, NY.

Creswell, J. W. (2005). *Educational research: Planning, conducting, and evaluating quantitative and qualitative research* (2nd ed.). New Jersey: Pearson Education.

Deltecs InfoTech Pvt Ltd. (2009). *Mobile training or Blackberry training-a new buzz word in corporate training programs.* Mambai, India. Retrieved August 2, 2009, from http://www.pressreleasepoint.com/ mobile-training-or-blackberry-training- new-buzz-word-corporate-training- programs

Doolittle, P., Lusk, D., Byrd, C., & Mariano, G. (2009). iPods as mobile multimedia learning environments. In H. Ryu & D. Parsons (Eds.). *Innovative mobile learning: Techniques and technologies*. (pp. 83-101). Hershey, PA: IGI Global.

Herrington, A. (2009). Using a smartphone to create digital teaching episodes as resources in adult education. In Herrington, J., Herrington, A., Mantei, J., Olney, I., & Ferry, B. (Eds.), *New technologies, new pedagogies: Mobile learning in higher education* (pp. 28–35). Wollongong, Australia: University of Wollongong.

Hramiak, A. (2005). A method for the analysis of data from online educational research. *Journal of Interactive Online Learning*, *4*(2), 82–93.

Kelly, F. S., McCain, T., & Jukes, I. (2009). *Teaching the digital generation: No more cookie-cutter high schools*. Thousand Oaks, CA: Corwin Press.

Kleijnen, M., Wetzels, M., & de Ruyter, K. (2004). Consumer acceptance of wireless finance. *Journal of Financial Services Marketing*, *8*(3), 206–217. doi:10.1057/palgrave.fsm.4770120

Kroeker, P. P. (2005). Technology to facilitate interaction in mobile learning. Unpublished master's thesis, Athabasca University.

Kumar, S., & Zahn, C. (2003). Mobile communications: Evolution and impact on business operations. *Technovation*, *23*, 515–520. doi:10.1016/S0166-4972(02)00120-7

Laouris, Y., & Eteokleous, N. (2005). *We need an educationally relevant definition of mobile learning.* Paper presented at the 4th World Conference on Mobile Learning. Cape Town, South Africa. Retrieved on August 12, 2009, from http://www.mlearn.org.za/CD/papers /Laouris2.pdf

Lefoe, G., Olney, I., Wright, R., & Herrington, A. (2009). Faculty development for new technologies: Putting mobile learning in the hands of the teachers. In Herrington, J., Herrington, A., Mantei, J., Olney, I., & Ferry, B. (Eds.), *New technologies, new pedagogies: Mobile learning in higher education* (pp. 15–27). Wollongong, Australia: University of Wollongong.

Lu, H., Lin, J., Lin, C., & Su, K. (2007). *A study of the construction of a mobile learning oriented mathematics learning activity.* 6th WSEAS International Conference on Applied Computer Science, Hangzhou, China. Retrieved March 27, 2008, from http://74.125.155.132/scholar?q=cache: HQ0M6UAVrVYJ:scholar.google.com/ &hl=en&as_sdt=2000

Luo, J. (2004). Portable computing in psychiatry. *Canadian Journal of Psychiatry*, *49*(1), 24–30.

Maag, M. (2006). iPod, uPod? An emerging mobile learning tool in nursing education and students' satisfaction. *Proceedings of the 23rd Annual ASCILITE Conference: Who's Learning? Whose Technology?* University of Sydney. Retrieved November 20, 2009 from http://www.ascilite.org.au/conferences/sydney06/proceeding/pdf_papers /p92.pdf

Manhattan Research. (2008). *Health in the palm of your hand.* Retrieved August 18, 2009, from http://www.manhattanresearch.com/ newsroom/Press_Releases/ physician-mobile-trends- release.aspx

Mauch, J. E., & Park, N. (2003). *Guide to the successful thesis and dissertation: A handbook for students and faculty* (5th ed.). New York, NY: Dekker.

Metcalf, D. S., & De Marco, J. (2006). *M-learning: Mobile learning and performance in the palm of your hand.* Amherst, MA: HRD Press.

Moura, A., & Carvalho, A. (2008). Mobile learning: Teaching and learning with mobile phones and podcasts. *Proceedings of the 2008 Eighth IEEE International Conference on Advanced Learning Technologies*. Retrieved January 28, 2010, from http://adelinamouravitae.com.sapo.pt/ mobilespainfinal.pdf

Osborne, H. (2008). *In other words…using text messages to improve medication adherence.* Retrieved January 8, 2010, from http://www.healthliteracy.com/ article.asp?PageID=7987

Palfrey, J., & Gasser, U. (2008). *Born digital: Understanding the first generation of digital natives*. New York, NY: Basic Books.

Pinkwart, N., Hoppe, H. U., Milrad, M., & Perez, J. (2003). Educational scenarios for the cooperative use of personal digital assistants. *Journal of Computer Assisted Learning*, *19*(3), 383–391. doi:10.1046/j.0266-4909.2003.00039.x

Pozzi, F. (2007). The impact of m-learning in school contexts: An inclusive perspective. In Stephanidis, C. (Ed.), *Universal access in Human-Computer Interaction: Applications and services* (pp. 748–755). Berlin/Heidelberg, Germany: Springer-Verlag. doi:10.1007/978-3-540-73283-9_81

Prensky, M. (2001). Digital natives, digital immigrants. *On the Horizon, 9*(5). MCB University Press. Retrieved November 2, 2009, from http://www.marcprensky.com/writing/ Prensky%20 -%20Digital%20Natives,%20 Digital%20Immigrants%20-%20Part1.pdf

Prensky, M. (2005). What can you learn from a cell phone? Almost anything! *Journal of Online Education, 1*(5). Fischler School of Education and Human Services at Nova Southeastern University. Retrieved on March 6, 2008, from http://www.innovateonline.info/index.php?view=article&id=83

Prensky, M. (2010). *Teaching digital natives: Partnering for real learning*. Thousand Oaks, CA: Corwin Press.

Quinn, C. (2000). M-learning: Mobile, wireless, in-your-pocket learning. *LINE Zine*. Retrieved August 18, 2009, from http://www.linezine.com/2.1/features/cqmmwiyp.htm

Rheingold, H. (2002). *Smart mobs, the next social revolution: Transforming cultures and communities in the age of instant access*. Cambridge, MA: Basic Books.

Scherr, D., Zweiker, R., Kollmann, A., Kastner, P., Schreier, G., & Fruhwald, F. M. (2006). Mobile phone-based surveillance of cardiac patients at home. *Journal of Telemedicine and Telecare*, *12*, 255–261. doi:10.1258/135763306777889046

Suri, H. (1998). *A critique of contemporary methods of research synthesis*. Paper presented at the Conference of the Association for Active Educational Researchers, Adelaide, Australia. Retrieved on February 4, 2009, from www.aare.edu.au/98pap/sur98250.htm

Tapscott, D. (2009). *Grown up digital: How the net generation is changing your world.* New York, NY: McGraw-Hill.

Thornton, P., & Houser, C. (2005). Using mobile phones in English education in Japan. *Journal of Computer Assisted Learning*, *21*, 217–228. doi:10.1111/j.1365-2729.2005.00129.x

Traxler, J. (2005). Defining mobile learning. *Proceedings IADIS International Conference Mobile Learning 2005*, Malta, (pp. 261-266).

Trifonova, A. (2003). *Mobile learning: Review of the literature.* (Technical report DIT-03-009). Trento, Italy: University of Trento, Department of Information and Telecommunications. Retrieved July 6, 2008, from http://eprints.biblio.unitn.it/archive/00000359/

Vavoula, G., Pachler, N., & Kukulska-Hulme, A. (Eds.). (2009). *Researching mobile learning: Frameworks, tools, and research designs*. New York, NY: Peter Lang.

Wagner, E., & Wilson, P. (2005). Why learning professionals need to care about mobile learning. *Training & Development*, 40–43.

ADDITIONAL READING

Adhikari, R. (2009, November 12). Teaching mature markets new m-commerce tricks. *E-Commerce Times*. Retrieved April 4, 2010, from http://www.ecommercetimes.com/story/ 68626.html

Ardito, C., Buono, P., Costabile, M. F., Lanzilotti, R., & Pederson, T. (2007). Re-experiencing history in archaeological parks by playing a mobile augmented reality game. In Proceedings of the OTM MONET'07 Workshop, Springer LNCS 4805, pp. 357-366. Retrieved May 5, 2008, from http://ftp.cs.umu.se/~top/pubs/MONET2007%20 Arditoetal_w_confinfo.pdf

Babcock, C. (2010). *Management strategies for the cloud revolution*. Toronto: McGraw-Hill.

Baron, N. S. (2008). *Always on: Language in an online and mobile world*. New York: Oxford University Press.

Cochrane, T., & Bateman, R. (2009). Transforming pedagogy using mobile web 2.0. *International Journal of Mobile and Blended Learning*, *1*(4), 56–83. doi:10.4018/jmbl.2009090804

Collins, A., & Halverson, R. (2006). *Rethinking education in the age of technology: The digital revolution and schooling in America*. New York: Teachers College.

Kaptelinin, V., & Nardi, B. (1997). Activity theory: Basic concepts and applications. In *CHI 97 Electronic Publications*: Tutorials. Retrieved February 5, 2010, from http://www.sigchi.org/chi97/proceedings/ tutorial/bn.htm

Kitchenham, A. D. (2006). Teachers and technology: A transformative journey. *Journal of Transformative Education*, *4*(3), 202–225. doi:10.1177/1541344606290947

Kitchenham, A. D. (2008). One-to-one computing and teacher transformation. In Hansson, T. (Ed.), *Handbook of research on digital information technologies: Innovations, methods, and ethical issues* (pp. 82–101). Hershey, PA: Information Science Reference.

Kitchenham, A. D. (2009a). Laptops and teacher transformation. In Rogers, P., Berg, G., Boettcher, J., Howard, C., Justice, L., & Schenk, K. (Eds.), *Encyclopedia of Distance Learning* (pp. 1313–1318). Hershey, PA: IGI Global.

Kitchenham, A. D. (2009b). E-professional development and rural teachers: Finding the blend. *International Journal of Mobile and Blended Learning*, *1*(3), 70–85. doi:10.4018/jmbl.2009092204

Kukulska-Hulme, A., & Traxler, J. (Eds.). (2007). *Mobile learning: A handbook for educators and trainers*. New York: Taylor & Francis.

Macdonald, J., & Creanor, L. (2010). *Learning with online and mobile technologies: A student survival guide*. Burlington, VT: Gower Publishing.

Montgomery, K. C. (2007). *Generation digital: Politics, commerce, and childhood in the age of the Internet*. Cambridge, MS: MIT Press.

Palfrey, J., & Gasser, U. (2008). *Born digital:Understanding the first generation of digital natives. NewYork*. Basic Books.

Prensky, M. (2010). *Teaching digital natives: Partnering for real learning*. Thousand Oaks, CA: Corwin Press.

Ryu, H., & Parsons, D. (Eds.). (2009). *Innovative mobile learning: Techniques and technologies*. Hershey, PA: IGI Global.

Sharma, S., & Kitchens, F. (2004). Web services architecture for m-learning. *Electronic Journal of e-Learning, 2,* Article 2. Retrieved March 4, 2008, from http://www.ejel.org/volume-2/vol2-issue1/issue1-art2.htm

Small, G., & Vorgan, G. (2008). *iBrain: Surviving the technological alteration of the modern mind.* Toronto, ON: Harper.

Tapscott, D. (1998). *Growing up digital: The rise of the Net Generation.* Toronto: McGraw-Hill.

Thomas, M. (2006). Book Review-mLearning: Mobile Learning and Performance in the Palm of your Hand. Retrieved January 30, 2008 from http://kt.flexiblelearning.net.au/tkt2006/ edition-9/book-reviewmlearning-mobile- learning-and-performance-in-the- palm-of-your-hand

Tomei, L. (2009). (Ed.). *Information Communication Technologies for enhanced education and learning: Advanced applications and developments*. Hershey, PA: IGI Global.

United Nations Department of Economic and Social Affairs. (2008, August). *UN Millennial Goals Report*. Retrieved January 20, 2010 from http://mdgs.un.org/unsd/mdg/Resources/ Static/Products/Progress2008/MDG_Report _2008_En.pdf

Webster, J., & Watson, R. T. (2002). Analyzing the past to prepare for the future: Writing a literature review [Electronic version]. *Management Information Systems Quarterly*, *26*(2), xiii–xxiii.

KEY TERMS AND DEFINITIONS

Bluetooth: A radio technology enabling devices such as mobile phones, laptop computers and printers to communicate with each other.

G3: High speed mobile communications network which allows for mobile video.

General Packet Radio Service (GPRS): A high speed form of wireless communication.

Global Positioning System (GPS): Based on satellite systems to provide positioning and navigation information.

Internet Explorer Mobile: Microsoft's version of Internet Explorer for cell phones and PDAs.

iPod: A small and portable media player, capable of playing digital audio, and with some versions, digital video.

iTunes: A digital media player application used for playing and organizing digital music and video files.

M-Learning: The use of mobile devices such as cell phones, PDAs, and iPods as educational tools. Also known as mLearning, mobile learning.

MMS: Multimedia messaging service used to exchange multimedia content such as photographs and videos between mobile devices.

MP3: A compressed digital music format for the transfer and playback of music on small digital music players such as iPods.

Personal Digital Assistant (PDA): Such as Palm Treo or Pocket PC. Some PDAs are now known as smartphones, meaning that the lines are blurring between PDAs and cellphones.

Podcast: A series of audio or video digital media files which are downloaded from the internet.

Real Simple Syndication (RSS): Which is a web feed format to publish frequently updated works such as blogs and podcasts.

Short Message Service (SMS): Also known as texting.

Smartphone: A cell phone with extended capabilities conducive to m-learning, such as high speed Wi-Fi for internet browsing and email.

Wireless Application Protocol (WAP): The wireless communication environment used for mobile web accessed from a mobile phone or PDA.

Wi-Fi: Wireless technology used for cell phones and PDAs.

Windows Mobile: A compact operating system designed for use with mobile devices such as smartphones and PDAs. Included with Windows Mobile can be Office Mobile, which is a suite of mobile versions of Word, Excel, PowerPoint, and Outlook

APPENDIX A. NUMERICAL LISTING OF M-LEARNING ARTICLES

Table 3.

	Author	Title	Field	Country	Hardware	Software
1	Moura, A. & Carvalho, A.	Mobile Learning: teaching and learning with mobile phones and podcasts	Education	Portugal	Mobile phone Mp3 player	MP3 software
2	Manhattan Research	Health in the Palm of Your Hand	Medicine	U.S.	PDA Smartphone	Epocrates MerckManual 5-Minute Clinical Consult
3	Thornton, P. & Houser, C.	Using mobile phones in English education in Japan	Education	Japan	Mobile phone PDA	Email Web Video
4	Sharma, S. & Kitchens, F.	Web services architecture for m-learning	Education	U.S.	Mobile phone PDA	Email Web Video
5	Lu, H., Lin, J., Lin, C., & Su, K.	A study of the construction of a mobile learning oriented mathematics learning activity	Education	Taiwan	Smartphone PDA	Web Arcade game
6	Trifonova, A.	Mobile learning: review of the literature	Education	U.K. Finland U.S. Taiwan	Mobile phone PDA	SMS MMS Web
7	Prensky, M.	What can you learn from a cell phone? Almost anything!	Education	U.S. Japan Europe	Mobile phone	Web GPS SMS
8	Metcalf, D. & De Marco, J.	mLearning: Mobile learning and performance in the palm of your hand	Business	U.S.	Smartphone PDA	Game software Web Email
9	Osborne, H.	In other words...Using text messages to improve medication adherence	Medicine	U.S.	Mobile phone	SMS
10	Kumar, S. and Zahn, C.	Mobile communications: evolution and impact on business operations	Business	Japan, U.S.	Mobile phone	Email Web SMS

continued on following page

Table 3. continued

	Author	Title	Field	Country	Hardware	Software
11	Kleijnen, M., Wetzels, M., & de Ruyter, K.	Consumer acceptance of wireless finance	Business	The Netherlands	Mobile phone	Web
12	Boardman, D., Casalegno, F., McMurray, B., & Pomeroy, S.	Rethinking the mobile social shopping experience	Business	Spain	Smartphone	Facebook ToT software
13	Wagner, E. & Wilson, P.	Why learning professionals need to care about mobile learning	Business	U.S.	Smartphone iPod	Web
14	Adatia, F. & Bedard, P.	Palm reading: 2. Handheld software for physicians	Medicine	Canada	PDA Smartphone Pocket PC	Web Email Various types of medical software Microsoft Palm OS
15	Luo, J.	Portable Computing in Psychiatry	Medicine	Canada	PDA	Web Various types of medical software Microsoft Pocket PC Palm OS Adobe PDF Reader
16	Maag, M.	iPod, uPod? An emerging mobile learning tool in nursing education and students' satisfaction	Medicine	U.S.	iPod	iTunes PowerPoint Web RSS feeds
17	Scherr, D., Zweiker, R., Kollmann, A., Kastner, P., Schreier, G., and Fruhwald, F. M.	Mobile phone-based surveillance of cardiac patients at home	Medicine	Austria	Mobile phone	Web Email SMS

Chapter 2
Producing Generic Principles and Pedagogies for Mobile Learning:
A Rigorous Five Part model

Davina Calbraith
Anglia Ruskin University, UK

Reg Dennick
University of Nottingham, UK

ABSTRACT

This chapter outlines how innovative research methods were developed (Calbraith, 2010), and how the model described in this chapter was based on (and adapted from) comprehensive research concerning learning objects (Calbraith & Dennick, 2009). It describes how the model was designed and developed to create a robust foundation on which to build rigorous research-based content for mobile learning. Taking a step-by-step approach it describes how reliable pedagogies were formed, how subsequent research testing distilled factors noted from this method into both unique and generic pedagogical principles, and how the principles formed can be used in any context or discipline to produce effective and enjoyable learning. The authors include analysis of a worked example using this approach (in this instance from Nursing) in order to illustrate how each stage of the model may be performed, and to make clear how the process may be replicated and incorporated into many different settings.

INTRODUCTION

This new mobile learning model was developed using an original systematic review method, grounded theory to develop emergent theory, and comprehensive research/usability testing on seven specially designed Learning Objects.

Using a step-by-step guide, the aim of this chapter is to practically highlight how this model was formed, the methods used, and most importantly

DOI: 10.4018/978-1-60960-511-7.ch002

how it can be replicated. It may be used either in part (to produce effective Learning Objects for mobile devices), or in full (to produce rigorous teaching and learning pedagogies within mobile learning environments). To this end, a practical example has been taken from an Intensive Care Nursing (ICU) setting from one of the author's pool of research studies. This ICU example is an exemplar in that it illustrates how the process may be applied. Although this is taken from ICU, generic parallels and implications for this model's use within other disciplines/settings are possible, and are therefore outlined and discussed. (N.B. All research studies used in the development of this model and the ICU example have full ethics approval from two universities and two Health-care Trusts in the UK. Further details available on request).

Peters (2007) defines mobile learning (or 'MLearning') as learning performed on handheld and desktop devices that are portable, interactive, connected and individual. This therefore includes Personal Digital Assistants (PDAs), laptops, mobile phones, smartphones, iPhones, and iPad. Kukulsha-Hulme and Traxler (2007) describe mobile learning as 'learning that is not time or space dependent' and note that it can be "informal, unobtrusive, ubiquitous and disruptive" (p1492) Kukulsha-Hulme and Petit (2008) therefore underline the need for the definition to take into account social and philosophical dimensions. For the purposes of this model, the working definition of mobile learning will therefore be 'any type of learning performed on a mobile device unconstrained by time or place'.

The term PDA (personal digital assistant), first used in 1992, is defined as a mobile device which functions as a personal information manager with the ability to connect with the internet. Many employ touch screen technology.

Nusca (2009) defines 'Smartphones' as "mobile phones that offer more advanced computing ability and connectivity than a basic feature phone" (i.e. they run complete operating system software and provide a platform for application developers - PC Magazine, 2010). The first smartphone (with touch screen and predictive text) was released publicly in 1993. The first phone to call itself a smartphone was the Ericsson R380 in 2000. Most smartphones (especially HTC ones) now have touch screen and/or stylus, 3G, windows media player, and mobile phones capabilities (Microsoft, 2010).

The term 'iPhone' (first used in 2007) is 'an internet and multimedia enabled smartphone'. Designed and marketed by Apple Inc, they use a multi-touch screen, allow third party applications with diverse functionalities. They have been described as 'PDA/cell phone hybrids' (CNET, 2010). The 'Blackberry', introduced in 2002, was the first smartphone optimised for wireless e-mail (CRN, 2010). iPad is the latest addition (released 2010) which is a 'smartphone/laptop hybrid' (Paczkowski, 2010, Arrington, 2010). It has Wi-Fi, 3G and 'multi-touch' capabilities i.e. finger-tip sensitive LCD (Martin 2010, Eaton 2010, Topolsky 2010)

For the purposes of this chapter the focus will be on Learning Objects *only* (within mobile learning) using PDAs, Smartphones and iPhones as described above. The working definition of a learning object will be a digital learning package "that addresses one clearly identifiable topic or learning outcome and has the potential to be reused in different contexts" (Weller et al 2003).

In short, this chapter describes how educators can build a process that ensures effective learning takes place anytime and anywhere using one clearly identifiable topic or learning outcome.

BACKGROUND

Mobile learning has enjoyed rapid expansion over the last few years and advancement of new technologies has undoubtedly influenced this. The United Nations predicted that the world's population would reach 6.3 billion by the end of

2009, and that 41% would be carrying mobile phones (Hall, 2009). Although ownership of mobile devices does not necessarily mean that mobile learning automatically follows, it does mean that the opportunity for it to happen exists. Given the rapid adoption of mobile technology in our everyday lives it is hardly surprising that mobile technology has 'spilled over' into learning environments.

In recent years diverse (and often dichotomous) information and practice has emerged concerning mobile learning which has 'complicated' emerging potential learning object (LO) pedagogies – i.e. some educators have taken up the challenge to develop LO pedagogies by either (i) developing the pedagogy as they go (constructivist-type approaches) or (ii) deconstructing existing effective pedagogical strategies with high 'face validity'. Unfortunately both approaches have met with limited success (with regard to the integration of desired pedagogies with effective evaluation strategies) due to 'pedagogical barriers' (Felix, 2005; Calbraith & Dennick, 2009). This has left educators wondering (i) why their learning packages are difficult or impossible to evaluate, and (ii) asking the question 'Where did I go wrong?" (Calbraith, 2009). Traxler (2009, p19) recognises the difficulty in developing evaluation strategies and states: "There is a need for a more comprehensive, eclectic, and structured approach to evaluation based on sound and transparent principles". The proposed model rectifies these problems as it allows pedagogy and evaluation strategies to develop 'hand-in-hand', and develops principles.

Some educators have recognized that a more systematic approach to mobile education is required (Neumeier, 2005), and specifically that "mobile devices have not yet been fully exploited nor their limitations addressed" (Cooper & Shufflebotham, 1995, p1). Others agree: "While the use of mobile devices has sparked the interest of an increasing number of researchers in recent years our knowledge of learners' preferences for the mobile platform and their usage patterns remains limited" (Stockwell, 2008, p253). In an effort to gain further insight, several usability studies and cohorts of user interviews were incorporated into the model's development.

THE MAIN FOCUS OF THE CHAPTER

A brief history and explanation of how the process was initially constructed is outlined, followed by the model's 5 component parts. For ease of understanding the reader will be taken through these sequentially. This is then followed by a description of how pedagogies and evaluation strategies were developed, how generic principles were formed, and how they may be used for new mobile learning contexts. Alongside these components parts, constant comparisons will be taken from an ICU learning example in order for the reader to see how their own discipline/area of interest may be substituted and the model applied.

How the Process was Constructed

The model (presented later in this chapter) is based on learning object research undertaken between 2004 to 2009. This was developed by means of

1. comprehensive research/testing (usability studies, computer package testing on seven specially designed Learning Objects, observation, questionnaire and interview techniques performed on several randomised samples of students and staff from a variety of disciplines. Research from each of these is reported elsewhere in detail (Calbraith, 2010). Data from each pedagogical and evaluative approach discovered during this process was analyzed, and an evaluation made as to whether the emergent strategies 'held true' across different disciplines, levels of practice, gender, location, and type of use. Generic principles were found and developed further using different disciplines.

2. a grounded theory approach (to develop emergent theory on every pedagogy/evaluation strategy found to be reliable in the review - i.e. those that were robust in creation/application and seen to be effective in practice); and
3. a new systematic review method with meta-analysis (developed according to the level of academic maturity found within this topic). It was gathered across a variety of disciplines to uncover the most reliable pedagogy/evaluation strategies and practice when using Learning Objects within mobile learning as a whole.

This Chapter's Method and Model

This model has been built using the above method and has five component parts:

Part 1: Usability testing
Part 2: Grounded theory
Part 3: Systematic review of mobile learning research literature relevant to area in question. For the purposes of this chapter, both generic and Intensive Care Nursing (ICU) mobile learning literature will be used.
Part 4: Grounded theory hypothesis testing
Part 5: The formation of generic principles

Each part will now be described in detail followed by the model in order to practically illustrate how new principles formed may be used.

Part 1: Usability Testing

As no usability questionnaire specifically established for mobile learning existed, a questionnaire based on components deemed to be important for all types of e-learning (according to Krug, 2006; The TLT Group, 2010; and Calbraith, 2010) was drawn up and the authors' research questions incorporated. The usability study consisted of observation, questionnaire, and interview.

During development, the model was tested on a randomly selected and blinded population of 57 participants drawn from nursing, medicine, science, and IT backgrounds and 'lay people' (those with no experience of these fields). This included adults aged 18-58 years from all disciplines and levels of learning. Field notes were taken from participant observations regarding 7 different Learning Objects (LOs) and collated with participants' verbatim comments.

General findings were that all participants (i) felt that navigation through LOs were good, (ii) liked the overall look of the LOs, (iii) felt the content was clear & well-organised, and (iv) felt that images loaded well. 12.5% wanted extra links. Differences found in the 'Lay people' group were that they (i) were the most critical group regarding the LOs, (ii) noticed pictures first, (iii) Men 'hopped around' the LOs, whereas women took a more linear route. In the Science teacher group: (i) Titles were noticed first, and (ii) there were less comments on the LO pedagogy. There were no specific differences in the IT group but the nursing group noticed images first and 25% suggested the inclusion of links to white papers, etc.

Differences between disciplines were due to learning preferences not pedagogy/evaluation issues and therefore deemed negligible. Negative points raised were minor and easily remedied (i.e. size of text in some places, etc). Apart from these, the LOs were deemed to be of sufficient quality to test evaluation methods adequately providing that consideration for specific relevance to specific subjects was incorporated. It was already clear at this point that there may be some generic principles at work as the main findings were similar across different disciplines, location, and in most cases, gender.

A usability study is necessary to ensure that (i) mobile learning is appropriate for the required context, and (ii) to gain data required for hypothesis testing for the next step. It should be noted that if a shorter version of this model is required, (i.e. the user already has usability data) it is possible to start at this point of the process having answered all the systematic review questions in the model.

Part 2: Grounded Theory

Participants' statements were taken from the usability studies and were put through Glaser and Strauss' grounded theory approach using the constant comparative method. The authors' findings included the following:

- 70% of participants preferred information presented as images/animations whilst the remaining 30% preferred written/text elements (summarised information, simple terminology/ definitions);
- 97% of participants liked 2 or more elements per screen to help focus learning and prevent boredom;
- 50% of participants preferred the inclusion of test elements, the remaining 50% appeared to favour any method that aided the practical application of learning;
- Minor changes wanted in elements were equally distributed between visual (i.e. larger pictures) and written elements (i.e. less/more text, more labelling);
- Most participants felt navigation through the learning material on-screen was good. Remaining negative comments were minor;
- Comments about content of diagrams were all positive - a small percentage of participants (3%) wanted slightly larger images.

The final grounded theory stages incorporate 'hypothesis testing' - i.e. (i) Projects with large numbers and same hypothesis; (ii) Confirming instances and their conditions; (iii) Disproving instances and their conditions; (iv) Central propositions, variables, & dimensions; and (v) Situations that push variables to their limit (i.e. do original effects hold true?). Hypotheses produced by the authors' usability testing included:

- Mixed elements within learning objects increases knowledge, interest and achieves a good 'element-interaction' balance;
- Layout that is not in line with participant learning preferences results in section links not being detected;
- Learning objectives that can be used to judge the level of learning attained result in participants believing the information is of good quality and that they have progressed;
- Labels giving the right amount of information result in clarity and simplicity that aids navigation and does not overload the participant.

When the data was broken down (into disciplines, location and gender) the above hypotheses remained true in most cases. Projects with large numbers and similar hypotheses were sought but not found, however some confirming instances and their conditions were found e.g. (i) It has been well documented that multi-modal learning within e-learning increases knowledge retention, interaction and participant interest which appears to support the 'Mixed elements' finding; (ii) No literature has reported 'section links not being detected due to learning preferences' in mobile learning, however it is tentatively suggested that inserting specific page positions on each page may circumvent the need for all learning styles to be catered for; (iii) Although few papers comment specifically on mobile learning objectives, it is well documented that good, appropriate and measurable learning objectives result in effective learning; (iv) 'Good labels do not overload' – No specific nursing examples were found, however Steve Krug's publication 'Don't make me think' highlights this exact hypothesis and suggests that intuitive e-learning designs provide the most effective usability testing (Krug, 2006).

In summary, full hypothesis development was not possible with the ICU example as there were few specific mobile learning instances recorded with which to provide further confirming/disproving instances/conditions. Calbraith & Dennick's (2009) work was deemed to be sufficient as secure basis on which to build pedagogical and evaluation model for mobile learning because: (i) similar instances in mobile learning research mirror e-learning LO research findings; and (ii) the rigor/reliability of the mobile learning systematic review findings have obvious parallels.

Part 3: The Systematic Review Method

A systematic review of both generic and ICU specific mobile learning literature was undertaken. This model can be performed for use in any discipline due to the generic literature – however, model users should substitute their own specific mobile learning literature for the ICU literature added to the generic basis given here. Model users can utilise/adapt the systematic review protocols, data extraction sheets, and methodological rigor sheets developed for this method (available from the contact author on request). These documents were developed during extensive learning object research to allow a more comprehensive type of data collection, and a greater evaluation of systematic features than is usual in systematic review protocols. They were reviewed by a specialist panel, deemed to be appropriate, and adapted by the researcher to include specific aspects of mobile learning. Of course, most educators do not have the time to undertake a systematic review of their area therefore a literature review plus application of the systematic review's important generic factors and generic principles (outlined later in the chapter) may suffice.

Inclusion criteria incorporated 'any research study that could provide evaluative or pedagogical information on mobile learning'. Inclusion criteria were deliberately wide in order for the authors to extract generic principles for use in the final model. Consequently, material from any discipline, any age of participants, any location, any type/part of course, and all types and years of all journals were eligible for inclusion - providing that descriptions of advantages, disadvantages and effects of mobile learning were reported.

A general electronic search was made for all studies that contained mobile learning research including the terms 'Mobile learning', 'Mlearning', 'PDA's, 'laptops', 'handheld devices', 'mobile devices', 'Smartphones', 'mobile phones', and 'iPhones'. This was repeated in CINAHL. Plumbed, Medline, and BNI databases for specific ICU literature (Model users should substitute their specific database searches here). These searches immediately showed a paucity of published literature so a further search of 'grey' literature was performed. 41 studies were highlighted and full text versions gathered where possible. Of these, 22 studies were found to contain mobile learning research, but only 15 had the required inclusion criteria. A check concerning the effectiveness of literature searching and protocol rigor was performed (using a peer-reviewed custom-built rigor assessment sheet with criteria similar to SCIE's Knowledge Object rigor criteria). This was judged to be adequate taking into account the level of academic maturity present and the large narrative nature of the proposed data. These 15 studies were put forward for data extraction.

Preliminary Analysis

Not surprisingly, no Randomised or Case Controlled Trials were found. 4 observational studies were located, along with 3 Controlled 'before-and-after' studies (1 contemporaneous, 1 case-controlled, 1 site-controlled), 3 qualitative studies, 2 cohort studies, 1 review, 1 focus group report, and 1 audit. Baseline measurement and/or control group performance were not reported in many cases. Outcome measures included level of collaboration; transition of passive to active learning; demoting factors for mobile adoption; increase in knowledge; user learning roles; and the benefits and limitations of mobile technology.

Review findings are discussed in detail from this point on because they form part of the model illustrated later in this chapter. To demonstrate how these findings can be used to formulate tailor-made strategies/principles for blended learning in any environment, an ICU setting example is used. It is suggested that the reader may substitute their own research/context for this ICU example - the model is readily adaptable and will accommodate a range of contexts.

Discussion of Review Findings: Important Issues

Analysis of the systematic review revealed several generic factors that appeared to influence effective mobile learning: The importance of quick information, the importance of timely access to resources at the actual point of need, the changing role of educators, the changing role of students, potential changes to the delivery of training, constraints, broadening educational goals, and the increase of negotiated curricula. Each factor is discussed below - giving summaries of the general mobile learning literature, specific (ICU) mobile learning literature, and a discussion of how each applies to the chosen example (N.B. This example was based around i/smartphones. Although PDAs may not historically form part of mobile learning in some disciplines, they will be mentioned here as this is relevant historically to ICU mobile learning). Model users may work through these factors (set out below) substituting ICU for their own examples:

i. ***The importance of quick information***: Brandon Hall (2009) state that the workforce has become more mobile which has influenced how learning is delivered (e.g. quick information for travelling staff). Information sent can be viewed whilst waiting or travelling thereby making use of what would otherwise be 'dead' time.
   - Although smartphone examples are not available yet, use of PDAs since 2001 has been shown to enhance learning in the clinical environment by the rapid and effective acquisition of relevant information (White et al., 2005). This was obviously a very useful aspect in the intensive care setting where quick access to information is always paramount – i.e. checking procedures/drugs for a patient whose condition may deteriorate extremely quickly. The author's research appeared to support previous PDA findings.
   - Users of this model should therefore ask themselves – Exactly how important is quick information for my setting? This point lays hand-in-hand with

ii. ***The importance of timely access to resources at the actual point of need***: Learners in their everyday lives are becoming accustomed to constant 'information connection' via PDAs, mobile/smart phones and laptops (Alexander 2004, Farrell & Rose, 2008) - allowing access anytime and almost anywhere (Holzinger et al., 2005). Mobile devices have the ability to combine distinct functionalities in one device that is versatile, customizable and portable (Baumgart, 2005). Research has shown potential for increases in leadership skills and professional confidence by providing students with "reinforcement of core knowledge and evidence-based information in real time as required by the student" (White et al., 2005, p150). However, few studies describe the implementation of mobile devices within nursing practice and even fewer examine the effects on nursing education (Farrell & Rose, 2008).
   - PDAs in healthcare settings have, to date, provided resources at the patients' bedside, during patient consultation in clinic, and provided quick answers to practitioner queries. Specifically,

Farrell and Rose (2008) describe 'e-tensive care units' where students gained timely e-learning resources at the 'point of care'. No ICU studies report the use of smartphones, iPhones or iPads.

- The ICU nurses in this example were excited at the prospect of having timely on-the-spot resources and immediately saw the potential work benefits. They reported that there was often a queue for computer terminals which could only be accessed when the patient's condition and nursing 'cover' for their patient/bed space allowed and considered mobile devices a great advantage.
- Users of this model should therefore ask themselves – Exactly how important is timely access to resources at the actual point of need? This will differ according to each discipline and the circumstances. In certain instances timely access to resources is not just helpful but crucial. Clearly, permanent adoption of mobile technologies in any environment would require several changes – one of which may be the role of educators;

iii. ***The changing role of educators***: Since 2005, Nurse Educators have been using a variety of new learning technologies that increasingly focus on students (i.e. user-centred teaching). As a result educators have taken a much greater role in course design, and aiding students' collection and application of information in the workplace (Billings, 2005). Educators' responsibilities, according to Pachler et al (2010), include taking greater responsibility for helping students to get the most out of new technologies, e.g. reflective practices. Not surprisingly, as this is still a new concept in healthcare training/learning, educators in the ICU example were unaware of how their role may change. However, they were readily able to highlight potential practical issues concerning 'confidentiality' and 'security of information'. Obviously, cultural changes will be necessary in any context (Dearnley et al., 2008), but encouragingly, Miller et al (2005) were pleased to note that established clinical nurses (after having initial doubts) gave increasing support to students once they understood that mobile devices could be used to access drug information in real time.

- Users of this model should therefore ask themselves – Exactly how should the role of educator change in my context? In the ICU example, educators would first need to adopt a cultural change, embed this into teaching practices, then teach staff/students how to use mobile devices to access appropriate software (thereby taking on a larger facilitating role);

iv. ***The changing role of students***: Mobile devices allow students to "instantly construct their own learning for immediate application in real-world contexts" (Billings, 2005). Farrell and Rose (2008) undertook a pilot study to discover whether mobile devices would enhance students' pharmacological/clinical contextual knowledge, and to identify the effects of PDA use in clinical practice (N.B. this is the first example of published smartphone research available in this area). Here role of the learner changed due to having to choose (i) the most appropriate time and place to access mobile information, and (ii) the amount of times the same information was accessed. In short, it placed a greater emphasis on the degree of student choice and responsibility for learning.

- Participants in the ICU example loved being able to choose their own subject and time of learning - they saw this as an enormous asset.

- Users of this model should therefore ask themselves – How necessary is it and exactly how should the role of students change in my context?

v. ***Potential changes to the delivery of training***: It has been suggested that mobile devices have the potential to change how training is delivered. However, handheld devices have been evaluated in a variety of clinical environments since the 1990s (Farrell & Rose, 2008) but have *not* yet shown a dramatic or widespread change as to how nursing is practiced - despite the expectation in 2001 that handheld computers would "dramatically change the way physicians practice medicine" (De Ville, 2008, p385). Perhaps one reason for this may be explained by Neumeier's findings: "In the course of designing CAL-supported material, it had become evident that a systematic investigation into the factors that shape mobile and blended learning experience in the context of language learning and teaching was missing and urgently needed" (Neumeier 2005, p163). It was not just language education (nor indeed medical education) that this applied to, but education per se (Prosser, 2007). Mobile learning was also in need of rapid and systematic investigation of factors - the authors' research and this model seek to provide this.
   - In health care, real-time access to information at the bedside has "the potential to improve the quality and safety of care, thus reducing adverse events and improving patient health outcomes" (Farrell & Rose, 2008). However, most examples that detail changes to training delivery entail 'location' and do not involve people at all.
   - In the ICU example, nurses could see that there was potential for positive changes in training delivery providing that funding for devices was secured.
   - Users of this model should therefore ask themselves – Exactly how should the delivery of training change? What needs to be put in place to make this happen? This touches on other aspects that have an impact on the adoption of mobile learning, e.g. constraints;

vi. ***Constraints***: "The adoption of mobile learning has been constrained by slow networks, limited services, anaemic devices, and a hesitancy by organisations to purchase hardware that could soon be obsolete" (Hall, 2009). All of these constraints were found in the ICU example, and cost/updating of devices was highlighted as a concern regarding mobile learning becoming a long-term/permanent fixture. Buying mobile devices was not considered by the unit; however ICU nurses felt comfortable at the thought of using their own mobile/smart phones providing they had been professionally approved for use on the unit.
   - Users of this model should therefore ask themselves – What constraints are there to the implementation of mobile learning in my setting? How can they be solved?;

vii. ***Broadening educational goals***: Felix (2005) believes broadening educational goals (i.e. lifelong learning concepts) have played a part in the impetus and interest regarding adoption of mobile learning. Felix (2005) also believes that this inevitably leads to a 'social/cognitive constructivist paradigm' for learning and teaching, and highlights the pedagogical dilemmas present. In practice the authors have witnessed the difficulties that the development of pure constructivist approaches bring and have therefore sought to provide greater workable solutions. Both (i) deconstructing pedagogical e-learning approaches that work in practice and (ii) building pedagogical approaches in a linear fashion (i.e. developing them as you go along), have met with limited results and pedagogical barriers preventing further development (Calbraith & Dennick 2009). In the field of mobile learning it is therefore suggested that pedagogy should be developed *in conjunction* with the evaluation processes required instead of evaluation and pedagogy being developed or appraised separately, thereby 'ironing out' any pedagogical problems early in development (Calbraith & Dennick, 2009).

- In the ICU example, broadening educational goals were thought to bring both advantages and disadvantages i.e. greater choice of relevant learning but also greater insecurity on whether learning choices were really appropriate, whether they would replace well-used traditional methods, and whether bought-in learning would provide value for money.
- Users of this model should therefore ask themselves – How do broadening educational goals (i.e. lifelong learning ideals) affect my setting? What needs to be considered?

viii. ***The increase of negotiated curricula***: Kulusha-Hulme and Shield (2008) note that learning is no longer 'solely and carefully crafted by lecturers' due to learners being more mobile and motivated by their own learning needs. During the development of the research techniques and Learning Objects (LOs) for this model, a great deal of interaction and consultation took part with stakeholders, educators, staff and students regarding content to ensure it was highly relevant to the setting. A pilot study using participants'' learning preferences (Calbraith, 2010) also fed into this process. ICU nurses were consulted about whether they would use these types of LO and many participants responded in the following ways: 'This is a great way to learn'; 'It made you excited and *want* to know the answer'; 'It made me want to see the next bit of information'. When asked why they felt this way they said that it was because of (i) how the LOs were constructed, (ii) the use of immediate feedback which guided the user toward reasoned clinical decisions and motivated them to want to know more, and (iii) the fact that they could choose what were the most relevant aspects to learn within any given subject. They also liked the way the Scenarios developed further reasoning and application skills (Scenarios had been specifically designed to aid 'transferral of learned skills' to new contexts).

- Users of this model should therefore ask themselves – How does the increase of negotiated curricula affect my setting? What do stakeholders or users think? What needs to be considered?
- At this point, model users may wish to evaluate whether findings from their practice correlate well to these multi-discipline systematic review findings. Clearly the ICU practice examples (outlined above) correlated well. If the model users' setting does not correlate with the above factors a specific literature/systematic review in their setting may be required.

Discussion of Review Findings: Problems with Mobile Learning

During the course of the review several *disadvantages* of mobile learning were noted (and can be used to shape learning when applying the model to the model user's context):

i. Waycott and Kukulsha-Hulme (2003) found several limitations when using Palm m105 PDAs: small screen size for 'scan reading', new/difficult text navigation requirements, and awkward methods for taking notes i.e. no full keyboard. In particular, users found flipping between reading documents and writing notes cumbersome - having to perform these actions consecutively rather than simultaneously. Chehimi et al (2006) found small screen sizes problematic.

- In the randomised ICU example, this was evident. Some participants were concerned that night use may be difficult due to decreased lighting conditions, and some thought that new text navigation requirements may be difficult at times. Back in 1995, authors highlighted difficulties with screen size – "the small amount of visible text means that the reader loses track of where they are in the document" (Bartlette 1995, this is available at: www.research.digital.com/wrl/techreports/abstracts/TN-46.html) and the potential size of nested lists are therefore limited (Cooper & Shufflebotham 1995, p3). There has since been considerable changes in technology (e.g. text input can now be done via a virtual keyboard where letters on screen are tapped, or by external keyboard connected via USB, IR or Bluetooth, or by letter or word recognition where letters/words are put onto screen and then translated into letters currently activated in the field, or by stroke recognition - e.g. Palm's 'Graffiti' - using a set of predefined strokes represent various characters used in the input). However, screen/text/size remained a small concern with ICU participants but they felt that once they had got used to the new text navigation this would become easier and perhaps cease to be a problem at all.
- ICU nurses also found navigation to be a slight concern - several of them suggested that they would prefer a 'site map indication' somewhere on the screen. As long as this potential problem is born in mind when creating the information to be accessed (i.e. adding 'page 2 of 4' for example) this should not be a major problem. Cooper & Shufflebotham (1995) suggest 3 further ways to solve this problem – (i) Text can be replaced with 'StretchText' (text grows/shrinks according to user preference/use); (ii) text 'folding' (where more information is 'hidden', 'collapsed' or 'nested' underneath heading texts); and (iii) screen rotation (rotating the device by 90° changes the page orientation from landscape to portrait and vice versa) which is an option with some devices (e.g. iPad). iPhone development has added further 'remedies' - icons grow bigger on screen to show which 'app' has been selected, and 'pinch' techniques can be employed to navigate quicker;

ii. Goth et al (2006) report mobile device users 'ignoring' the environment. Kristofferson & Ljungberg (1999) believe 'focus' and 'attention' can be potentially problematic in mobile learning. Ignoring the environment in ICU could be catastrophic for patient care. However, at no time during testing of this model did the focus of the ICU nurse get 'stuck' with the learning device. Each bell, buzzer and change in patient status was picked up and acted on immediately. This may be explained by the following possible reasons:

a. mobile device use was not yet an accepted norm thereby users may regard it as 'having to slip it in where possible' during their work. If this is the case, this attitude may change if mobile device use was operating as part of a permanent and accepted practice;

b. an unspoken discomfort about using mobile devices in the ICU context – Wishart (2008, p359) noted that language students felt that they "could not disrupt the established practice with the novel technology" due to the socio-cultural environment of the placements. Stockwell (2008) too highlights some effects on the establishment that mobile learning brings indicating that *support* in the workplace may be paramount for successful use/implementation;

c. training/practice development on the unit encouraged a particularly good model of ICU nursing practice – i.e. finely-tuned recognition and response to patient needs; or

d. Goth et al's findings may not be generalisable to this context. Further research is required.

iii. Wishart (2008) acknowledges that trainees did not fully explore mobile device potential because they 'were not yet confident in their pedagogical identities'. This hints that 'standard pedagogies' (if they exist) would be useful and may increase user confidence. In ICU, there was a reticence to explore mobile device capabilities beyond what they had been told it could do and how it should be used for the purposes of the research;

iv. Chehimi et al (2006) found that mobile devices are limited to primitive battery power. Wishart (2008) found that some PDAs lose stored content/applications when the battery runs down thereby requiring the re-installation of applications (e.g. Toshiba Pocket PC). Wishart therefore suggests one hour maximum usage when being deployed in wireless environments thus limiting continuous use (Ganger & Jackson, 2003). Since 2008 improvements to battery life have been made, therefore more reliable devices may be used to get around this problem. Alternatively, devices in ICU could be used for limited periods (between 5-30 minutes at a time) and batteries could be recharged on a constant basis using recharging 'cradles' (like ICU current practice for other well-used devices);

v. Cooper and Shufflebotham (1995) noted that bandwidth may limit speed/access of information retrieval. Kukulska-Hulme and Pettit (2007a) state that WiFi connection can sometimes be difficult if not impossible. Whilst acknowledged to be true, the ICU example did not have a problem with speed or access using laptops but some did using mobile phones. It should be noted that in Miller et al's (2005) study speed of information access and readability were factors that *increased student satisfaction*:

vi. Chehimi et al., (2006) stated that mobile mediums are limited to diminutive processing power and parameterised memory. In this example, this was not a problem, however mobile devices can be plugged into other equipment allowing 'higher specification peripherals' if the learning content requires greater power - e.g. using serial ports and/or USB cables (Cooper & Shufflebotham, 1995). Conversely, an extension card with an Ethernet port and/or RJ-45 adaptor can be used (Wikipedia 2010). Extending the capabilities of desktop computers using pen drives has become normal practice - the only difference is that mobile learning devices are not yet 'everyday equipment' for most so the practice is not yet familiar;

vii. Finally, and most importantly, there is a potential for security/privacy issues to arise between users (i.e. confidential patient information stored) or if using wireless (i.e. due to exchange of data transmitted between client and WAP). This can be overcome by ensuring a private connection wireless link (Ganger & Jackson, 2003). Although none of these potential problems were actual problems in ICU it is noted that using other material may have presented such problems. Teaching staff quickly identified that security/confidentiality issues would have to be adequate - to deal with the new technology to prevent breaches in care, cheating during on-line tests, or staff attainment of competencies. Wilkinson et al (2006) have an easy way to prevent such problems - make a database of the IP addresses all of computers in the required setting that will be used to perform the required function (i.e. test/exam), and combine this with a computer system where the server is able to reject log-in requests from unauthorised persons or prevent re-entry to an exam paper once the page is 'exited'.

There was one specific ICU concern not found in the generic literature i.e. some nurses were initially concerned about whether mobile devices could potentially interfere with pumps, monitors and equipment. It was explained that compatibility of equipment was paramount for patient safety and advice from a unit technician would be sought (and therefore would be assured for this type of learning). For healthcare settings it is suggested that a formal risk assessment should be made each time a different use or make of mobile device is required.

Discussion of Review Findings: Advantages of Mobile Learning

During the course of the review several *advantages* of mobile learning were noted:

i. Waycott, J. (2002) Found that viewing applications at the touch of a finger/stylus was quick and easy for users. ICU directorate nurses all used keyboards but were asked about their thoughts concerning finger/stylus use. Even nurses that had not previously used Smartphones felt that choosing 'apps' and web-links with fingers/stylus would be relatively easy (providing they had received a session on how to use them);

ii. Some devices can communicate with other local PDAs/computers by 'beaming' or 'synchronising' data (Cooper & Shufflebotham, 1995). Synchronisation software can be used – e.g. iSync can be used on MacOS X, HotSynch Manager (for Palm OS Handhelds), and Microsoft ActiveSync for Windows XP (or Windows mobile device centre for Windows Vista) when using windows mobile or pocket PC. Information can be stored using Microsoft Outlook or 'ACT!'. The ICU nurses were familiar with similar methods for other contexts – e.g. BARS system for blood labels. However, it should be noted that if the device uses bluetooth technology, bluetooth compatibility should be checked and maintained if it is not to become a major barrier;

iii. When used in language teaching, Nah, White and Sussex (2008) found that mobile phones enhanced listening skills and encouraged students to actively engage and participate in the learning. The ICU nurses were actively engaged - it is suggested that any method that promotes active engagement has merits;

iv. Waycott, J. (2002) found that the portability of mobile devices was a great asset. Although the ICU example did not use PDAs, they did use mobile phones, Smartphones and laptops in the course of the research and felt that portability would allow bedside use, outside work, and whilst travelling if necessary. Indeed, these devices "are constantly available to their users due to them being 'personal' and constantly to hand" (Cooper & Shufflebotham, 1995);

v. The main way of navigation is for mobile learners to select the links they are interested in, this said, "the simple interface does not appear to get in the way of the task" (Gessler

& Kotulla, 1994 available at: http://www.ncsa.uiuc.edu/SDG/IT94/Proceedings/DDay/gessler/wwwpda.html). Kukulska-Hulme and Pettit (2007) describe the main advantage of Smartphones as being 'self-service' education – i.e. led by students' own learning needs with tutor support available. Cooper & Shufflebotham (1995) believe this is particularly true if the device "is set up to perform certain actions for its user without the need for prompting or filling in information before access to learning material is gained on behalf of the user";

vi. Brandon Hall (2009) suggest that most learners already own mobile phones so there is already a psychological advantage for them regarding 'ownership' of the learning – i.e. users see themselves as directing their own learning. In the ICU example the users expressed great enthusiasm, however further debate/research is required to establish whether a greater level of ownership of learning is gained using 'self-owned' versus 'given/loaned for specific purpose' devices, and whether users have similar ownership of learning when the same learning choices are presented on laptops and computers;

vii. Chehimi et al (2006) talk about 3D applications for Smartphones. It is easy to envisage how 3D graphics could aid healthcare learning, particularly for anatomy and physiology. This would have undoubtedly enhanced LO images used in the ICU example;

viii. Wishart (2006) says that all handheld recording methods were popular, and students particularly valued the ability to capture 'on-the-spot' events and reflections through video recording. This facility was not used in the ICU example permitting no direct comparison. However, it is easy to see how consented recording of patient assessments could be used as evidence of competencies gained. It is suggested that ethical consideration of patients' feelings and dignity should be considered at length if this kind of use is chosen as many patients or relatives may find this type of 'care' a little too intrusive during their time of crisis.

Recommendations

In conclusion – taking the important issues highlighted from the systematic review (and both disadvantages and advantages) into account - it is clear that there should not be major problems with mobile learning ***providing that:***

i. the devices, software, text size and content are all carefully considered and specifically designed for mobile use;

ii. content is chosen according to the specific intended use;

iii. the devices are installed, monitored and serviced according to health and safety regulations (and in healthcare settings patients/relatives should give informed consent).

Indeed, many of the potential problems listed here did not become problems in practice when researched. Clearly mobile device advantages have not yet been fully exploited either. Unexploited possibilities undoubtedly have the potential to change dramatically the way tools are used (Waycott 2002).

Part 4: Grounded Theory Hypothesis Testing

Having followed previous parts of this model, this final part of the model provides tailor-made, reliable and context-relevant principles with which to create effective, practice-based mobile learning packages for the required learning context.

It was important to see whether each hypothesis gathered from the usability study held true. Hence, the two top performing learning approaches (from Part 2) were taken together with two low rigor approaches (i.e. Qualitative studies using 'experiential' and 'just in time' learning) and were each given a null hypothesis – i.e. there will be no difference between the 4 approaches (i.e. levels of practice, disciplines, gender, location). In this example, scenarios were aligned with the LOs (Borau & Ullrich 2009) and specifically designed to help users transfer learned skills into different clinical contexts. Carroll (2000, p87) states that "scenarios are central for understanding people's needs and the ways they use the technology". Romero and Wareham (2009) highlight the question we should be asking 'What type of learning do we want from mobile technologies?' They compare 'permanent behavioural change' versus 'speedy problem-solving'. It is suggested that ICU environments require both – speedy knowledge to deal with immediate problems but new information should then become part of practitioners' established practice.

35 ICU nurses were randomly selected from 110 to participate in the study – 31 agreed. Each participant was blinded and randomised to one of the 4 approaches. Lesser degrees of student learning satisfaction was noted on the 2 low rigor approaches, and knowledge mean score ratings were 30-40% higher when using the higher performing approaches. All hypotheses were compared and were found to link very accurately between systematic review findings and usability studies.

Part 5: How Generic Principles were Formed

All factors from the systematic review were considered and fed into the development of an original principles. Romiszowski (1992) describes three types of communication within *The Instructional Process*: (i) From the Instructional system to the learner (i.e. information to be learned); (ii) From learner to the Instructional system (information on the learning progress); (iii) from the Instructional system to the learner (i.e. feedback information or correction). However, on the grounds of experience, the authors would argue that a fourth type is becoming increasingly relevent: i.e. information on students learning experience (From learner to the Instructional system). Laurillard outlines similar thoughts (HEA 2005). By its very nature, the instructional process has to be informed by practice if it is to be used to its best advantage (i.e. observing how LOs work in practice, what the tutors using them think, what learning package designers think, and what students think). The evaluation framework must show that it is both rooted and functional in practice. The authors' primary aim, therefore, is for mobile learning to be 'practice-driven' i.e. to draw on and be effective in practice.

Generic principles were distilled directly from good pedagogy and evaluation as discovered by the previous parts of this model. E.g. Participants thought good pedagogy (i) "makes you prioritise your care and use it like in real life", (ii) "is clever because combines scientific knowledge and nursing practice", (iii) "helps you know what you've learned", (iv) "reminds you of what you've forgotten and helps you re-apply it", and (v) "is interesting and keeps your attention active which is easier than learning from books". Participants thought good evaluation was when (i) both coursework and summative assessment are used as "this is good balance for assessment", (ii) when the evaluation includes questions as "this retains knowledge", (iii) when the evaluation guides users towards reasoned answers "as the information given can be used practically", (iv) when the evaluation has MCQs "as this requires reasoned answers which can be built upon", (v) when the evaluation shows that new learning has built upon existing knowledge; (vi) when the evaluation itself aids application of the new knowledge, and (vii) when evaluation is set out using various different contexts/methods as "it focuses the attention on important parts of the learning". Evaluation and pedagogical principles were formed directly from these main findings as follows:

Effective mobile evaluation includes *(Generic principles):*

- Both formative and summative assessment to achieve a good balance
- Questions to aid knowledge retention
- Guidance of users towards reasoned answers
- A demonstration that new learning has built upon existing knowledge
- The easy application of new knowledge
- Focused attention on the important parts of the learning

Effective mobile pedagogy *(Generic principles):*

- Reminds the user what they have learned and how to apply it
- Keeps attention active

(Principles unique to discipline – in this example, Nursing):

- Encourages the same prioritisation of care as in real-life
- Combines scientific knowledge and nursing practice

THE MODEL

In 2007, the authors felt that usability and the observation of technology interaction was an integral part of any mobile/e-learning model. Waycott et al (2005) found that some users adapt tools to their everyday practice and preferences. Clough et al (2007) stress the importance of incorporating this into the design of mobile learning. The model was designed with this in mind and tested thoroughly in 2009 (See Figure 1).

USING THE MODEL

As already indicated, several uses of this model are possible. If the creation of mobile learning principles and model for new contexts is required it is suggested that the full process (i.e. all 5 parts) should be performed using either a systematic/narrative review or literature search. If usability testing has already been performed (and this usability study addresses similar questions to regarding navigation, platform, aesthetics, and student preferences) the user can start using the model from Part 2 onwards.

When using this model for other contexts the following are useful tips:

- Evaluate strengths/weaknesses of each framework/conceptual approach and define (where possible) what conditions are necessary for them to be effective;
- Establish whether the approaches/frameworks can be used in entirity or only in part;
- Establish a rationale for use;
- Consider the following – (i) How flexible does the approach/framework need to be? Under what circumstances? (ii) Are any reviews or models currently available? If they exist, how well do they perform? What is lacking? (N.B. Sheard & Markham 2005 have stated the need for evaluation of *any* web-based learning evaluation to encompass both the educational process and that associated with the functional usability of the technology).
- What can previous research show concerning effective evaluation/pedagogical frameworks that is applicable to the new context?
- Do certain frameworks/conceptual approaches seem to be effective?
- How do the answers to the above questions inform the development of a useful evaluation/pedaogical framework in the new context?
- How do the answers to the above questions inform the theory and practice of mobile learning as a whole?

Figure 1.

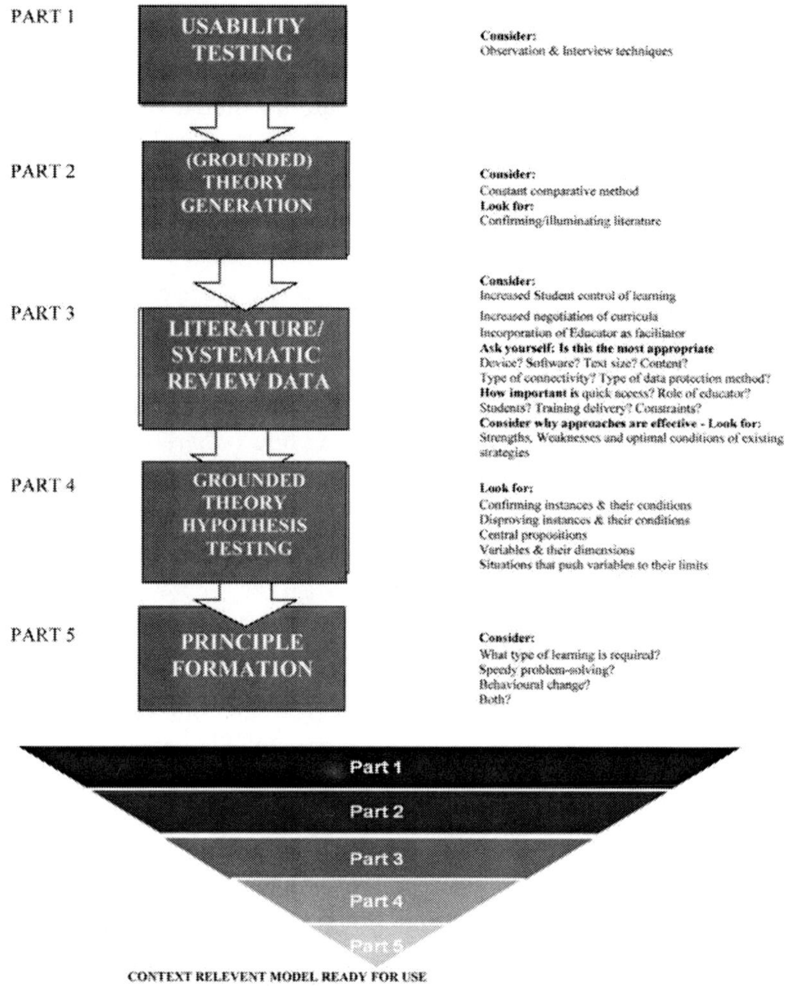

In this example, the 5 point approach resulted in 2 high scoring strategies – 'collaborative' learning and 'constructivist' learning. When these approaches were researched with different evaluation strategies and disciplines they continued to score well and were therefore chosen to be used for the next stage of this model. Model users should choose strategies which consistently score well in their discipline across different mobile contexts using this approach.

The systematic review had highlighted ***which*** pedagogical and evaluation strategies were performing well in practice and discovered some caveats for effective use, however, as expected it had not discovered exactly ***why*** these approaches were successful. Hence, an ICU usability study was created to observe how participants used devices/laptops, to ascertain their thought processes, and to check that operational features would not confuse the research with confounding variables. A grounded theory approach was employed to develop findings further.

Once generic principles are gained, this model may be used (irrespective of the mobile device) for multiple contexts providing that the objective is the same: student computer sessions, via projector as part of a lecture, on-line as part of student exam revision or distance education module, asynchronised as part of continued professional development, or as quick competency guides for the rapid integration of new members of staff. An illustration, of how principles may be used in an asynchronised distance education package, for example, is as follows:

- Create the learning package and (i) draw attention to important parts of the learning (using flashing lights, bright colour, underlined text, etc); (ii) Guide users towards reasoned answers (using feedback, giving direct information, etc); (iii) keep attention active (using change of stimuli, etc);
- Tell students that (i) working through the on-line package will provide formative self-assessment/ evaluation due to the questions/test elements included which aid knowledge retention; (ii) they can do this package as many times as they want in preparation for summative evaluation;
- Tell the students they will be expected to partake in an on-line scenario (that combines scientific knowledge and nursing practice) in which they will have to prioritise care using knowledge presented in the package.

FUTURE RESEARCH DIRECTIONS

Despite all provisions made during the research to ensure generic principles are effective in most cases, it is anticipated that 'individual-institution mismatches' are still possible when using the model. Pachler et al (2010) warn about potential disconnection between the way students live their lives and the way educational institutions interact with them. With this in mind, several 'blended' directions are suggested:

i. Students may be familiar with 'ask-the-audience' questions on TV quiz shows. PDAs could be employed during lectures for students to answer multiple choice questions (Ganger and Jackson 2003) and results collated and projected in real time for all to see. Lecturers could respond immediately to general areas of weakness, give feedback, or use this method to formatively assess. Shared views, attitudes, and misconceptions could be discussed;
ii. iPods, PDAs, mobile phones and MP3 players (with playback facility), could be used to listen to missed lectures (Ferneley 2010);
iii. Downloads of videoed cases, scenarios, skill demonstrations, exemplars/guidelines could be available for students to access/watch (via podcast etc);
iv. Robinson et al (2010) have devised a way to combine automatic monitoring with Java Smartphones for real-time monitoring of chemistry experiments whilst away from the lab. Similar monitoring could be set up – i.e. remote real-time monitoring of 'Sim-Man' patients;
v. PDAs could be used in remote locations for exam taking - Ganger & Jackson (2003) used such a method where E-mails of student performance were sent to tutors automatically (making immediate feedback to students possible). Smartphones could also be used by connecting to a specified secure web-page/ site to answer questions;
vi. Bridge & Ginsburg (2001) discuss the use of PDAs to evaluate students during their clinical placements which are then downloaded via 'cradles'. Students could use PDAs on clinical practice then download the data back at university. Providing mentors agree and witness the learning, this could be used as an original way to provide evidence of student practice, or act as a competency 'sign-off sheet' which could then be collated by tutors;
vii. Researched scenarios in this model were very effective and students really enjoyed them. Further scenarios could be offered where SMS text messaging takes place

as part of the learning pedagogy - E.g. on top of a mountain during an earthquake. Information, medical histories and instructions about patients could texted in stages following real life examples.

The possibilities appear endless. Blended learning factors found to date generally complement rather than replace existing resources when using mobile/Smartphones and PDAs (Waycott, 2002). No published research concerning iPhones or iPads is currently available. Billings (2005) believes the real issue revolves around how learning technologies can be used to *improve* student learning and academic program outcomes. Adams et al (2009, p5) warn "good teaching and engaged learning should not be determined by the use of certain instructional tools but by the guiding principle that learning is an active and recursive process where knowledge must be contextualised to be relevant to the learner". In the ICU sample, learners said that the Learning Objects created in this way motivated them to see and learn more, and made them excited about learning. The most important point is that the pedagogy used for each piece of mobile learning should not only deliver all that it needs to with regard to information but it should be relevant, immediately useful to the learner, and guide their learning using informed reasoning.

CONCLUSION

The challenge of the next decade and beyond is huge. The key is to find methods and models such as these on which to systematically assess practical research, to provide a robust and practical basis against which approaches can be evaluated and developed, and to find approaches that build a pedagogically secure foundation for the new e-learning curricula of the future. The ability of these methods to adapt to the ever-changing pedagogy (that new ways of learning will undoubtedly require) is paramount.

REFERENCES

Adams, N. B., DeVaney, T. A., & Sawyer, S. G. (2009). *Measuring conditions conducive to knowledge development in virtual learning environments: Initial development of a model-based survey*. (pp. 1-23). Retrieved July 2, 2010, from www.jtla.org

Arrington, M. (2010). The unauthorized TechCrunch iPad review. *TechCrunch,* Retrieved July 2, 2010, from http://techcrunch.com/2010/04/02/the-unauthorized-techcrunch-review/

Bartlett, J. F. (1995). *Experience with wireless World-Wide Web client*. (Technical report, note TN-46). Digital Western Research Lab. Retrieved July 2, 2010, from www.research.digital.com/wrl/techreports /abstracts/TN-46.html

Baumgart, D. C. (2005). Personal digital assistants in healthcare: Experienced clinicians in the palm of your hand? *Lancet*, *366*, 1210–1222. doi:. doi:10.1016/S0140-6736(05)67484-3

Billings, D. M. (2005). From teaching to learning in a mobile, wireless world. *The Journal of Nursing Education*, *44*(8), 343.

Borau, K., & Ullrich, C. (2009). Mobile learning with open learning environments at Shanghai Jiao Tong University, China. *Learning Technology Newsletter*, *11*(1-2), 7–8.

Brandon Hall Research. (2009). *Mobile learning comes of age: How and why organizations are moving to learning on mobile devices*. Retrieved July 2, 2010, from www.brandon-hall.com/publications/mobilelearning/mobilelearning- comesofage.shtml

Bridge, P. D., & Ginsburg, K. A. (2001). An integrated approach for evaluating students' achievements of clinical objectives. *Medical Education Online*, *6*(9), 1–6.

Calbraith, D. (2010). *Towards an effective pedagogy and evaluation framework for learning objects in medical education and healthcare*. Unpublished doctoral thesis, Nottingham University.

Calbraith, D., & Dennick, R. (2009). *Pedagogical strategies for learning objects*. Paper presented at the ICERI international conference 2009, Madrid, Spain, 16-18th November. Full paper published by IATED. Retrieved July 2, 2010, from www.iated.org/iceri2009/index.php? section=19

Carroll, J. M. (2000). *Making use: Scenario-based design of human-computer interactions*. Retrieved July 2, 2010, from www.mitpress.mit.edu/catalog/item/default .asp?ttype=2&tid=4114

Chehimi, F., Coulton, P., & Edwards, R. (2006). Delivering 3D advertising to mobile phone. *Proceedings of the IEEE International conference on consumer electronics*, Las Vegas, January 2006. Retrieved July 2, 2010, from http://eprints.lancs.ac.uk/22928/1/ getPDF17.pdf

Clough, G., Jones, A. C., McAndrew, P., & Scanlon, E. (2007). Informal learning with PDAs and smartphones. *Journal of Computer Assisted Learning, 24*(5), 359–371. doi:.doi:10.1111/j.1365-2729.2007.00268.x

CNET. (2010). Best smartphones. Retrieved July 2, 2010, from http://reviews.cnet.com/best-smartphones

Cooper, I., & Shufflebotham, R. (1995). *PDA Web browsers: Implementation issues*. Retrieved July 2, 2010, from http://alethea.ukc.ac.uk/wp?94ihc

CRN. (2010). *Mobile learning*. Retrieved from July 2, 2010, from http://www.crn.com.mobile/222002587;jsessionid=DTFGA1N1W4DPVQE1GPDA

De Ville, K. (2008). The cure is in hand? The brave new world of handheld computers in medicine. *Cambridge Quarterly of Healthcare Ethics, 17*, 385–400. doi:.doi:10.1017/S0963180108080523

Dearnley, C., Haigh, J., & Fairhall, J. (2008). Using mobile technologies for assessment and learning in practice settings: A case study. *Nurse Education in Practice, 8*(3), 197–204. doi:.doi:10.1016/j.nepr.2007.07.003

Eaton, K. (2010). The iPad's biggest innovation: Its $500 price. *Fast Company*. Retrieved July 2, 2010, from http://fastcompany.com/article/apples-tablet- introduced?page=0%2C0

Farrell, M. J., & Rose, L. (2008). Use of handheld computers in clinical nursing education. *The Journal of Nursing Education, 47*(1), 13–19. doi:.doi:10.3928/01484834-20080101-03

Felix, U. (2005). E-learning pedagogy in the third millennium: The need for combining social and cognitive constructivist approaches. *ReCALL, 17*(1), 85–100. doi:.doi:10.1017/S0958344005000716

Ferneley, E. (2010). *Feasibility study into MP3 and MP4 assisted learning*. Retrieved July 2, 2010, from www.edu.salford.ac.uk/scd/tlqis/ abstracts/ferneley0506.rtf

Ganger, A. C., & Jackson, M. (2003). Wireless handheld computers in the preclinical undergraduate curriculum, *Medical Education Online, 8*(3). Retrieved July 2, 2010, from www.med-ed-online.org

Gessler, S., & Kotulla, A. (1994). *PDAs as mobile WWW browsers*. In The Second International WWW Conference '94: Mosaic and the Web, Chicago, USA, October 1994, retrieved from http://www.ncsa.uiuc.edu/SDG/IT94/ Proceedings/DDay/gessler/www pda.html

Group, T. L. T. (Teaching Learning and Technology Group). (2010). *Questionnaire*. Retrieved July 2, 2010, from www.tltgroup.org

Krug, S. (2006). *Don't make me think: A common sense approach to Web usability* (2nd ed.). Berkeley, CA: New Riders Publishing.

Kukulsha-Hulme, A., & Petit, J. (2008). Practitioners as educators: Emergent practice in personal mobile teaching, learning, work, and leisure. In Ally, M. (Ed.), *Mobile learning: Transforming the delivery of education & training* (pp. 135–155). Athabasca, Canada: Athabasca University.

Kukulska-Hulme, A., & Pettit, J. (2007a). *Self-service education: Smartphones as a catalyst for informal collective and individual learning.* In Mlearn '07 Mobile Learning Conference, 16-19th October 2007, Melbourne, Australia. Retrieved July 2, 2010, from http://oro.open.ac.uk/16621/

Kukulska-Hulme, A., & Pettit, J. (2007b). *Do smart devices make smart learners?* In MLearn '07 Mobile Learning Conference, 16-19th October 2007, Melbourne, Australia, pp.1-9. Retrieved July 2, 2010 from http://oro.open.ac.uk/16621/

Laurillard, D. (2005). Presentation given at the HEA 2005 conference: *Harnessing technology to improve the learning experience.* Higher Education Academy.

Magazine, P. C. (2010). Smartphone definition. Retrieved June 13, 2010, from http://pcmag.com/encyclopedia_term .php?gid=310

Martin, D. (2010). *Apple releases iPhone OS SDK 3.2 Beta 2 for iPad.* Retrieved July 2, 2010, from the http://reviews.cnet.com/8301-19512_7-10450636-233.html

Microsoft. (2010). *Smartphones.* Retrieved July 2, 2010, from http://www.microsoft.com/windosmobile/ engb/devices/smartphones.mspx?13=&14=&31.35=1&24=1&56=1&25=1

Miller, J., Shaw-Kokot, J. R., Arnold, M. S., Boggin, T., Crowell, K. E., & Allegri, F. (2005). A study of personal digital assistants to enhance undergraduate nursing education. *The Journal of Nursing Education*, *44*, 19–26.

Nah, K. C., White, P., & Sussex, R. (2008). The potential for using a mobile phone to access the Internet for learning EFL listening skills within a Korean context. *ReCALL*, *20*(3), 331–347. doi:. doi:10.1017/S0958344008000633

Neumeier, P. (2005). A closer look at blended learning–parameters for a blended learning environment for language teaching and learning. *ReCALL*, *17*(2), 163–178. doi:.doi:10.1017/S0958344005000224

Nusca, A. (2009). Smartphone vs. feature phone arms race heats up: Which did you buy? Retrieved August 20, 2009 from http://www.zdnet.com/blog/gadgetreviews /smartphone-vs-feature-phone-arms-race- heats-up-which-did-you-buy/6836

O'Malley, C., Vavoula, G., Glew, J. P., Taylor, J., Sharples, M., & Lefrere, P. (2003). *Guidelines for learning/teaching/tutoring in a mobile environment.* Retrieved July 2, 2010, from www.mobilearn.org/results/results.htm

Pachler, N., Bachmair, B., & Cook, J. (2010). *Mobile learning: Structures, agency, practices.* London, UK: Springer.

Paczkowski. J. (2010). *Initial iPad demand greater than initial iPhone demand.* All things digital. Retrieved July 2, 2010, from http://digitaldaily.allthingsd.com/ 20100223/initial-ipad-demand-greater-than- iphone-demand/

Peters, K. (2007). M-learning: Positioning educators for a mobile connected future. *International Review of Research in Open and Distance Learning*, *8*(2).

Robinson, J. M., Frey, J. G., Stanford-Clark, A. J., Reynolds, A. D., & Bedi, B. V. (2010). Chemistry by mobile phone (or how to justify more time at the bar). Retrieved July 2, 2010, from http://eprints.soton.ac.uk/28632/01/481.pdf

Romero, M., & Wareham, J. (2009). Just in time mobile learning model based on context awareness information. *Learning Technology Newsletter*, *11*(1 & 2), 4–6.

Romiszowski, A. J. (1992). *The selection and use of instructional media*. London, UK: Kogan Page.

Schraefel, M. C., Andre, P., & Van Kllek, M. (2010). This time it's personal: From PIM to the perfect digital assistant. Retrieved July 2, 2010, from http://eprints.lancs.ac.uk/22928/1/getPDF.pdf

Stockwell, G. (2008). Investigating learner preparedness for and usage patterns of mobile learning. *ReCALL*, *20*(3), 253–270. doi:.doi:10.1017/S0958344008000232

Topolsky, J. (2010). *Apple iPad: The definitive guide (so far)*. Retrieved July 2, 2010, from http://engadget.com/2010/01/29/apple- ipad-the-definitive-guide-so-far/

Traxler, J. (2009). The current state of mobile learning. In M. Ally, M. (Ed.), *Mobile learning: Transforming the delivery of education* (pp. 9-25). Athabasca, Canada: Athabasca University Press.

Waycott, J. (2002). Reading with new tools: An evaluation of personal digital assistants as tools for reading course materials. *ALT-J*, *10*(2), 38–50. doi:.doi:10.1080/0968776020100205

Waycott, J., & Kukulsha-Hulme, A. (2003). Students' experiences with PDAs for reading course materials. *Personal and Ubiquitous Computing*, *7*(1), 30–43. doi:.doi:10.1007/s00779-002-0211-x

White, A., Allen, P., Goodwin, L., Breckinridge, D., Dowell, J., & Garvy, R. (2005). Infusing PDA technology into nursing education. *Nurse Educator*, *30*, 150–154. doi:.doi:10.1097/00006223-200507000-00006

Wikipedia. (2010a). *iPhone*. Retrieved July 2, 2010, from http://en.wikipedia.org/wiki/IPhone

Wikipedia. (2010b). *PDA*. Retrieved July 2, 2010, from http://wikipedia.org/wiki/PDA

Wilkinson, S., Dennick, R., & Rae, H. (2006). A Touchstone for online summative assessment. *Issues and News on Learning and Teaching in Medicine, Dentistry and Veterinary Medicine*, 32-33.

Wishart, J. (2006). *Personal digital assistants–teachers prefer the personal*. Paper given at Mlearn'06, 5th World Conference on Mobile Learning, October 22-26th 2006, Banff, Ab, Canada, 15.

Wishart, J. (2008). Challenges faced by modern foreign language teacher trainees in using handheld pocket PCs (PDAs) to support their teaching and learning. *ReCALL*, *20*(3), 348–360. doi:.doi:10.1017/S0958344008000736

ADDITIONAL READING

Adams, N. B., DeVaney, T. A., & Sawyer, S. G. (2009) *Measuring conditions conducive to knowledge development in virtual learning environments: initial development of a model-based survey*, pp1-23, available from www.jtla.org

Alexander, R. (2004). Going Nomadic: Mobile learning in Higher Education. *Educause*, *39*(5), 29–35.

Calbraith, D., & Dennick, R. (2009) *Pedagogical strategies for learning objects*, oral paper given at the *ICERI international conference 2009*, Madrid, Spain, 16-18th November, full paper published by IATED – also available at www.iated.org/iceri2009/index.php? section=19

Carroll, J. M. (2000) *Making use: Scenario-based design of human-computer interactions*, 6 x 9, p87 also available at: www.mitpress.mit.edu/catalog/item/default .asp?ttype=2&tid=4114

Carroll, J. M. (2000) *Making use: Scenario-based design of human-computer interactions,* 6 x 9, p87 also available at: www.mitpress.mit.edu/catalog/item/default .asp?ttype=2&tid=4114

Clough, G., Jones, A. C., McAndrew, P., & Scanlon, E. (2007). Informal learning with PDAs and Smartphones. *Journal of Computer Assisted Learning, 24*(5), 359–371. doi:.doi:10.1111/j.1365-2729.2007.00268.x

Edirisingha, P., & Salmon, G. (2007) *Pedagogical models for podcasts in higher education*, paper presented at the EDEN conference, 13-16th June 2007, BDRA conference papers and presentations – available at: http://hdl.handle.net/2381/405

Felix, U. (2005). E-learning pedagogy in the third Millennium: the need for combining social and cognitive constructivist approaches. *ReCALL, 17*, 85–100. doi:.doi:10.1017/S0958344005000716

Goth, C., Frohberg, D., & Schwabe, G. (2006) *The focus problem in mobile learning,* Proceedings IEEE 4th International workshop on wireless, mobile and ubiquitous technologies in Education (WMTE'06), Athens, pp.153-160

Holzinger, A. Nischelwitzer, A., & Meisenberger, M. (2005) Lifelong learning support by M-learning: example scenarios, *Association of Computing Machinery*, ACM eLearn Magazine, 5, p12

Krug, S. (2006). *Don't make me think: A common sense approach to web usability* (2nd ed.). California: New Riders Publishing.

Kukulska-Hulme, A., & Pettit, J. (2007) *Self-service Education: Smartphones as a catalyst for informal collective and individual learning,* in Mlearn '07 mobile learning conference, 16-19th October 2007, Melbourne, Australia available at: http://oro.open.ac.uk/16621/

O'Malley, C., Vavoula, G., Glew, J. P., Taylor, J., Sharples, M., & Lefrere, P. (2003) *Guidelines for learning/teaching/tutoring in a mobile environment,* available at: www.mobilearn.org/results/results.htm

Romero, M., & Wareham, J. (2009). Just in time mobile learning model based on context awareness information, *Learning Technology Newsletter*, Vol.11, Issue 1 & 2 January – April 2009 IEEE. *Computers & Society*, 4–6.

Romiszowski, A. J. (1992). *The Selection and Use of Instructional Media* (2nd ed.). London: Kogan Page.

Stockwell, G. (2008). Investigating learner preparedness for and usage patterns of mobile learning [Cambridge University Press.]. *ReCALL, 20*, 253–270.

Waycott, J., Jones, A., & Scanlon, E. (2005). PDAs as lifelong learning tools: an activity theory-based analysis. *Learning, Media and Technology, 30*(2), 107–130. doi:.doi:10.1080/17439880500093513

KEY TERMS AND DEFINITIONS

Generic Principles: Principles which can be used in a variety of contexts without compromising their integrity.

Learning Object: A learning object addresses one clearly identifiable topic or learning outcome and has the potential to be reused in different contexts (Weller et al 2003).

Mobile Learning: Any learning that incorporates the use of mobile devices (PDAs, Smartphones, mobile phones, laptops).

Chapter 3
But Do They Want Us in "Their" World?
Evaluating the Types of Academic Information Students Want through Mobile and Social Media

Tim Brown
University of Central Florida, USA

Amanda Groff
University of Central Florida, USA

ABSTRACT

The growth of social media and mobile communication provides educators with an opportunity to transmit course-related information to students in new ways. But are students willing to accept course information through those channels, typically seen as "fun" and "social?" The study in this chapter examines the reasons that students use different types of personal media and how appropriate certain types of communication channels are for academic information. Results show that students prefer to get their academic information through "official" channels, such as email and course management systems. However, they are willing to accept certain types of information through social channels (mobile devices, social networking), as long as they do not have to share personal information.

INTRODUCTION

Social networking sites continue to grow in popularity. Facebook, Twitter, YouTube and others have changed the way that many people communicate on the Web. This growth has led educators to experiment with different ways to use these Web 2.0 tools to transmit content to their students. But while some experiments have focused on differ-

DOI: 10.4018/978-1-60960-511-7.ch003

ent ways to use these tools (Young, 2004; Young, 2009), very little scholarship has examined the motivations people have for using these tools. Without a clear understanding of why people use their social networking tools, educators may be spending energy on experiments that may have little chance of success.

This study examines the motivations that college students have for using different social networking sites and mobile media. By grounding this study in the uses and gratifications paradigm, we can come to a more complete understanding as to the reasons that people use these tools. In addition, by incorporating conceptual aspects of the Technology Acceptance Model into this study, we can also see how useful students see these tools and whether students are ready to accept academic content through these very social media. If they aren't, then educators are trying to force students to think of "work" (their classes) in a communication channel that they may be using primarily for play. A better understanding of what students are looking for in social and mobile media in the classroom – and what they don't want – can help educators convey academic information more effectively and more efficiently.

BACKGROUND

In order to examine what students want out of their social media and mobile tools in the classroom, we're going to use two fundamental theoretical foundations: uses and gratifications, and the technology acceptance model (TAM). The uses and gratifications approach helps us examine the motivations, or reasons, that people use different types of media, whether it's television, radio, the Internet, or mobile phones. The technology acceptance model (TAM) introduces the concepts of usefulness and ease of use; in other words, does the user think the technology or media is useful and easy to use. We'll examine the specific questions a bit later, but first some necessary background on these theories and the work that has brought us to this point.

Uses and Gratifications

The study of motivations for using different forms of media has a long history in communications research (Katz, 1959; Katz, Gurevitch, & Haas, 1973; Ruggiero, 2000), focusing primarily on traditional mass media such as television and radio. Rubin and Perse (1987) noted that television news viewers tended to become more involved with the news if they watched for a specific reason, rather than just out of habit. Diddi and LaRose (2006) noted that "escape" motivations, or passing the time, were likely to drive college students to view news across many different platforms rather than just television or radio. The approach has also been used to study personal media; Dimmick, Sikand and Patterson (1994) determined that sociability was a major factor in using the telephone, but coordination of social events was a subfactor. More recent research has examined the motivations for using newer media, such as cell phones and pagers (Leung & Wei, 1998; Wei & Lo, 2006), home computers (Perse & Dunn, 1998), email (Dimmick, Kline, & Stafford, 2000) and satellite radio (Lin, 2006). Lin (2006) noted that just because a person is drawn to one type of new media (online radio) doesn't necessarily indicate their acceptance of another, similar media (satellite radio).

As that research has grown, so has the notion that even as the motivations for using media are different, the motivations can differ within the media, rather than just the need. Matthews and Schrum (2003) noted that college students living in dorms found the convenience of high speed access important for social communication through email and IM, but a distraction when trying to study. More recent research has examined how and why college students use social networking sites (Raacke & Bonds-Raacke, 2008; Ray, 2007). Sheldon (2008) noted that college students use Facebook in large part to maintain existing relationships instead of finding new ones. Ray (2007) found that in addition to relationship management, social networking sites also allowed users to meet multiple types of needs, on multiple levels. Other research has examined how users of Twitter may start with social motivations, but over time, determine they are finding more satisfaction of their information seeking motives (Dong, 2008; Johnson & Yang, 2009).

Research into communication tools and education has also employed the uses and gratifications approach. As that research has developed, it's become clear that students are becoming more adept at determining just how they want to use newer media. Ebersole (2000) noted early on in Internet gratifications research that middle school aged students were more likely to use the World Wide Web for entertainment purposes, while older students were likely to use it for research and academic reasons. Brown (2007) noted that when students were presented with new media options for what they were doing in old media, the students looked for specific functions of the new media before using it. Others have examined student attitudes toward course websites and laptops (Bonds-Raacke & Raacke, 2008). Bonds-Raacke and Raacke (2008) found that students liked how Tablet PC's made the classroom more engaging, even though the faculty may not have determined exactly how to use them. In other words, the students were beginning to determine their own uses for the technology, separate from the faculty.

Technology Acceptance Model

In examining why students use different technologies in class, Park and colleagues (Park, 2005; Park, Kwan, & Cheong, 2007; Park, Lee, Jae, & Roman, 2007) have incorporated the technology acceptance model, or TAM. Developed by Davis and colleagues (Davis & Venkatesh, 2004; Venkatesh & Davis, 1996; Venkatesh, Morris, Davis, & Davis, 2003) partly as an extension of Rogers' Diffusion of Innovation theory (Rogers, 2003), TAM looks at the influences of perceived usefulness (PU) and perceived ease of use (PEOU) on a person's intention to adopt new technology and use it. Zhou (2008) examined how journalists in China viewed using the Internet based on whether they were forced by their organizations to adopt it or if they had previously adopted it before their employer. Park and colleagues examine not just whether students are ready to accept new technology in the classroom (2005), but whether faculty are ready to use it (2007) and the impact that has on the learning process.

Social Networking, Mobile Media and the Classroom

Research into mobile media and education has increased as the technology has increased as well. Studies examining course websites (Bonds-Raacke, 2006) have shown that students seem to gravitate toward mobility and ease of access. Early studies of podcasting (*Seven things you should know about podcasting,* 2005; Chapin, 2005; GCSU, 2006) examined their convenience for students, primarily as ways to provide lecture notes and supplemental material. Other work (Bonk & Zhang, 2006; Brown, 2006) has shown that students will often be drawn to the convenient nature of the technology, but will use it only as far as it's convenient for them. Brown (2006) saw students gravitate toward using AOL's instant messenger service for online office hours because it was convenient in two ways: (1) they were already using the technology, and (2) it allowed them to avoid meeting their professors face-to-face. Huntsberger and Stavitsky (2007) found that students enjoyed the mobility of podcasts, but also enjoyed being a part of "review session" podcasts. If this is the type of trend that is to continue (convenience), it is certainly worth exploring what information students want out of their communication tools, and just how much they are willing to accept.

Research Questions and Hypotheses

In summary, by using the theoretical foundations of uses and gratifications we've pointed out that people have different reasons for using different types of media. We've also established, through the technology acceptance model (TAM), that people adopt technology because of how useful they perceive it to be and how easy it is to use. By examining research with those two foundations, we've found that college students often have specific reasons for using different types of communication channels, specifically social networking tools. But what we need to know is specifically how college students evaluate the kind of information they get through different communication channels, and how that might impact their decision to use those channels (social networking, mobile phones, iPods, etc.). If we understand how students evaluate the type of communication channel and if we know what relationship exists between their motivations for using social networking and for using those appropriate communication channels, we can get a better understanding as to what they think of academic content they receive through social networking.

Therefore, it's worth it to ask:

RQ1: How do students evaluate the appropriateness of certain academic information communicated across various platforms, such as email, text messaging, and social networking?

- As noted above, the uses and gratifications approach provides an excellent framework from which to examine the motivations that students say drive them to use certain media. If students say they go to social networking sites to relax or catch up with friends, will they be inclined to get "course related" information from those same sites? Therefore:

RQ2a: How do motivations for social networking use relate to the evaluation of the appropriateness of academic information through various platforms for communication?

- This is valuable to find out because there may be differences in what drives students to use social media as opposed to the other, personal media that they use (cell phones, etc.).
- As mentioned above, the technology acceptance model provides a framework which can help educators understand whether students perceive a form of new media is useful in accomplishing their goals and if it is too complex to use to be of any value. Following in the work of Park, et al:

RQ2b: How do the factors of perceived usefulness (PU) and perceived ease of use (PEOU) of both social networking and course management sites impact how students evaluate the appropriateness of academic information transmitted through certain communication channels?

- As mobile media use continues to grow and it becomes more and more a part of our daily lives (Horrigan, 2009), it is natural for educators to find ways to use that mobile technology for educational purposes. But will students be willing to use that technology? Consistent with previous research into the use of technology hardware on campus;

RQ3: What university tools do students say they will use through mobile media?

RQ4: Are there differences among those who would use the mobile university tools based on their evaluation of different levels of appropriateness for communication tools?

METHODOLOGY

Students from various universities around the United States were invited to take part in the survey, although the majority of the respondents came from one large southern university. The questionnaire was "live" (available on a secure website) for one week, and only one reminder was sent out to the students.

Only some of the instructors of the classes from which the sample was taken use any type of social networking in their classes. The use varied between the instructors: one used Twitter as a way to send out emergency notifications (i.e., "class cancelled today" or "instructor delayed by 15 minutes"), the other used a Twitter widget on the front page of the course management system (CMS) sites as regular announcements. Neither required students to obtain Twitter accounts.

Participants were asked several questions about their use of various communication tools and how appropriate (1=extremely inappropriate, 5=extremely appropriate) they believed certain types of communication from their professors would be through four different channels: email, text messages, posts on the students' personal social networking pages and posts on a class page set up by the instructor (e.g., "Please indicate how appropriate you feel it is to get the following through ____: reminders of assignments, change in class schedule, links to content used for class, grades on assignments, grades on tests, comments on assignments, emergency university information, general class announcements, confirmation of office visit, reminder of tests.). Students were also asked about their use of different social networking tools, including Twitter. Students who indicated they had a Twitter account were asked questions consistent with uses and gratifications research to determine motivations for using media; students who indicated they did not have a Twitter account were not asked those questions. All students were directed to items which asked about their reasons for using the course management system at the university and whether they felt it was easy to use and useful in an academic setting (1=low, 5=high). Items about Twitter were modified from Johnson and Yang (2009); items eliciting information about motives for the course management system in general were modified to reflect the use of the course management system as opposed to using Twitter. Items indicating level of agreement with whether Twitter and the course management system were easy to use (Perceived Ease of Use, PEOU) and useful (Perceived Usefulness, PU) were modified from previous studies using the technology acceptance model (TAM) that involved both voluntary and involuntary adoption of new technologies (e.g., Zhou, 2008). Principle component factor analysis with varimax rotation was used to determine the relationship among items within a communication tool. It should be pointed out that even though the motivations for use of Twitter are modified from Johnson and Yang, principle component factor analysis rather than confirmatory factor analysis was used because the items were modified to the degree that differences would be more apparent in a principle factor analysis.

RESULTS

Respondents

Seven hundred and seventy eight students responded to the questionnaire. Because the questionnaire was also sent to students at other universities, it is difficult to determine an overall response rate; however the response rate at the sponsoring university was 55%. The gender breakdown of the respondent sample was much more female than male -69% to 31%. However, many of the classes from which the sample was taken are communication classes, which have traditionally attracted females more than males. With regard to race, 71.2 percent of respondents reported being Caucasian; 7.2 percent reported being African-American; 13 percent reported as Hispanic and 2.9% reported as Asian-American. In addition, 5.1 percent of the respondents indicated they were multi-racial. With regard to age, the median age of the sample was 19 years old, with most respondents indicating they had been online at least ten years (or more).

Table 1. Means, standard deviations, and α of appropriate communication channels

Scale	M	SD	α
Email Information	4.42	.75	.91
Email Grades	3.83	1.24	.94
Text Emergency Information	4.01	1.11	.65
Text Class Info	2.9	1.21	.91
Text Grades	2.26	1.13	.92
Personal Page Info	2.58	1.26	.96
Personal Page Grades	1.8	1.07	.95
Class Page Info	4.18	1.10	.98
Class Page Grades	3.08	1.43	.94

Note: 1 = not at all appropriate, 5 = completely appropriate

Just over 20% of students reported having a Twitter account; however, more than 90% reported having a page on a social networking site, and even more reported using a course management system at least once in a college class. While much larger numbers of students answered questions about the CMS and social networking as a whole, only those students who reported having a Twitter account were asked questions about their use of Twitter.

RQ Results

RQ1 asked, "How do students evaluate the appropriateness of certain academic information communicated across various platforms, such as email, text messaging, and social networking?" In order to answer that question, scales were created based on the principle component factor analysis of the ten items for each channel (email, text message, personal social networking page, and class page on a traditional social networking site). Table 1 reports the scales and their means, standard deviations and Chronbach's alpha scores for each scale. In general, the items factored out with great consistency: items related to general course information content tended to group together and items relating to the reporting of grades tended to group together, developing two factors per communication channel. One exception was in the "text message" channel, where a third factor developed – emergency university information.

Two items of interest should be pointed out here. First, the high mean (4.42) and relatively low standard deviation of the Email information scale seems to indicate that students evaluate communicating and receiving almost all types of information are appropriate through that channel. Even though the grades questions for email factored out as a separate variable, those items were highly ranked as well (M=3.08); however, that measure had the highest standard deviation, which seems to indicate that there are more people who would prefer a more secure method of sharing grades than through email. Second, while one might think that text messaging would be seen as a less secure route to report grades, it appears that it is seen as more private than a personal social networking page; more people are willing to receive their grades via text (M=2.26) than through a posting on a personal networking page (M=1.81).

So we see that these students said they wanted specific, private information (such as grades) sent through what they saw as a more private, secure communication channel (email). We also saw that the students did not like the idea of their grades being posted on their personal social networking page; that shows us that they are very much interested in keeping that kind of information private. Students seem to be willing to *receive* general information on their personal pages, but they are reluctant to receive personal grade information that might be shared with their friends, thereby making it public. That seems to indicate that students see social networking sites as less private that other forms of communication. The next step in this process, then, is to determine what relationships – if any – exist between how students evaluate the appropriateness of transmitting academic information through these channels and why they choose to use social networking sites. If we can understand that relationship, then we come closer to knowing what specific types

Table 2. Means, standard deviations, and α of motivations scales

Scale	M	SD	α
Twitter Relaxation	3.78	1.02	.93
Twitter Comm –Utility	1.91	.96	.86
Twitter Comm – Info seek	3.48	1.11	.77
Twitter Comm – Social	3.27	1.26	.91
CMS Relax	1.61	.75	.96
CMS Comm – Utility	4.51	.69	.70
CMS Comm - Info seek	3.42	1.03	.83
CMS Comm – Social	3.11	1.06	.72
SNS Relaxation	4.32	.63	.89
SNS Comm – Utility	1.83	.78	.82
SNS Comm – Info seek	3.32	.85	.77

Note: 1 = low motivation, 5 = high motivation

of information they are willing to accept through these social networking sites.

RQ2a asked, "How do motivations for social networking use relate to the evaluation of the appropriateness of academic information through various platforms for communication?" This question (and RQ2b) was answered by developing scales on motivations for media use and comparing them through correlational analysis with the scales created for communication appropriateness. Much as with RQ1, principle component factor analysis was used on several items examining the motivations of use for Twitter, the school course management system (CMS) and social networking sites in general. Table 2 lists the scales and their means, standard deviations, and Chronbach's alpha scores. In general, four factors emerged from each communication channel: relaxation (e.g.,"to relax," " to be entertained"); communication for utilitarian purposes for school (e.g., "to get announcements," "to communicate with my professor"); communication for information seeking purposes (e.g., "to meet new people"); and communication for social integration (e.g., "I like to keep up with family and friends"). Bivariate correlation was then used to determine the relationship among the social networking scales and the communication platform scales; Table 3 reports the results.

The same analysis was used to determine both innovativeness factors (utility and fashion/status) and to determine factors related to the technology

Table 3. Pearson's r values for correlations of communication appropriateness measures and motivations for using Twitter and other social networking sites

	Twitter Relax	Twitter Comm Utility	Twitter Comm Info Seek	Twitter Comm Social	SNS Relax	SNS Comm Utility		SNS Comm Info seek
Email Information	.00	.04	-.06	-.06	.22**	-.02		.07
Email Grades	-.09	.15	-.08	-.12	.11**	.04		.06
Text information	.02	.31**	.02	.08	.07	.21**		.12**
Text grades	-.38	.43**	.04	.02	-.02	.30**		.13**
Text Emergency Information	.00	.18*	.00	-.03	.23**	.08**		.15**
Personal Page information	.10	.24**	.11	.03	.10*	.28**		.17**
Personal Page Grades	.03	.33**	.04	.03	-.42	.35**		.16**
Class Page information	.24**	.08	.09	-.01	.18**	.02		.11**
Class Page Grades	.10	.25**	.02	.00	.02	.20**		.11**

** $p<.01$ * $p<.05$

Table 4. Means, standard deviation and α of PU and PEOU scales

Scale	M	SD	α
Innovativeness – Fashion	3.22	.93	.79
Innovativeness – Utility	4.28	.66	.80
Twitter PU	1.89	.94	.93
Twitter PEOU	4.0	.74	.86
CMS PU	3.20	.97	.85
CMS PEOU	4.03	.74	.89
SNS PU	*	*	*
SNS PEOU	3.33	.85	.57

Note: 1 = low, 5 = high *SNS Perceived Usefulness did not factor out of items

acceptance model (perceived usefulness, or PU; perceived ease of use, PEOU). Table 4 reports the means, standard deviations, and Chronbach's alpha scores of those scales. Note that no factor accounting for perceived usefulness of social networking sites emerged from the rotation – this is likely due to a change in the items for that communication tool. Table 5 reports the results of bi-variate correlation tests involving the innovativeness scales and the perceived usefulness and perceived ease of use of social networking sites in general (and Twitter specifically) and the evaluation of appropriate communication tools.

It is perhaps useful at this point to discuss the results that have been reported so far, in order to better understand the results of the remaining research questions. The means for the "information" channels (sending class information through a channel) are highest for email (M=4.42) and for a class social networking page (M=4.18); by contrast, students seem to say that they don't think it's appropriate to send class information through personal social networking pages (M=2.58) or through text messages (M=2.9). This would seem to indicate a preference on the part of students to receive information from faculty through either "professional" channels (email), or through a channel that is designated a clearinghouse for information for the entire class (a class page on a traditional social networking site). Indeed, post-

Table 5. Pearson's r values for correlations between communication appropriateness measures and PU and PEOU of social networking sites and innovativeness factors

	Twitter PU	Twitter PEOU	SNS PEOU	Innovative- fashion	Innovative- utility
Email Information	.04	.06	-.02	.06	.21**
Email Grades	.13	-.02	.07	.15**	.13**
Text information	.26**	.11	.10*	.14**	.12*
Text grades	.40**	-.06	.05	.16**	.03
Text Emergency Information	.18	.13	.20**	.10**	.17*
Personal Page information	.32**	.08	.15**	.03	.07
Personal Page Grades	.41**	-.09	.08*	.11**	-.01
Class Page information	.20**	.13	.11**	-.06	.10**
Class Page Grades	.37**	-.06	.03	.03	-.01

** $p<.01$ * $p<.05$ NOTE: SNS Perceived Usefulness did not factor into a variable

hoc correlation analysis shows a significant – albeit low to moderate – correlation between those two factors (r=.27, p<.01). Conversely there is a moderate to strong correlation (r=.43, p<.01) between receiving class information through text messages and through a personal social networking page; because the means for each of those measures is low, there seems to be a reluctance on the part of students to get information from faculty through a "personal" channel (text through a phone, "Facebook" page that shares information with the professor). The means for communication motives through social networking pages (Table 2) also indicate that they seem more geared toward fun and relaxation (M=4.32) as opposed to school utility (M=1.83).

It begins to become even more clear when examining the correlations between social networking motivations and the measures for appropriate communication channels (Table 3). There are very few significant correlations among any of the motivations for using Twitter. Most of the significant correlations center on the motivation for using Twitter as a school utility communication tool (to get announcements from the professor, to communicate with the professor). The school/utility motivation correlates with many of the "personal" communication tools (text message, personal SNS page). Because the means for many of these communication tools are low, it is apparent that students who use Twitter are relatively consistent in their motivations for using (or not using) personal media channels for school work.

This trend holds when examining the relationships between the perceived usefulness and perceived ease of use of Twitter and other social networking sites for communicating course information and the communication channel measures, as well as innovativeness scales derived from previous research (Chang, Lee, & Kim, 2006).

So now we know that students don't think it's appropriate to send/receive personal information through certain channels that might be more public (personal social networking pages) or less secure (text messages), and that they see these social networking sites primarily as fun diversions or ways to communicate with their friends rather than their professors. Next, we want to know how students evaluate the usefulness and simplicity of both social networking sites and traditional course management systems. RQ2b asked, "How do the factors of perceived usefulness (PU) and perceived ease of use (PEOU) of both social networking and course management sites impact how students evaluate the appropriateness of academic information transmitted through certain communication channels?" By understanding these relationships, it can become more clear as to what channels of communication are seen as similar for different types of content (grades, course announcements, supplemental material).

As with the motivations for social media use, significant and moderate to strong correlations emerge between the perceived usefulness of Twitter and the "personal" communication tools of personal SNS pages and text messages. The lower means for each seem to indicate that students feel that personal communication channels are meant for personal media and information, not for school work. Each of the innovativeness scales, which measure students' willingness to try new things, show low to moderate correlations with some of the communication appropriateness measures. What is interesting is that the innovativeness utility scale, which measures the students' interest in using new technology for things that are important to them, has low correlations with items such as texting emergency information (r=.12) and even general information (r=.17). Because those means are closer together than other means (Innovative utility= 4.28, text emergency information= 4.01, text class information= 2.9), one would think they would have stronger relationships, that those who want the newest technology for information purposes would want general course-related information through text. However, it is possible that the mobile nature of different technologies (SNS, email) creates more options for those stu-

Figure 1. Percentage of users who say that they would use the following university tools if they had mobile access on their phones

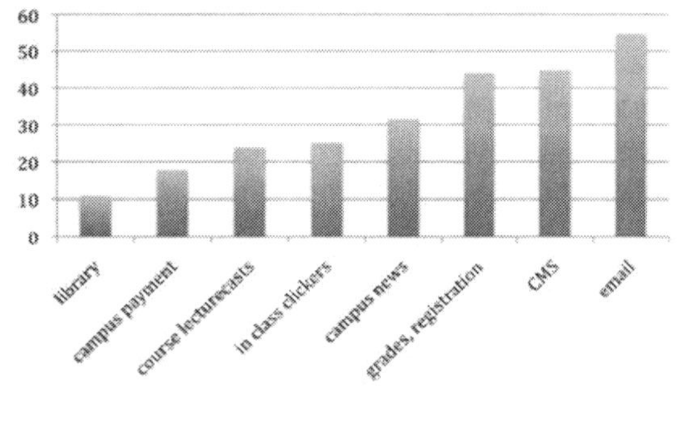

dents, which would mean they would be willing to accept information in a variety of formats, just not all the time.

RQ3 and RQ4

So now that there seems to be a trend in student distaste for sending and receiving certain types of information through "personal media" (text message, personal SNS pages), it is valuable to examine just what kind of information students would want through more mobile communication. Using a well established framework of education technology questions (Smith, Salaway, & Borreson Caruso, 2009), RQ3 asked, "What university tools do students say they will use through mobile media?" Students were asked to choose which of the following university services they would use through mobile internet connections on their personal mobile phone: grades and registration; library services, course management systems (CMS); university email; campus news; campus payment systems; streaming/listening to recorded course lectures; and clicker services for classrooms. Figure 1 displays the results.

That the students' university email system is the most popular answer should not be a surprise, given the nature of mobile communication. However, note that two services used primarily for utility – the school course management system (CMS) and grades and registration services – are the second most popular. This actually seems to make sense in light of the means for the class social networking site (SNS) page information measures (M=4.18). If students are given the option to *receive* information through their personal media, they may see it as convenience – recall that the mean for text emergency information measures was quite high (M=4.01). Students seem to be saying that they are willing to *get* certain types of information through personal channels, as long as they don't have to *reveal* any personal information, as they would have to with personal social networking pages.

But is there a difference in how students rate the appropriateness of the university tools on mobile devices based on their willingness to use certain tools through mobile media? In other words, even though *all* the students in the survey may have evaluated the appropriateness of communication in the aggregate, will those who are willing to use personal, mobile for school resources evaluate their communication channels differently than those who aren't? That's what RQ4 sought to determine: Are there differences among those who would use the mobile university tools based on their evaluation of different levels of appropriateness for communication tools? Based on independent sample t-tests, the answer seems to be somewhat mixed. The communication appropriateness measures were examined through a t-test, using the top four mobile university tools (grades/registration, campus news, CMS, and email) as grouping variables. Table 6 reports the results. In six of the nine communication measures, there is a significant difference between the means of those who say they would use a grade/registration service on their phone and those who would not. Those who say they would use the mobile grade service statistically significant means that were slightly higher than those who said they would not use the university's mobile grade service on their devices. The only other significant difference came with the mobile version of "campus news"

Table 6. Results of independent sample t-tests for communication appropriate measures with mobile uses as grouping variable

Scale	Grades	CMS	Email	Campus News
Email Information	$p<.01$	*	*	$p<.05$
Email Grades	$p<.01$	*	*	*
Text Emergency Information	$p<.05$	*	*	*
Text Class Info	*	*	*	*
Text Grades	$p<.01$	*	*	*
Personal Page Info	*	*	*	*
Personal Page Grades	*	*	*	*
Class Page Info	$p<.05$	*	*	*
Class Page Grades	$p<.05$	*	*	*

*no significance

and the email information scale – those who said they would use the mobile campus news service on their devices evaluated getting general class related information via email as more appropriate than those who said they would not use the mobile campus news service.

What is the importance of these findings? It adds further evidence to the idea that students are willing to use their personal media for *some* school activities, but not for others. The significant differences in the means of the communication tools indicate that students who say they would use mobile grades and registration tools are willing to do so mainly because it's information they *receive*, rather than *send*. Note how there are no significant differences in the means for the measures examining personal SNS page use. In those cases, using the personal SNS page for information, even receiving it, would require the student to provide some personal information to the sender of the message (status update, friend list, etc). It would appear that, because there is no difference in the reported statistical values, *all* students in this study feel relatively the same way. Note also that there are no significant differences among those students who say they would use mobile CMS and email applications. These are functions that require little personal information being shared; indeed, these may be considered more professional modes of communication for academic work by students.

SOLUTIONS AND RECOMMENDATIONS

So what do all of these findings mean? In summary, this study has provided evidence that students compartmentalize the use of certain communication tools for different reasons – social tools for social time, work tools for work time. In addition, students seem to be saying that they have limits as to what kind of academic information they want to receive through personal communication channels. The recognition that personal pages on social networking sites (SNS) would mean that faculty would be able to view students' *personal* information in addition to academic information does not sit well with the students in this survey. They seem to prefer to stay with formal, professional channels – such as email or course management systems – for school work in most cases.

There are, however, exceptions. Students are willing to *receive* course related information on their personal media (SNS, text, mobile phones) in certain situations, such as emergency information or a change in course schedule; or, in the case of mobile phones, email and CMS information, most likely because of their "official" designation from the university. There are also a few students who view potential benefits of the use of social networking sites in the classroom, specifically Twitter. However, they account for only about 20% of the students in this survey, and even their evaluation of that tool's use currently in the classroom is low.

So what does that mean for educators who want to incorporate different teaching tools? This study provides some mixed results. Students do see *some* value in using social networking sites to transmit information, but they have limits as to what they want through those channels. Faculty may want to focus on *general* course information through social networking, which would allow students to get the "utility" information in the same place they get social information about their friends. However, educators should be careful not to expect students to respond in too many ways to social networking in the classroom. They don't want to show their grades to other students, and they seem to be careful about sharing too much information with others in the class who might connect with them via social networking sites.

However, when it comes to getting emergency information (such as a class being cancelled), students seem to be more inclined to accept the information over their mobile devices. This seems logical, as the information is important to their daily schedules and is pushed to their mobile devices. But beyond that, it seems unlikely that students (at least the ones in this sample) would be willing to accept just any kind of course-related information sent to their mobile devices. That doesn't mean that faculty shouldn't try to craft information to send to mobile devices. Recall that students were ambivalent about getting general course information through text messages (M=2.9 on the 1-to-5 scale). What that tells us is that they're not closed to the idea to getting text or mobile messages related to class; faculty should make a point of keeping that information specific and easy to act on (reminders of homework or tests or notices that grades have been posted).

FUTURE RESEARCH DIRECTIONS

As scholars continue to examine the value of mobile and social media for education, much of their work is likely to be dictated by mobile devices and their functionality. Industry reports indicate that the use of smartphones in the United States continues to grow at a rapid pace – some estimate that smartphones will have accounted for more than 40% of the mobile phone market in the U.S. (Walsh, 2010). Mobile Internet use continues to grow in the U.S., and as standard university applications (course management system, email, grades/registration) become more usable on mobile devices, it is reasonable to assume that more students will not only own mobile Internet devices but also be willing to use them for these services. The trick for educators will be to make sure that they find those uses that students are willing to accept and make the most of them.

Among the limitations of this study is the construction of the sample. While the study aims to determine the motivations for using social networking tools in higher education, the sample is made up of students from several liberal arts and sciences classes, with very little representation from any technology or computer science classes. While it is possible that some students in those majors would be in the general education classes represented here, they would not be in the same numbers as if classes from those technology majors had been included. It is possible that students who are more versed in using different types of technology might find more uses for the different communication tools examined here, and therefore alter the findings. Another limitation is the small number of Twitter users; only 20% of those respondent indicated that they had their own Twitter accounts. Future research should make an effort to seek out more college students who use Twitter.

CONCLUSION

Educators face a growing battle over how to integrate new technology into their classrooms. The battle usually centers around whether to use the technology because it's available and the students have access to it, or to use it sparingly until their respective institutions push it from above. What is key in either case is to determine what material is best suited for each specific communication channel, be it mobile phones, social networking or course management systems. In addition, knowing what students are willing to accept through those communication channels is important; if students aren't willing to share course information on their personal social networking pages, then educators may be spending a lot of time developing material with little hope of a return on their investment.

This study has provided some insight into what students feel is appropriate when it comes to academic material transmitted through different communication channels. Students in this study seem to separate their communication preferences: they choose their personal/social channels for themselves, and prefer the "official" channels such as email and course management systems for "official" school/academic information. The direction for educators to take is one of cautious experimentation: educators should be ready to use new communication technologies to transmit course information, but they should keep in mind that not all channels are useful for all types of information. And if students aren't willing to accept certain types of information, then the effort is wasted. The more that educators can "dip their toes" in the water of new technologies, the more we can learn about how to use these tools most effectively.

REFERENCES

Bonds-Raacke, J. M. (2006). Students' attitudes toward the introduction of a course website. *Journal of Instructional Psychology*, *33*(4), 251–255.

Bonds-Raacke, J. M., & Raacke, J. D. (2008). Using tablet PCs in the classroom: An investigation of students' expectations and reactions. *Journal of Instructional Psychology*, *35*(3), 235–239.

Bonk, C. J., & Zhang, K. (2006). Introducing the R2D2 model: Online learning for the diverse learners of this world. *Distance Education*, *27*(2), 249–264. doi:10.1080/01587910600789670

Brown, T. (2006). Using instant messenger programs for virtual office hours. *Feedback*, *47*(1).

Brown, T. (2007). Gratifications of handheld computers. *Academic Exchange Quarterly*, *11*(4).

Chang, B., Lee, S., & Kim, B. (2006). Exploring factors affecting the adoption and continuance of online games among college students in South Korea: Integrating uses and gratification and diffusion of innovation approaches. *New Media & Society*, *8*(2), 295–319. doi:10.1177/1461444806059888

Chapin, A. (2005). *iPods and second language acquisition*.

Davis, F. D., & Venkatesh, V. (2004). Toward preprototype user acceptance testing of new Information Systems: Implications for software project management. *IEEE Transactions on Engineering Management*, *51*(1), 31–46. .doi:10.1109/TEM.2003.822468

Diddi, A., & LaRose, R. (2006). Getting hooked on news: Uses and gratifications and the formation of news habits among college students in an internet environment. *Journal of Broadcasting & Electronic Media*, *50*(2), 193–210. .doi:10.1207/s15506878jobem5002_2

Dimmick, J. W., Kline, S., & Stafford, L. (2000). The gratification niches of personal e-mail and the telephone. *Communication Research*, *27*(2), 227–248. doi:10.1177/009365000027002005

Dimmick, J. W., Sikand, J., & Patterson, S. J. (1994). The gratifications of the household telephone: Sociability, instrumentality, reassurance. *Communication Research*, *21*(5), 643–663. doi:10.1177/009365094021005005

Dong, Q. (2008). *Explaining why young adults use MySpace and Facebook through uses & gratification theory*. Conference Papers-National Communication Association.

Ebersole, S. (2000). Uses and gratifications of the Web among students. *Journal of Computer-Mediated Communication*, *6*(1).

EDUCAUSE. (2005). *Seven things you should know about podcasting.*

GCSU. (2006). *The GCSU iPod story*. Georgia College and State University.

Horrigan, J. (2009). *The mobile differenc*e. Pew Research Center report.

Huntsberger, M., & Stavitsky, A. (2007). The new podogogy: Incorporating podcasting into journalism education. *Journalism and Mass Communication Educator*, *61*(4), 397–410.

Johnson, P. A., & Yang, S. (2009). *Uses and gratifications of twitter: An examination of user motives and satisfaction of twitter use*. Association for Education in Journalism and Mass Communication.

Katz, E. (1959). Mass communication research and the study of popular culture. *Studies in Public Communication*, *2*(1), 1–6.

Katz, E., Gurevitch, M., & Haas, H. (1973). On the use of the mass media for important things. *American Sociological Review*, *38*, 164–181. doi:10.2307/2094393

Leung, L., & Wei, R. (1998). The gratifications of pager use: Sociability, information seeking, entertainment, utility, and fashion and status. *Telematics and Informatics*, *15*(4), 253–264. doi:10.1016/S0736-5853(98)00016-1

Lin, C. A. (2006). Predicting satellite radio adoption via listening motives, activity and format preference. *Journal of Broadcasting & Electronic Media*, *50*(1), 140–159. doi:10.1207/s15506878jobem5001_8

Matthews, D., & Schrum, L. (2003). High-speed Internet use and academic gratifications in the college residence. *The Internet and Higher Education*, *6*(2), 125–144. doi:10.1016/S1096-7516(03)00020-4

Park, N. (2005). *User acceptance of e-learning in higher education: An application of technology acceptance model*. International Communication Association.

Park, N., Kwan, M. L., & Cheong, P. H. (2007). University instructors' acceptance of electronic courseware: An application of the technology acceptance model. *Journal of Computer-Mediated Communication*, *13*(1), 163–186. .doi:10.1111/j.1083-6101.2007.00391.x

Park, N., Lee, S., Jae, E. C., & Roman, R. (2007). *User acceptance of a digital library system in developing countries: An application of the technology acceptance model.* Conference Papers-National Communication Association.

Perse, E. M., & Dunn, D. G. (1998). The utility of home computers and media use: Implications of multimedia and connectivity. *Journal of Broadcasting & Electronic Media*, *42*(4), 435–457.

Raacke, J., & Bonds-Raacke, J. M. (2008). MySpace and Facebook: Applying the uses and gratifications theory to exploring friend-networking sites. *Cyberpsychology & Behavior*, *11*(2), 169–174. doi:10.1089/cpb.2007.0056

Ray, M. B. (2007). *Needs, motives, and behaviors in computer-mediated communication: An inductive exploration of social networking websites.* International Communication Association.

Rogers, E. M. (2003). *Diffusion of innovations* (5th ed.). New York, NY: Free Press.

Rubin, A. M., & Perse, E. M. (1987). Audience activity and television news gratifications. *Communication Research, 14*(1), 58–84. doi:10.1177/009365087014001004

Ruggiero, T. E. (2000). Uses and gratifications theory in the 21st century. *Mass Communication & Society, 3*(1), 3–37. doi:10.1207/S15327825MCS0301_02

Sheldon, P. (2008). Student favorite: Facebook and motives for its use. *Southwestern Mass Communication Journal, 23*(2), 39–53.

Smith, S., Salaway, G., & Borreson Caruso, J. (2009). *Undergraduate students and information technology, 2009. No. Research Study (Vol. 6)*. Boulder, CO: Educause Center for Applied Research.

Venkatesh, V., & Davis, F. D. (1996). A model of the antecedents of perceived ease of use: Development and test. *Decision Sciences, 27*(3), 451–481. doi:10.1111/j.1540-5915.1996.tb01822.x

Venkatesh, V., Morris, M. G., Davis, G. B., & Davis, F. D. (2003). User acceptance of information technology: Toward a unified view. *Management Information Systems Quarterly, 27*(3), 425–478.

Walsh, M. (2010). Summer of smartphones II. *MobileMarketingDaily.*

Wei, R., & Lo, V. (2006). Staying connected while on the move: Cell phone use and social connectedness. *New Media & Society, 8*(1), 53–72. doi:10.1177/1461444806059870

Young, J. R. (2004, August 13). Students say technology has little impact on teaching. *The Chronicle of Higher Education, 50*, A28.

Young, J. R. (2009, November 22). Teaching with twitter: Not for the faint of heart. *The Chronicle of Higher Education*, 56.

Zhou, Y. (2008). Voluntary adopters versus forced adopters: Integrating the diffusion of innovation theory and the technology acceptance model to study intra-organizational adoption. *New Media & Society, 10*(3), 475. doi:10.1177/1461444807085382

ADDITIONAL READING

Bramble, N. (2009, Fifth period is Facebook. *Slate.Com,* Carta, D. (2008, Sept. 18). Emodo is a Twitter for education. *Mashable.Com*

Catone, J. (2009, July 24). In the future, the cost of education will be zero. *Mashable.Com*

Chapin, A. (2005). iPods and second language acquisition

Chu, C., & Lin, H. (2006). Factors influencing the adoption of wireless technologies on campus. *International Journal of Mobile Communications, 4*(3), 291–308.

Crofts, S., Dilley, J., Fox, M., Retsema, A., & Williams, B. (2005). Podcasting: A new technology in search of viable business models. *First Monday, 10*(9).

Debatin, B., Lovejoy, J. P., Horn, A., & Hughes, B. N. (2009). Facebook and online privacy: Attitudes, behaviors, and unintended consequences. *Journal of Computer-Mediated Communication, 15*(1), 83–108. .doi:10.1111/j.1083-6101.2009.01494.x

Dimmick, J., Chen, Y., & Li, Z. (2004). Competition between the Internet and traditional news media: The gratification-opportunities niche dimension. *Journal of Media Economics*, *17*(1), 19–33. doi:10.1207/s15327736me1701_2

E-learning and all that jazz. (2006). *Medical Teacher, 28*(4), 396.

Ellison, N. (2010). *Facebook use by undergraduates: An educational tool? Kelleher, T., O'Malley, M., & Oh, M. (2003). Adapting and extending the technology acceptance model in globally linked PR classes International Communication Association. doi:ica_proceeding_12240*. PDF.

Lee, M. J. W., & Chan, A. (2005). Exploring the potential of podcasting to deliver mobile ubiquitous learning in higher education. *Journal of Computing in Higher Education*, *18*(1), 94–114. doi:10.1007/BF03032726

Leung, L., & Wei, R. (1999). Seeking news via the pager: An expectancy-value study. *Journal of Broadcasting & Electronic Media*, *43*(3), 299–315.

Lin, C., Shih, H., & Sher, P. J. (2007). Integrating technology readiness into technology acceptance: The TRAM model. *Psychology and Marketing*, *24*(7), 641–657. doi:10.1002/mar.20177

Lowery, C. (2009). An explosion prompts rethinking of Twitter and Facebook. *Nieman Reports*, *63*(3), 32–34.

Lu, J., Lu-Zhuang Wang, & Chun-Sheng Yu. (2007). Is TAM for wireless mobile data services applicable in China? A survey report from Zhejiang, china. *International Journal of Mobile Communications*, *5*(1), 11–31. .doi:10.1504/IJMC.2007.011487

McEwen, H. (2006). At NKU, students turn in podcasts. *Cincinnati Enquirer/Cincinnati.Com*,, Nov. 6, 2006.

Miller, D. B. (2006, October 16). Podcasting at the University of Connecticut: Enhancing the educational experience. *Campus Technology,* Neff, J. (2009). School's out: Are students taking a pass on Facebook? *Advertising Age*, *80*(31), 43–43.

Palmgreen, P., & Rayburn, J. D. (1982). Gratifications sought and media exposure: An expectancy-value model. *Communication Research*, *9*(4), 561–580. doi:10.1177/009365082009004004

Perse, E. M., & Dunn, D. G. (1998). The utility of home computers and media use: Implications of multimedia and connectivity. *Journal of Broadcasting & Electronic Media*, *42*(4), 435–457.

Read, B. (2005, July 1). Duke U. assesses iPod experiment and finds it worked - in some courses. *The Chronicle of Higher Education*, *51*, A28–A28.

Read, B. (2005, March 18). Seriously, iPods are educational. *The Chronicle of Higher Education*, *51*, A30–A32.

Read, B. (2006, May 12). Duke stops giving students free iPods but will continue using them in classes. *The Chronicle of Higher Education*, *52*, A39–A39.

Rogers, E. M. (2003). Learning reconsidered: Education in the digital age. *Journalism and Mass Communication Educator*, *57*(4), 306–309.

Siegert, K. (2005). Podcasting in medicine. *Journal of Visual Communication in Medicine*, *28*(4), 176. doi:10.1080/01405110600560128

Sloan-C. (2006). Beyond lecturecasting: Using podcasts for discussion and student content generation. *SloanCWiki,*, Feb. 16, 2007.

Snell, J. (2007, January). College via podcast. *Macworld*, *24*, 128.

Souza, J., & Bingham, P. (2006). Integration of available and new technologies to raise student understanding and engagement. *Journal of Educational Technology Systems*, *34*(2), 189–198. doi:10.2190/79MV-UAKA-CK2M-KYBV

Sreebney, O. (2007). *Digital rendezvous: Social software in higher education*. Educause Center for Applied Research.

Stern, B. B., Royne, M. B., Stafford, T. F., & Bienstock, C. C. (2008). Consumer acceptance of online auctions: An extension and revision of the TAM. *Psychology and Marketing*, *25*(7), 619–636. doi:10.1002/mar.20228

Thoms, K. J. (2000). Integrating technology into the classroom. *6th Annual Mid-South Instructional Technology Conference,* Murfreesboro, TN.

Thornton, P., & Houser, C. (2005). Using mobile phones in English education in Japan. *Journal of Computer Assisted Learning*, *21*(3), 217–228. doi:10.1111/j.1365-2729.2005.00129.x

Trepte, S., Ranne`, N., & Becker, M. (2003). 'Personal digital assistants'- patterns of user gratifications. *The European Journal of Communication Research*, *28*(4), 457–474.

Trimmel, M., & Bachmann, J. (2004). Cognitive, social, motivational and health aspects of students in laptop classrooms. *Journal of Computer Assisted Learning*, *20*(2), 151–158. doi:10.1111/j.1365-2729.2004.00076.x

Vicario, T., Henninger, E., Austin, M., & Chambliss, C. (2002). *College students' responses to new communication technologies* (Reports-Research. Pennsylvania: Ursinus College. Virvou, M., & Alepis, E. (2004). Mobile educational features in authoring tools for personalised tutoring. *Computers & Education*, *44*(1), 53–68.

Voss, B. D. (2003). *The economic development case for investment in higher education information technology* (Research Bulletin. Boulder, CO: ECAR: Educause Center for Applied Research. Webster, J. G., & Lin, S. (2002). The Internet audience: Web use as mass behavior. *Journal of Broadcasting & Electronic Media*, *46*(1), 1–12.

Wei, R. (2001). From luxury to utility: A longitudinal analysis of cell phone laggards. *Journalism & Mass Communication Quarterly*, *78*(4), 702–719.

Zurita, G., & Nussbaum, M. (2004). Computer supported collaborative learning using wirelessly interconnected handheld computers. *Computers & Education*, *42*(3), 289–314. doi:10.1016/j.compedu.2003.08.005

KEY TERMS AND DEFINITIONS

Course Management System (CMS): Such as Blackboard or WebCT, which allows faculty to post assignments online.

Information Seeking Communication: Communication for the purpose of finding information about friends and family.

Innovativeness: A trait that refers to how "forward thinking" or adventurous a person may be when trying something new.

Social Networking Site (SNS): Such as Facebook or MySpace, designed to allow people to communicate with friends and meet new people.

Technology Acceptance Model (TAM): A theoretical approach that examines how different people evaluate the usefulness of a new technology before adopting it.

Twitter: A social networking site that allows people to "follow" the updates of others, increasingly used for both social and business purposes.

Uses and Gratifications: A theoretical approach that examines the reasons why people use certain media, such as watching the news to get specific information.

Utility Communication: Communication that holds a specific use for school or work.

Chapter 4
Standardized, Flexible Design of Electronic Learning Environments to Enhance Learning Efficiency and Effectiveness

Jennifer R. Banas
Northeastern Illinois University, USA

ABSTRACT

The demand for mobile, electronic learning environments has increased, but so, too, has the demand for performance-based outcomes. Within this context, efficiency and effectiveness have become the gold standards for quality. The design of electronic learning environments, both blended and fully mobile, requires unique considerations, particularly in regards to self-regulated learning, cognitive load, and learner characteristics. Repeated development of an effective and efficient electronic learning environment can be facilitated through the use of a standardized, flexible course design model. A sample course design model that promotes efficiency and effectiveness, while catering to the unique considerations for mobile learning in an electronic learning environment are presented along with suggestions for future conversations and research.

DOI: 10.4018/978-1-60960-511-7.ch004

INTRODUCTION

Henry Ford, American industrialist and pioneer of the automobile said, "A market is never saturated with a good product, but it is very quickly saturated with a bad one." With the current economic market's emphasis on improved performance comes the necessity to establish a trend towards improved the design of blended and full mobile electronic learning environments. Standardization may help.

The concept of standardization may bring about thoughts of Ford's best known invention, the assembly line. Instead of an automobile factory, though, it is a learning factory in which knowledge is the product of instructional components strung together like an anthology of widgets. This vision begets the question: "Can standardization generate a rich, relevant learning experience or does it lend to the diploma mill image too often associated with mobile learning?" The answer to this two-part question is "maybe" and "not necessarily," respectively. The answer is dependent on the structural design of the "assembly line" and the selection of "widgets." If the design is architecturally matched to how people learn and offers opportunities for meaningful learning experiences, then the possibilities for advantageous learning outcomes abound.

Contrary to negative connotations often associated with standardization, calibrating instructional design can actually facilitate and even ensure the development of a repeatedly rich, germane, electronic learning environment. When that environment is flexible and coupled with high quality, relevant learning objects, and features that promote self-regulated learning and reduced cognitive load, both instructors and learners can profit. Learning, however, cannot be the only measure of excellence. During an economic era in which greater pressure is being placed on individuals and organizations to improve performance, efficiency and effectiveness become the golden measure of quality.

This chapter will make a case for using a standardized, flexible course design model that supports learning objects to improve the efficiency and effectiveness of instruction. The discussion will begin with a look into the current demand for and the popular criticism of mobile learning. Within this context, the unique challenges and opportunities afforded by electronic learning environments (ELEs) will be examined from the perspective of efficiency and effectiveness. (ELEs, as a term, will be used synonymously throughout the chapter to refer to online, computer-based, or high technology learning forums). Considerations for human cognitive architecture, particularly cognitive load and self-regulated learning, will be part of this examination. Next, a standardized, flexible course design model will be proposed as a means to address these considerations and capitalize on these opportunities. Part of this defense will include a foundational review of the standardization of distance-based learning and how the principles relayed relate to the design of ELEs. Within this discourse, a sample course design model and other enhancements to improve ELEs will be presented. Finally, suggestions for future conversations and research will be shared.

BACKGROUND

The Growth and Criticism of Mobile Learning

Wiley (2000) stated that "Technology is an agent of change, and major technological innovations can result in entire paradigm shifts" (p. 2). This observation is particularly evident in the explosion of online courses, degree programs, and universities. According to the United States-based 2006 Sloan Consortium report, *Making the Grade – Online Education in the U.S.*, enrollment in one or more online courses increased from 1.6 million students in 2002 to 3.2 million students in 2005; the growth rate from 2004 to 2005 was thirty-five percent Using results from over 2,500 colleges and universities nationwide, this annual survey (a collaborative effort between the Babson Survey Research Group, the College Board and the Sloan Consortium) is the leading indicator of electronic learning in the United States. The 2009 (Allen and Seaman) report indicated growth from 2007 to 2008 was seventeen percent. While the growth rate was lower than some previous years, the increase in learners taking at least one course online from 1.6 million in 2002 to 4.6 million represented a compounded growth rate of nineteen percent!

Despite the demand, blended and fully mobile degree programs have not gone without criticism. Most arguments center on quality. Merrill pointedly indicated "easy to use tools and inexpensive availability of server hosting makes it possible for anyone with even minimal computer skills to uncritically shovel information onto the internet and call it instruction"(2007, p. 360). Consequently, some academics have yet to accept online education (Allen & Seaman, 2006). Inman, Kerwin and Mayes (1999) found that while higher education institutions are willing to teaching blended or fully mobile courses, they rated the quality as equal to or lower than on campus courses. Part of the disdain is due to the fact that much of the design and development of ELEs is left to faculty members who have had little to no training in best practices for online education (Moller, Foshay, & Huett, 2008). Hereto, faculty members fall back on a craft approach. Under the *craft approach*, instructors design and develop their online courses based on what has worked for them in the classroom (Moore & Kearsley, 1996). The problem with this classroom approach is that ELEs present their own set of unique opportunities and challenges.

ELECTRONIC LEARNING ENVIRONMENTS

Efficiency and Effectiveness in ELEs

The design of ELEs requires a consideration for human cognitive architecture and those factors relevant to learning at a distance. Failure to consider these factors and their influence on learning in an ELE could negatively impact the user's experience and/or result in a missed opportunity. These considerations will be explored in terms of self-regulated learning and cognitive load. A brief discussion about efficiency and effectiveness, within the context of ELE design, however, will take place first.

According to Hjeltnes and Hansson (2005), the focus on the flexibility of ELEs should never be at the sacrifice of quality. Technology without consideration for effectiveness and efficiency has the potential to make electronic learning very expensive. "Efficiency," stated Rumble (1997), "is ratio of output to input. A system increases its cost-efficiency when it maintains output with a less than proportional increase in inputs" (p. 120). From an ELE perspective, the time, money, and effort one puts into learning (or accessing learning for that matter), should not exceed the benefits of the learning and performance outcomes. Regarding effectiveness, a term pointedly borrowed from economics, Rumble (1997) indicated that an organization is effective when its outputs are relevant to and meet the needs and demands of its clients as set forth by an established set of criteria.

Efficiency and effectiveness in education should not be confused. An ELE can be effective, but not efficient. For example, an institution can put an exorbitant amount of money into designing and delivering education, but if the costs exceed the output, the ELE is not sustainable. Similarly, if an institution delivers a highly efficient ELE, but it is not well matched to the needs of its learners or the desired performance outcomes of the learner's sponsoring institution, the demand will become defunct. In order for an ELE to be both effective and efficient, designers and instructors must take into consideration several factors related to the cognitive architecture of the human brain which present themselves uniquely in such an environment. Self-regulated learning and cognitive load are two such factors.

Self-Regulated Learning and ELEs

Electronic learning environments are well-suited for (and actually require) self-regulated learning behaviors. *Self-regulated learning* refers to "learning that occurs largely from the influence of students' self-generated thoughts, feelings, strategies, and behaviors, which are orientated toward the attainment of goals" (Schunk & Zimmerman, 1998, p. viii).

Self-regulated learners, according to Schunk and Zimmermann (2008), are students who efficiently control their learning experiences in a variety of ways. These techniques include establishing a conducive work environment and using resources effectively; employing various cognitive and metacognitive learning strategies to understand course materials; regulating emotions during academic tasks; and adopting positive motivational beliefs about their own capabilities, what they are learning, and how they can improve upon their learning (Artino, 2008; Schunk & Zimmerman, 2008). From an academic standpoint, self-regulated learning has been studied to understand better how successful students adapt their cognition, motivation, and behavior to improve learning (Artino, 2009). The belief is that learners who use more self-regulated learning strategies are more like to outperform their less self-regulated counterparts (Pintrich, 1999).

In ELEs, learners are required to self-regulate their learning by way of directing, generating, and monitoring their own learning progress and performance outcomes. Each of these activities depends on the design efficiency and effectiveness of the learning environment. From an efficiency standpoint, learners must easily be able to schedule their learning, to locate and retrieve course materials, to submit work, and to complete performance assessments in a similar fashion each time. If enrolled into more than one course, module, training, or other type of learning unit, symmetrical design of the learning environments further adds to the efficiency of the experience. This same efficiency principle applies to the instructor and instructional designers. These course leaders should easily be able to navigate the ELE and to modify course content to improve relevancy. From an effectiveness standpoint, learners should be able to track and reflect their progress. They should also be presented with opportunities to apply new knowledge or practice new skills, without penalty. Features such as these, and the standardized inclusion of them, can all help to improve the efficiency and effectiveness of ELEs from a self-regulated learning standpoint. Cognitive load and the influence of memory are another consideration.

Memory, Cognitive Load, and ELEs

Knowledge and awareness of long-term memory, working memory, and cognitive load is essential for the effective and efficient design of ELEs. *Long-term memory* is the large, central location in which information is stored, managed and retrieved for later use; *working memory* is the conscious locus of short-term information processing; and *cognitive load* is the burden placed on working memory during instruction (Sweller, 2003, 2004). Cognitive load can be affected by the mental effort necessary to process new information (intrinsic cognitive load), the manner in which the material is presented (extraneous cognitive load), and the effort required for activating schema (germane cognitive load) (Sweller, 1988). Comprehending the interplay between long-term and working memory helps to justify why a standardized course design model is best for both improved efficiency and effectiveness. Foundational research regarding these concepts and the design implications for ELEs are presented next.

Much knowledge regarding long-term and working memory is derived from the foundational works of De Groot (1965) and Chase and Simon (1973) and their observation of chess players. These researchers found that much of expert player's success could be attributed to the phenomenon of long-term memory. Repeated playing of the game lead to memory storage of winning chess moves such that they did not have to "think" about their next move, they just "knew" it, much like a violinist does not thinking about playing each chord in a song, he/she just knows it. Consequently, their working memory was not overburdened in the way a novice player who has to approach each move as a relatively new experience. This means "that expertise....in any area, is at least heavily dependent and possibly solely dependent on knowledge held in long-term memory" (Sweller, 2004, p. 11).

Cognitive load influences both the efficiency and effectiveness of one's learning experience in an ELE. "Limited working memory is one of the defining aspects of human cognitive architecture, and accordingly, all instructional designs should be analyzed from a cognitive load perspective" (Sweller, Van Merriënboer & Paas, 1998, p. 262). According to Sweller (2004), information can enter working memory via one of two routes: from long-term memory as previously learned material, or as new information via one's sensory. In a multimedia environment, "a number of task characteristics, including format, complexity, use of multimedia, time pressure, and pacing of instruction.....have been identified by CLT research to influence cognitive load" (Paas, Ayres, & Pachman, 2008) .

From an ELE design standpoint, a standardized course design model would lessen the load placed on working memory because learners would not have to relearn how to navigate the environment (including activities such as locating course materials, submitting assignments, completing assessments, and participating in discussion boards) for each new course, unit, or other learning module. In other words, by using the same course design, learners can more quickly become experts (as frequently repeated tasks are stored in long-term memory), thus lightening the load on working memory and improving the efficiency of the learning experience. With a lesser burden placed on working memory, learners are freed to devote more energy toward learning the course material, thus improving opportunities for deeper learning (i.e. effectiveness) and greater performance outcomes.

Another cognitive load consideration for the design of ELEs is the transitory nature and relatively small capacity of working memory. The transitory aspect was uncovered by Peterson and Peterson (1959) who found that people, when asked to remember an unfamiliar combination of letters, could only do so for a short period of time without rehearsing them. Without rehearsing, the information cannot pass from working to long-term memory. Applying this aspect to the design of ELEs, this means that the number of steps to complete a task, such as in submitting an assignment, retrieving an article, or posting in a weekly discussion board should be limited. It also means tasks steps are best standardized so as not to impose a new combination of steps to be learned each time the learner must complete a task. The idea here is to make processes as efficient as possible.

Regarding the capacity of working memory, Miller (1956) found that people could hold no more than 5-9 chunks/elements of previously unknown information. When asked to do work with those chunks, the load placed on working memory is even greater, and the number of chunks that can be held is less. For this reason, course material presented within an ELE's is best chunked into standardized units/modules, and sometimes even sub-units and sub-modules. The more complex the concepts or skills to be learned, the more chunking that needs to take place. Further discussion about how to organize course material is shared later in this chapter.

OTHER CONSIDERATIONS

Poor design, inattention to varying learning styles, lack of a support system, and failure to recognize the self-selecting content needs and socialization patterns of adult learners has turned some students away from mobile learning (Frontline Group, 2001). Additionally, despite the convenience that ELEs may offer, less technologically adept learners are often intimidated by idea of learning in a digital environment where they are isolated from a face-to-face support system. Knowing human cognitive architecture and its relationship to learning outcomes are rooted in one's ability to receive, process, and use of information, carefully assessing learners' characteristics is a critical step in the design process.

Paas et al. (2008) referred to the interactions between task and learner characteristics as *mental load*. Learner characteristics to consider in the design of (and tasks required in) ELEs include technology expertise level, age, and spatial ability. Of these characteristics, technology expertise deserves special mention. According to Lohr and Gall (2008), cognitive load will vary based on expertise. In light of the self-regulated learning required to perform successfully in an ELE, learners may lose the motivation and will to extend effort to complete a course if frustration or fear too often supersedes the intrinsic reward of new knowledge and completion. In this regard, the mental load (the interplay of task and learner characteristics) becomes too great. This means that ELEs, minimally, need to be designed in a way that supports the needs of the least technologically adept learners. The challenge of a learning unit should lie with *learning* the content, not the interaction with an ELE.

A final consideration not aforementioned is the cognitive load placed on ELE instructors. ELE instructors bring to the table their own set of learner characteristics which interact with task characteristics. Technology expertise, for example, can significantly impact their ability, and consequently, their desire to engage with learners, the course material, and course management tasks. Recall that cognitive load can be affected by the mental effort necessary to process new information (intrinsic cognitive load), the manner in which the material is presented (extraneous cognitive load), and the effort required for activating schema (germane cognitive load) (Sweller, 1988). If ELE instructors' level of expertise are poorly matched to the ELE or the ELE is poorly designed and inefficient, instructors may be unable or become unmotivated to perform necessary course functions due to the extra time commitment required to be successful. In such situations, both learners and the level of quality associated with the instructor's institution will suffer. Additionally, those instructors may choose not to return to teaching in ELEs. According to Green, Alejandro, and Brown (2009), the major factor discouraging teachers from teaching in a blended or fully mobile environment is time commitment. High turnover rate of online instructors and poor learner performance could possibility be avoided given effective and efficient course design.

For the all the considerations shared above –self-regulated learning, cognitive load, learner characteristics—a standardized, flexible course design model offers instructional designers a solution, and both learners and instructors the opportunity to focus on what is most important: learning and teaching.

ELE DESIGN

Tallent-Runnels, Thomas, Lan, Cooper, Ahern, and Shaw's (2006) review of mobile teaching, suggested that while convenience is important for students, the quality of the instructional design is the most vital element, even for the most focused and motivated student (p. 112). Artino (2008) found that, "perceived instructional quality was the strongest individual predictor of overall satisfaction" and that "students who felt the course utilized effective instructional methods were also more likely to be satisfied with their online learning experience" (p. 267). If the quality of a course is to be measured in terms of effectiveness and efficiency, a poorly designed course could lend itself to an unsatisfactory rating due to factors un-related to the course content.

The ability of a learner to navigate his/her way through the learning environment and any associated motivation he/she feels towards completing course requirements are critical considerations for ELE design. A well-designed course model can also lead to improved learning outcomes. Runyan (2007) found an increase in student achievement after redesigning a mobile course to reflect the standards of a high quality ELE as established by Quality Matters, an organization whose peer review process is designed to certify the quality of electronic learning environments. Some of these quality standards included, but were not limited to: (1) The tools and media support student engagement and guide the student to become an active learner; (2) Navigation throughout the online components of the course is logical, consistent, and efficient; (3) Students have ready access to the technologies required in the course; (4) The course components are compatible with current standards for delivery modes; and (5) The course design takes full advantage of available tools and media. These standards and that of similar organizations are representative of the growing trend to improve the quality of mobile learning. One other such organization is the Association for Educational Communications and Technology (AECT).

AECT is one of the leading and oldest organizations dedicated to improving instruction by way of technology. Members of this organization include those from the fields of educational technology design and development, distance learning, teacher education, school library media sciences, and information technology. Recently, the organization revised their definition of educational technology to read: "Educational technology is the study and ethical practice and facilitating learning and improving performance by creating, using, and managing appropriate technological processes and resources" (Januszeweski & Molenda, 2008, p.2). Taking into account the latter part of this definition, "creating, using, and managing appropriate technological processes and resources" and the unique consideration for learning in an ELE, a case firmly can be made for researching, testing, implementing ELE course design model to improve both the efficiency and effectiveness of instruction.

A HISTORICAL CASE FOR STANDARDIZING ELE DESIGN

Foundational thoughts about distance learning and the standardization of instruction can be traced back to Peters' (1988) research into distance teaching institutions in the 1960's. He proposed that distance education could be analyzed in a way similar to the industrial production of goods and evaluated using categories such as standardization, assembly line, planning, organization, mass production, and scientific control methods. Concluding that the process of teaching as a whole could be improved through mechanization and automation, he noted that the development of distance study courses was just as important as the preparatory work taking place prior to the production process; the effectiveness of the teaching process was particularly dependent on planning and organization; and that courses must be formalized and expectations from students standardized. With the exception of students' expectations needing to being standardized (a stark contrast to constructivist approaches to instruction) there is some valuable insight to be taken from these principles.

The first principle to be taken away from Peters' research is that the design of an ELE requires considerate planning and organization in addition to the development of the course content. As mentioned earlier, a craft approach in which instructors simply design their online courses to reflect their teaching practices in the classroom fails to recognize the different challenges and rewards that accompany blended and fully mobile learning, for both the instructor and the learner. Interactions in an ELE require designers to plan ahead for procedures such as assignment and assessment submission, document retrieval, discussion board postings, and course communications. Each of these tasks and the steps required to complete them must be planned ahead and clearly communicated to the learner.

A second principle to take away from Peters' research is that courses should be formalized. In other words, legitimate (i.e. research-driven) methods should be applied consistently to the design of ELEs. Thoughtful consideration for cognitive load, self-regulated learning, and other factors previously mentioned also should be included into the design features. This includes manageable and appropriate packaging of learning units/modules, frequent opportunities for assessment and self-reflection, meaningful and relevant applications, and other features proven to be effective instructional elements (Merrill, 2002).

Moller, Foshay, and Huett's (2008) contended that if electronic learning is to become mainstream (in other words, accepted into the global communities of business and academia as a viable and superior means of learning) then there is a need to develop and use a research-based, standardized, flexible course design models. These models should allow instructors and/or instructional designers to efficiently tailor and deliver course content in ELEs that effectively meet the needs of the learner audience. What these models should look like, the means to evaluate their quality, and the directions for future research comprise the remaining sections of this chapter.

IMPROVING EFFICIENCY AND EFFECTIVENESS BY DESIGN

"Failure is simply the opportunity to begin again, this time more intelligently" – Henry Ford

According to Meyer (2003), electronic learning has brought with it a "renewed focus on pedagogy and instructional design" (p. 20). Within this discourse, an emphasis has been placed on efficiency and effectiveness. Such has been the driving force behind some institutions to develop a standardized, flexible course design model for degree programs that efficiently and effectively incorporates known instructional design theories and principles. A standardized, flexible course design model is one that can be used over and over again to build different modules, courses and even entire programs, with interchangeable, relevant learning objects. When a standardized, course model does not exist, the user may be able to find an alternative way to reach the desired outcome, but unnecessary effort, time, or money may be expended in the process; and worse yet, the quality of the desired outcome may suffer or be rendered unachievable.

Having served both full-time and part-time at fully mobile and blended online institutions in multiple capacities, including course designer, professor, and dean, this author argues that despite the criticism, superior learning and instruction can feasibly take place in an ELE when that environment is well-designed. This design includes a high-quality, standardized, flexible course design model that facilitates the use of appropriate learning objects; motivational enhancements to support self-regulated learning; frequent progress monitors and formal assessments that provide feedback to both the learner and instructor; meaningful and relevant assignments to practice new skills and integrate new knowledge into current practice; and technological support devices to reduce fear and lessen cognitive load while navigating the ELE. These design components will be addressed from the standpoint of efficiency and effectiveness respectively.

OPPORTUNITIES FOR IMPROVED EFFICIENCY

From an efficiency standpoint, both instructors and learners benefit from the use of a standardized, flexible course design model. Instructors benefit from a course development stand point; learners benefit from a cognitive load standpoint. Each of these efficiency opportunities will be described next.

Opportunities for instructor efficiency: Taking into account the benefits of planning, organization and formalization uncovered in Peter's (1988) foundational research into distant learning, providing an instructor or instructional designer with a standardized, flexible course design model is much like providing the skeletal outline of English composition essay to either an experienced or novice writer. The experienced writer benefits from the outline in that he/she is able to focus efforts on and to devote time to developing other writing techniques and deepening existing skills. The novice writer benefits from an outline in that he/she is given a minimum standard of excellence from which to build, or in the cases of some novices, insight into what such an essay even looks like in the first place. Similarly, an instructor or instructional who is given a standardized, but flexible course design model is gifted with more time to develop course content rather than using energies to develop a template or organizational scheme for each course.

A standardized, flexible course design model permits the instructor or instructional designer to easily customize the ELE to match the desired learning outcomes. Consider the customization offered with a video game avatar or a car. Each of these items has a predictable model (or framework) one has come to expect. They both also have certain parts/objects placed into that framework that can be tailored. The car, for example, usually has seats, doors, a stereo, and floor mats that allow for customization. Similarly, the avatar permits a customized name, hair, clothing, and voice. Together, the framework and its customized components work together efficiently to deliver a personalized experience. These customized components in an ELE are referred to as learning objects.

Learning objects are any unit, digital or non-digital, which can be used, re-used or referenced during technology supported-learning (LOM, 2000). Learning objects vary in type and size; they can be as small as a single image, a 30-second cameo video, or an exam question; or large and complex as 20-30 minute video lectures, a series of multi-media slides, an interactive tutorial, or a collection of readings (McGreal, 2004). Some learning objects may be used one, twice, or many times in different courses or modules, or in different programs. A standardized, flexible course design model makes it easy for instructional designers or instructors of ELEs to "slide" relevant learning objects in and outdated or irrelevant learning objects out of the course without needing to rebuild the learning management system that accommodates them. It is this customization feature that also helps the instructor or instructional designer to improve the effectiveness of the ELE.

Opportunities for learner efficiency: From a learner standpoint, a standardized, flexible course design model can help to reduce cognitive load and anxieties associated with navigating an unfamiliar learning environment from one course to the next. When the format remains the same, learners efficiently can go about the task of learning the course content. This is particularly of benefit to those new to ELE learning or who are less technologically adept. Merrill (2006) reminded us that for most young adults, computers, electronic games, and other digital devices are a way of life for them, but for older learners, computers often cause anxiety. Such is the difference between today's digital natives and yesterday's digital immigrants. Terms coined by Prensky (2001, 2005), a *digital native* is someone for whom digital technologies already existed when they were born; a *digital immigrant* is someone who grew up without digital technology and adopted it later.

In an ELE, digital immigrants may need to extend an additional working memory to technological tasks more commonplace for the digital native. For example, this author, at her former online academic institution, often found herself coaching new ELE learners in basic "housekeeping" (i.e. course procedures) tasks such as attaching documents to emails, copying/pasting text, downloading and saving course documents to one's computer, and uploading assignments. Additionally, these coaching events were often accompanied by emails in which learners shared feelings of being overwhelmed with the course content. Based on cognitive load theory, these feelings make sense. These learners, much like De Groot's (1965) and Chase and Simon's (1973) novice chess players, were expending more working memory to learning the environment and the types of "moves" to be made than embodying the content. After learning and practicing these navigation skills in the first course, learners went on to complete more quickly these tasks without support and frustration in their next courses. Questions in emails received from these "expert" students related more to interpretation of the course content than how to perform procedural tasks.

Part of learners' decrease in technology frustration and increased attention to content learning at this author's former institution could be attributed to learners having mastered basic technology skills; but, other credit must be given to the standardized design used in all courses. Each course within a given degree program utilized the same course design model. The course syllabus, assignments, supporting course materials, and assessments could be found in the same place each time. Additionally, procedures for interacting with these components, as well as the discussion boards, were identical from one course to the next. Learners never needed to devote energies after their first course to locating and using these items. Finally, the number of assignments, assessments, and discussion board postings remained constant. Such standardization of design of procedures benefited the digital native as well as the immigrant.

OPPORTUNITIES FOR IMPROVED EFFECTIVENESS

Merrill (2000) commented on a sign he had read on the wall of an instructional design company: "If a product does not teach, it has no value" (p.1). In other words, a product is only effective to the extent that it meets the needs of its targeted audience. As discussed earlier, efficiency does not beget effectiveness; therefore, the effectiveness of a course design model must be addressed for its own unique contribution to the quality of the learning experience. A standardized, flexible course design model not only buttresses improved instructional and learning efficiency, but also provides opportunities to improve course effectiveness. Being able to make course content more relevant by way of interchangeable learning objects is one such opportunity. A second is providing a structure into which motivational enhancements that foster self-regulated learning practices can be built. A third is freed time for learners to further develop technology literacy. And finally, a fourth is providing a structure into which research-driven effective instructional design principles can be consistently applied. Each of these factors will be described respectively in the context of effectiveness.

Opportunities to build in relevancy - A standardized, flexible course design permits instructional designers and instructors of ELEs to modify the content of the course to include relevant videos, readings, discussion post topics, and audio files that are a best match to the audience without changing the framework of the course. In an ELE where instructors and learners are separated by physical distance, it is even more important to find ways to make the course content relevant. The verbal and non-verbal feedback obtained by instructors in a traditional classroom to modify instructional examples and course materials to make them more relevant is less frequent in ELEs.

Relevancy is important in that it can be tied to course effectiveness. Banas (2008, 2009) found task relevancy and task attractiveness to be positively correlated with learning intentions and perceived effort given. This is because relevancy leads to an optimistic task- appraisal and consequently, an increase in learners' commitment to give effort while performing a given task (Boekaerts, 1992, 1996). Banas (2008, 2009) also uncovered a positive relationship between task attractiveness and task relevancy with cognitive performance. These findings are consistent with Bee and Bjorkland (2004) who found that adult learners are less likely to consider information that is not relevant. If information is not considered, it cannot be learned. Knowing that relevancy could improve learning outcomes makes the practice of instilling relevant learning objects into a standardized, flexible course design model worthwhile.

Opportunities to foster self-regulated learning - In addition to facilitating the development of a relevant learning environment, a standardized, flexible course design model provides a structure that promotes self-regulated learning. According Howland and Moore (2002), self-regulated learning behaviors are essential in ELEs. The socio-emotional support systems that typically accompany face-to-face instruction and help to the keep the learner both motivated and on track are less readily available in ELEs where instructors or classmates are not "there" and class is not "meeting." Consequently, the ELE needs to be designed in a way that promotes self-regulated learning behaviors. A standardized, flexible course design model can provide the forum for these behaviors by way of course content organized into manageable "chunks" and frequent opportunities to monitor progress and performance.

Regarding the organization of course content and promoting self-regulated learning behaviors, Artino (2008) suggested breaking assignments into multiple stages. This also should apply to the organization and chunking of course content. When assignments and course content are broken into smaller parts, the load placed on working memory is reduced and learning tasks become more manageable. This is much like the principle of breaking a larger goal into a series of smaller goals so that individuals can be motivated by the small achievements they make along the way. To do this in an ELE, a workable set number of learning modules, assignments, and assessments should be set for each course. What is workable will depend on the length of the course and the knowledge/skill levels of the learners. Process and impact evaluations (discussed later in the chapter), can help to uncover what learners feel is manageable and at what point performance is maximized.

In addition to chunking course materials, frequent opportunities to assess progress and performance can promote self-regulated learning. When learners know how they are doing and how far they have come along, they are more motivated to persevere (Artino, 2008). Progress monitors can include brief non-credit, multiple-choice, checks for understanding after readings/lectures and quizzes after learning modules, as well as unit tests. If the course is built into a learning management system with an online grade book, learners can monitor their progress from both a completion and performance standpoint, both of which generate the self-regulated learning behaviors associated with improved learning outcomes (Artino, 2008). If an online grade book is not available, a form into which learners can check off their progress in the course and enter scores for assignments/assessments can be provided.

A final feature that promotes self-regulated learning is naturally inherent to a standardized, course design model. When learners know what will be expected of them in terms of learning course content, assignments, and assessments from one course to the next, they can schedule and manage better their learning time in advance. Being able to do this not only reduces the anxiety that can sometimes accompany unknown expectations, but also promotes the self-regulated learning behaviors needed to be successful in an ELE when the learner is separated from his or her instructor.

Intentional and inherent features such as chunking, frequent opportunities to assess progress and performance, and predictability each help to promote self-regulated learning and increase the likelihood for improved learning and performance outcomes. They also leave the room for the development of technology literacy.

Opportunities to develop technology literacy-Development of technology literacy skills is another benefit afforded by a standardized course design model. When the format for a course remains the same from one to the next, learners can devote more time to deeper learning. One layer of deep learning that naturally complements learning in an ELE is technology literacy. According to Judson (2010), "While students today are certainly far more comfortable and confident approaching technological tools than students 20 years past, this poise does not necessarily translate into being literate in technology" (p. 272). They may be comfortable with the end products of technology such as video games and websites, but they not have yet learned how to use technology to solve problems, analyze information, and model complex ideas. These abilities are characteristics of a learner who is technologically literate as indicated by the National Education Technology Standards (NET-S) and Performance Indicators for Students (ISTE 2007). Such individuals are also able to construct knowledge and develop innovative products and processes. By standardizing the course design model, the working memory of the digital immigrant is freed to apply his/her efforts to learning the course content and becoming more technologically literate learner. Consequently the learner is not only learning from technology, but also *with* technology.

Opportunities to apply research driven practices – A fourth benefit of a standardized course design model is that it provides a consistent structure to apply research-driven, design techniques known to improve instructional effectiveness. One such set of techniques are Merrill's First Principles of Instruction (2002).

Merrills' First Principles of Instruction (2002) are the product of a systematic review of instructional design theories, models and research. From these sources Merrill abstracted a set of interrelated prescriptive instructional design principles. To be included into this set, the principle frequently had to appear in most instructional design theories; had to promote effective, efficient, or engaging learning; and had to be both design-orientated and supported by research. According to the First Principles of Instruction (Merrill, 2002), effective learning occurs when: instruction is presented in the context of real world problems; instruction attempts to activate relevant prior knowledge or experience; instruction demonstrates what is to be learned rather than merely tell information about what is to be learned; learners have an opportunity to practice and apply their newly acquired knowledge or skill; and instruction provides techniques that encourage learners to integrate the new knowledge or skill into their everyday life. Merrill's (2002) first principles of instruction offer much in the way of developing an effective standardized course design model. They are described in greater detail next.

The first principle reminds instructors and instructional designers that primary desired outcomes should be identified in the context of real world problems. According to Mendenhall, Buhanan, Suhaka, Mills, Gibson, and Merrill (2006), real world tasks effectively motivate learners to retrieve and apply acquired skills and knowledge. This could accomplished via a video case study, a newspaper article showcasing how the desired outcome relates to a current event, or requiring learners to identify the desired outcome in their own environment.

According to the second principle, instruction should activate relevant prior knowledge or experience. The importance of relevancy, particularly in terms of motivating self-regulated learning behaviors that lead to effective learning outcomes was discussed earlier in this section. Also discussed was how a standard, flexible course design model efficiently facilitates the use and reuse of relevant learning objects. In an ELE, discussion board questions that prompt the sharing of past experiences and self-assessments can also help to activate prior knowledge.

The third and fourth principles are closely tied. The third principle indicates that instruction should *demonstrate* what is to be learned rather than merely telling learners them; the fourth principle indicates that instruction should provide students with opportunities to practice parts of the whole task in assignments. This means that instruction should include provide examples of the final product or performance. In an ELE, this demonstration could take place in the form of videos, podcasts, or sample written works. Following the example, learners should have opportunities to practice and then apply fully their new knowledge and/or skills.

The fifth and final principle indicates that instruction should encourage learners to integrate their new knowledge and skills into their everyday life. This aligns with Knowles, Holton, and Swanson (2005) who indicated adult learners are motivated to learn that which satisfies needs and interests, is life-centered, and allows them to be self-directed. It also aligns with the concept of relevancy earlier discussed in this section. Instructors and instructional designers should provide learners with the opportunity to practice and apply real and relevant learning tasks within the confines of the ELE or in their own environment.

Abiding by these principles help to avoid what Wiggins and McTighe (2005) refer to as the twin sins of curricular design. These sins are *activity focused teaching* and *coverage focused-teaching*. Respectively, too often instructors fall back on favorite activities that lack an explicit focus on important or appropriate evidence of learning; or they feel pressure to cover an extensive amount of factual material, far beyond what a learner could ever retain beyond the length of the course or use in the "real world." So what might a standardized course design model that incorporates these principles and other elements explored look like?

A SAMPLE STANDARDIZED, FLEXIBLE COURSE DESIGN MODEL

Figure 1 presents a sample standardized, flexible course design model for an 8-week long blended or fully mobile course. Week 1 provides an opportunity for the learner to become comfortable with the ELE. During this time, the learner familiarizes him/herself with course expectations, policies and procedures, and the technology to be used. Information about how to contact technical support, technology FAQs, a site map and a course content map should be included into or introduced by this point because they work to reduce anxiety and/or the load placed on working memory so that the learner can go about doing what is most important in the course: learning. Site maps, course content maps and their primary features are described in greater detail next.

Site maps are commonly found on web sites to assist users in locating specific topics or to provide insight into the type of information offered. An ELE site map would point out how to navigate the environment and to complete common course activities. These activities may include but are not limited to locating readings, submitting assignments, participating in discussion boards, downloading the video plug in, checking grades, and communicating with their instructor. To do this, the ELE instructor or instructional designer could take a screen shot of the course home page and add text boxes which point out the primary components with which learners will need to interact. Taking this map a step further, the instructor could number these components and provide an annotated description for each.

Course contents map provide a navigating function similar to the site map, but rather help

Figure 1. Sample standardized, flexible course design model for a blended or full mobile ELE

Week 1	Week 2	Week 3	Week 4	Week 5	Week 6	Week 7	Week 8
Course intro: syllabus, policies, procedures, technology support	Self Assessment (knowledge and skill pre-test)						Self Assessment (knowledge and skill post-test)
Task in context (reading, video, or field experience)	Whole task demo (live, reading, video, or field experience)	Lecture (live, reading or video)	Lecture (live, reading or video)	Lecture (live, reading or video)	Lecture (live, reading or video)	Lecture (live, reading or video)	Lecture (live, reading or video)
Readings	Readings	Readings	Readings	Readings	Readings	Readings	Readings
	Discussion post	Discussion post	Discussion post	Discussion post	Discussion post	Discussion post	
	Review & progress monitor	Review & progress monitor	Review & progress monitor	Midterm	Review & progress monitor	Review & progress monitor	Final
	Task needs assessment	Partial task assignment Pt1	Partial task assignment Pt2	Partial task assignment Pt3	Peer review of partial tasks	Complete task submission	Task Reflection

the learner to navigate principal *understandings*. A course content map highlights the course modules, primary concepts and how they relate to one another. This content map could look like the Community Health Education course presented in Figure 2 or like a traditional concept map with primary components written into various shapes and connected by arrows. Regardless of the format chosen, the purpose is to provide learners with a visual representation of the logic behind and ideas driving the course. Repeated reference to the map also facilitates the storage of these ideas into long term memory, much like the expert chess player who develops a picture in his/her mind of the board and possible moves to be played.

Beyond Week 1, each element relates directly back to Merrill's First Principles of Instruction

Figure 2. Sample course content map

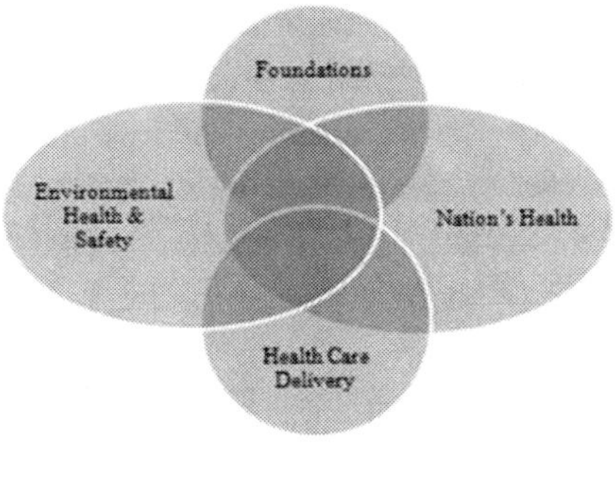

(2005) and motivational enhancements that support self-regulated learning. For example, the self-assessments, non-credit progress monitors, and exams provide learners with the opportunity to assess their gains in knowledge and skills over the length of the course. Week 2 provides learners with a complete task demonstration or example, and a need assessment that could require learners to consider the task in the context of their own setting or a given case study. Weeks 3 through 5 require learners to complete lectures and readings that will help prepare them to complete manageably-sized portions of the whole task. Peer review during Week 6 provides learners with the opportunity to receive non-graded feedback on their work before final submission. Week 8 asks learners to reflect on the task after completion and could ask learners to indicate how they will apply their new knowledge and skills beyond the course so that they make a connection to their everyday life. Weekly discussion boards throughout the course provide a forum in which learners can connect with other learners, share ideas, and reflect more deeply on concepts learned.

The elements shared above are only one way to organize course components into a standardized, course design model. The flexibility of such a design allows for the model to be supported with relevant learning objects in the form of case study videos or written stories, news media podcasts, research articles about best practices, videos clips of practitioners, or video lectures delivered by content expert. However learning objects are used, instructors and instructional designers are cautioned to monitor the number of objects used. Too many objects inserted into a given course increases the risk of overtaxing the cognitive load placed on the learner in the way of the number of items to click on, download, view, etc. Sometimes less is more. If undecided about which learning objects to use, remember that relevancy supersedes novelty.

A RECAP ON EFFICIENCY AND EFFECTIVENESS

This section shared many ideas regarding how a standardized, flexible course design model can efficiently and effectively help to deliver instruction via an ELE. Cognitive load, self-regulated learning, first principles of instruction, learning objects and other concepts were presented as elements to consider in the design. Recall that a system is efficient when it maintains output with a less than proportional increase in inputs and is effective when its outputs are relevant to and meet the needs and demands of its clients as set forth by an established set of criteria (Rumble, 1997). Constant evaluation of this dynamic balance is critical to the success of an ELE.

Making Sure the Design Fits

Due to the physical separation between the instructor and the learner in an ELE, design feedback in regards to pace, clarity, level of difficulty, access to materials, or other issues may be delayed or go unnoted. By instilling process and impact evaluations into the course at regular intervals, learners can more frequently contemplate their experiences, assess their progress, and provide feedback to their instructors, particularly in regards to efficiency and effectiveness.

Process evaluations can provide feedback on course material and activities, including access and appropriateness. Questions in a process evaluation may include: "Were you able to access all course readings? If not, which ones could you not obtain and why?" or "Were there any parts of the Week 3 or 4 assignments for which you would have liked greater clarification on expectations? If yes, which ones?" or "How much time did it take you to respond to the Week 2 discussion question?" Information from such analyses can help the instructor better to design future courses, or to make critical adjustments to the current course. These evaluations also help to get learners thinking more about their interaction with the course materials, particularly items that they may be underutilizing, or in some cases, that they didn't know were there!

Impact evaluations study the immediate or direct effects of the course on learner knowledge, attitude, or skills. For example, in the 4th week of an 8-week course about budgets and budgets management, can learners identify common expense and income categories and line items, or justify expenses? This type of evaluation can take the form of a quiz (in which points are scored) or a progress monitor (for which points are not score, or only participation points for completion are given). Unlike the final assessment, these assessments can be used to gauge learner comprehension and skills levels before they get behind.

Instructors and instructional designers can use feedback from process and impact evaluation to design better future courses or to make critical adjustments to the current ones. Other methods of evaluation and gaps in research regarding the evaluation of ELEs are described in the next and last section.

FUTURE DIRECTIONS

The goal for blended and mobile ELEs should be to deliver an instructional experience as rich as or richer than that received by the learner sitting in a classroom at his or her desk. According to Schlosser and Simonson (2006), "The more equivalent the experiences of distant students are to that of local students, the more equivalent will be the outcomes of the learning experience" (p. 25). This belief is representative of equivalency theory. Equivalency theory (Schlosser & Simonson, 2006) states that local and mobile learners have fundamentally different environments in which to learn, and it is up to the educator to design instruction that provides experiences with equal value for learners.

To be taken from equivalency theory is the continued need to establish standards for ELEs and comprehensive tools by which to measure those standards. Tools exist to measure motivation, self-regulated learning practices, as well as both skill and knowledge-based achievement, but very few, outside of the Quality Matters rating system, provide a comprehensive review of efficiency and effectiveness. With more and more institutions turning to ELEs to deliver blended or fully mobile learning, quality control measures are imperative to ensure desirable performance based outcomes. Efficiency related items could evaluate cognitive load in terms of visual literacy skills and technology skills required; access in terms of universal design components, technology requirements, and ease of use; and learner support in terms of technology aids. Effectiveness related items could evaluate content in terms of overall course goals and course objectives, authenticity of tasks and associated assessment, learner engagement, and opportunities for self-assessment and reflection. This list of items points to some of the unique aspects of ELEs, as well as course content in general. What the list does not include are items related to instructor effectiveness.

Much like the need to develop tools to evaluate blended and fully mobile ELEs, additional research and development is needed into the evaluation of ELE instructors and their knowledge of best practices. Questions to ask instructional designers and instructors before they set out to build and teach in ELEs should include: What kinds of activities best stimulate self-regulated learning? What means of instructor-learner communication work best? What are other means of reducing learner cognitive load? What visual literacy skills are required in electronic learning environments and how can the learning environment best support those requirements? What adaptations are necessary in terms of universal design? How can a community of learners be built when learners and their instructor are physically separated? How does one best support and engage a learner audience made up of both digital natives and digital immigrants? Which instructional design theories best support unique content needs? Questions such as these can drive the processes to design and deliver a more equivalent instructional experience for all learners. They also help to drive the topics for training new instructors and instructional designers of blended or fully mobile ELEs.

CONCLUSION

"Failure is simply the opportunity to begin again, this time more intelligently" – Henry Ford

With the continued trend towards blended and mobile learning in ELEs comes the opportunity to redefine the standards of quality instruction. This chapter made a case for using a standardized, flexible course design model as means to improve upon the efficiency and effectiveness of ELE instruction. Efficiency and effectiveness, as measures of quality, are particularly relevant and significant given today's economical pressures to improve both individual and organizational performance.

Despite existing criticisms of blended and fully mobile instruction, superior learning can take place given a course design model that takes into consideration the unique aspects of learning in ELE. The considerations include cognitive load, self-regulated learning behaviors, and learner characteristics. ELEs can include features to address each of these considerations while also facilitating instructor or instructional designer's development of the learning environment.

A standardized, flexible course design model permits instructors and instructional designers to use, reuse, and remove learning objects that help to customize a course or module and to make instruction more meaningful. Application of basic principles of instruction can help instructors and instructional designers design an ELE that is most relevant to their learner's needs and institution's desired performance based outcomes.

The discussions had and the questions asked in this chapter are only a starting point. Henry Ford, inventor of the assembly line and one of the Industrial Era's most successful businessman said, "Coming together is a beginning; keeping together is progress; working together is success."There are many challenges yet to be faced in the design of equitable, efficient, and effective blended and mobile electronic learning environments, but given the current demand, so to ares the opportunities.

REFERENCES

Allen, I., & Seaman, J. (2006). *Making the grade: Online education in the United States, 2006. Sloan Consortium Group*. Needham, MA: Sloan-C.

Allen, I., & Seaman, J. (2009). *Learning on demand: Online education in the United States, 2006. Sloan Consortium Group*. Needham, MA: Sloan-C.

Artino, A. R. (2008). Promoting academic motivation and self-regulation: Practical guidelines for online instructors. *TechTrends*, *52*(3), 37–45. doi:10.1007/s11528-008-0153-x

Baddeley, A. D. (1986). *Working memory*. New York, NY: Oxford University Press.

Banas, J. (2008). Behavioral construct tailing: A motivational design technique for electronic learning environments. In M. Simonson (Ed.), *31st Annual Proceedings: Selected Research and Development Papers at the Annual Convention of the Association for Educational Communications and Technology* (pp. 23-38). Orlando, FL: Nova Southeastern University.

Banas, J. (2009). Generating motivation to learn via tailored lesson introductions. *International Journal of Information and Communication Technology Education*, *5*(4), 1–17. doi:10.4018/jicte.2009041001

Bee, H. L., & Bjorkland, B. R. (2004). *The journey to adulthood* (5th ed.). Englewood Cliffs, NJ: Prentice Hall.

Chase, W. G., & Simon, H. A. (1973). Perception in chess. *Cognitive Psychology*, *4*, 55–81. doi:10.1016/0010-0285(73)90004-2

De Groot, A. (1965). *Thought and choice in chess*. The Hague, Netherlands: Mouton.

Frontline Group. (2001). *Building a culture that supports and encourages electronic learning*. Paper presented at the Technologies for Learning Conference 2001. Retrieved June 15, 2010, from http://www.frontline-group.com/ news/news_081301.html

Green, T., Alejandro, J., & Brown, A. (2009, June 10). The retention of experienced faculty in online distance education programs: Understanding factors that impact their involvement. *The International Review of Research in Open and Distance Learning North America*. Retrieved June 4, 2010, from http://www.irrodl.org/index.php /irrodl/article/view/683/1279

Hjeltnes, T. A., & Hansson, B. (2005). *Cost-effectiveness and cost-efficiency in e-learning*. Norway: TISIP Research Foundation.

Inman, E., Kerwin, M., & Mayes, L. (1999). Instructor and student attitudes toward distance learning. *Community College Journal of Research and Practice*, *23*, 581–591. doi:10.1080/106689299264594

ISTE. (2007). *National Education Technology Standards (NET-S) and performance indicators for students*. Retrieved June 9, 2010, from http://www.iste.org/content/navigationmenu / nets/forstudents/2007standards/nets_for_ students_2007_standards.pdf

Januszewski, A., & Molenda, M. (2008). *Educational technology: A definition with commentary*. New York, NY: Lawrence Erlbaum Associate & Taylor and Francis Group.

Judson, E. (2010). Improving technology literacy: Does it open doors to traditional content? *Educational Technology Research and Development*, *58*(3), 271–284. doi:10.1007/s11423-009-9135-8

Kayuga, S. (2007). Enhancing instructional efficiency of interactive e-learning environments: A cognitive load perspective. *Educational Psychology Review*, *19*, 387–399. doi:10.1007/s10648-007-9051-6

Knowles, M., Holton, E. F. III, & Swanson, R. A. (2005). *The adult learner: The definitive classic in adult education and human resource development* (6th ed.). Burlington, MA: Elsevier.

Lohr, L. L., & Gall, J. E. (2008). Representation strategies. In Spector, J., Merrill, M., Van Merrienboer, J., & Driscoll, M. (Eds.), *Handbook of research on educational communications and technology* (3rd ed., pp. 85–96). New York, NY: Lawrence Erlbaum Associates.

McGreal, R. (Ed.). (2004). *Online education using learning objects*. London, UK: Routledge Falmer.

Merrill, M. D. (2002). First principles of instruction. *Educational Technology Research and Development*, *50*(3), 43–59. doi:10.1007/BF02505024

Merrill, M. D. (2006). Levels of instructional strategy. *Educational Technology*, *46*(4), 5–10.

Merrill, M. D. (2007). A task-centered instructional strategy. *Journal of Research on Technology in Education*, *40*(1), 5–22.

Meyer, K. A. (2003). Face-to-face versus threaded discussions: The role of time and higher-order thinking. *Journal of Asynchronous Learning Networks*, *7*(3), 55–65.

Miller, G. A. (1956). The magical number seven, plus or minus two: Some limits on our capacity for processing information. *Psychological Review*, *63*, 81–97. doi:10.1037/h0043158

Moller, L., Foshay, W., & Huett, J. (2008). The evolution of distance education: Implications for instructional design on the potential of the Web (Part 2: Higher education). *TechTrends*, *52*(4), 66–70. doi:10.1007/s11528-008-0179-0

Moore, M. G., & Kearsley, G. (1996). *Distance education*. Wadsworth Publishing Company.

Paas, F., Ayres, P., & Pachman, M. (2008). Assessment of cognitive load in multimedia learning: Theory, methods, and applications. In Robinson, D. H., & Schraw, G. (Eds.), *Recent innovations in educational technology that facilitates student learning* (pp. 11–35). Charlotte, NC: Information Age Publishing.

Peters, O. (1988). Distance teaching and industrial production: A comparative interpretation in outline. In Sewart, D., Keegan, D., & Holmberg, B. (Eds.), *Distance education: International perspectives* (pp. 95–111). London, UK/New York, NY: Croom Helm/St. Martin's Press.

Peterson, L., & Peterson, M. (1959). Short-term retention of individual verbal items. *Journal of Experimental Psychology*, *58*, 193–198. doi:10.1037/h0049234

Pintrich, P. R. (1999). The role of motivation in promoting and sustaining self-regulated learning. *International Journal of Educational Research*, *31*, 459–470. doi:10.1016/S0883-0355(99)00015-4

Prensky, M. (2001). Digital natives, digital immigrants. *Horizon*, *9*(5), 1–6. doi:10.1108/10748120110424816

Prensky, M. (2005). Listen to the natives. *Educational Leadership*, *63*(4), 8–13.

Quality Matters. (2008-2010). *Quality Matters rubric standards 2008-2010 with assigned point values*. Retrieved June 5, 2010, from http://qminstitute.org/home/Public%20 Library/About%20 QM/RubricStandards 2008-2010.pdf

Rumble, G. (1997). *The costs and economics of open and distance learning*. London, UK: Kogan Page.

Runyon, J. (2007). Quality in course design and impact on student achievement. Paper presented at the 104th Sloan-C International Conference on Online Learning. Retrieved May 13, 2010, from http://www.sloan-c.org/effective_practices/ quality-course-design-amp-impact-student- achievement

Scholosser, L. A., & Simonson, M. (2006). *Distance education: Definition and glossary of terms* (3rd ed.). United States: Information Age Publishing.

Schunk, D. H., & Zimmerman, B. J. (Eds.). (1998). *Self-regulated learning: From teaching to self-reflective practice*. New York, NY: Guilford Press.

Simonson, M., Schlosser, L. A., & Hanson, D. (1999). Theory and distance education: A new discussion. *American Journal of Distance Education*, *13*(1), 60–75. doi:10.1080/08923649909527014

Sweller, J. (1988). Cognitive load during problem solving: Effects on learning. *Cognitive Science*, *12*(2), 257–285. doi:10.1207/s15516709cog1202_4

Sweller, J. (2003). Evolution of human cognitive architecture. In Ross, B. (Ed.), *The psychology of learning and motivation* (pp. 215–266). San Diego, CA: Academic Press.

Sweller, J. (2004). Instructional design consequences of an analogy between evolution by natural selection and human cognitive architecture. *Instructional Science*, *32*, 9–31. doi:10.1023/B:TRUC.0000021808.72598.4d

Sweller, J., Van Merriënboer, J., & Paas, F. (1998). Cognitive architecture and instructional design. *Educational Psychology Review*, *10*, 251–296. doi:10.1023/A:1022193728205

Tallent-Runnels, M. K., Thomas, J. A., Lan, W. Y., Cooper, S., Ahern, T. C., Shaw, S. M., & Liu, X. (2006). Teaching courses online: A review of the research. *Review of Educational Research*, *76*(1), 93–135. doi:10.3102/00346543076001093

Wiggins, G., & McTighe, J. (2005). *Understanding by design*. Alexandria, VA: Association for Supervision & Curriculum Development.

Wiley, D. A. (2000). Connecting learning objects to instructional design theory: A definition, a metaphor, and a taxonomy. In D. A. Wiley (Ed.), *The instructional use of learning objects*. Retrieved from http://reusability.org/read/

ADDITIONAL READING

Bransford, J. D., Brown, A. L., & Cocking, R. R. (Eds.). (2000). *How people learn: brain, mind, experience* (2nd ed.). Washington, D.C: National Academy Press.

Clark, R. C., & Mayer, R. E. (2007). *Electronic learning and the science of instruction: Proven guidelines for consumers and designers of multimedia learning*. San Francisco, CA: John Wiley & Sons.

Garmire, E., & Pearson, G. (Eds.). (2006). *Tech tally. Approaches to assessing technological literacy. Washington, DC: The National Academies Press. Hannafin, M. & Spector, J. M. Educational Technology Research and Development*. New York, NY: Springer.

Hannafin, M., West, R., & Shepard, C. (2009). The cognitive demands of student-centered, web-based learning: Current and emerging perspectives. In Zheng, R. (Ed.), *Cognitive effects of multimedia learning* (pp. 194–216). New York: Information Science Reference.

Jonassen, D. H. (Ed.). (2003). *Handbook of research on educational communications and technology* (2nd ed.). Mahwah, NJ: Lawrence Erlbaum.

Jonassen, D. H. (2005). *Modeling with technology: Mind tools for conceptual change* (3rd ed.). OH: Prentice Hall.

Joyce, B., Weil, M., & Calhoun, E. (2003). *Models of teaching* (7th ed.). Boston, MA: Allyn & Bacon.

Kauffman, D. (2004). Self-regulated learning in web-based environments: Instructional tools designed to facilitate cognitive strategy use, metacognitive processing, and motivational beliefs. *Journal of Educational Computing Research*, *30*(1/2), 115–137.

Mayer, R. (2005). Cognitive theory of multimedia learning. In Mayer, R. (Ed.), *The Cambridge handbook of multimedia learning* (pp. 31–36). New York: Cambridge University Press.

MERLOT. (2000). *Multimedia educational resource for learning and on-line teaching website* [On-line]. Available: http://www.merlot.org/

Merrill, M. D. (in press). Converting e3 - learning to e3 -learning: an alternative instructional design method. In Carliner, S., & Shank, P. (Eds.), *Electronic learning: Lessons learned, challenges ahead*. San Francisco, CA: Pfeiffer/Jossey-Bass.

Morrison, G. R., Ross, S. M., & Kemp, J. E. (2004). *Designing effective instruction* (4th ed.). San Francisco, CA: Wiley.

Reigeluth, C. M. (Ed.). (1983). *Instructional design theories and models: An overview of their current status*. Mahwah, NJ: Lawrence Erlbaum Associates.

Reigeluth, C. M. (Ed.). (1999). *Instructional-design theories and models: A new paradigm of Instructional theory*. Mahwah, NJ: Lawrence Erlbaum Associates.

Reiser, R. A., & Dempsey, J. V. (Eds.). (2006). *Trends and issues in instructional design and technology* (2nd ed.). OH: Prentice Hall.

Shapiro, A. (2008). Hypermedia design as learner scaffolding. *Educational Technology Research and Assessment*, *56*, 29–44.

Simonson, M. & Schlosser, C. *Quarterly Review of Distance Education*. Nova Southeastern University.

Spector, J. M., Merrill, M. D., Van Merrienboer, J., & Driscoll, M. P. (Eds.). (2007). *Handbook of research on educational communications and technology* (3rd ed.). Mahwah, NJ: Lawrence Erlbaum.

Uden, L., & Beaumont, C. (2005). *Technology and problem-based learning*. London: Information Science Publishing.

Wang, M. J. (2007). Designing online courses that effectively engage learners from diverse cultural backgrounds. *British Journal of Educational Technology*, *38*(2), 294–311. doi:10.1111/j.1467-8535.2006.00626.x

Wang, M. J., & Kang, J. (2006). Cybergogy of engaged learning through information and communication technology: A framework for creating learner engagement. In Hung, D., & Khine, M. S. (Eds.), *Engaged learning with emerging technologies* (pp. 225–253). New York: Springer Publishing. doi:10.1007/1-4020-3669-8_11

Waterhouse, S. A. (2004). *The essential guide for teaching in the digital age*. Boston, MA: Allyn & Bacon.

Wiggins, G., & McTighe, J. (2004). *Understanding by design workbook*. Alexandria, VA: Association for Supervision & Curriculum.

Wiley, D. A. (2000). Connecting learning objects to instructional design theory: A definition, a metaphor, and a taxonomy. In D. A. Wiley (Ed.), *The instructional use of learning objects: Online version*. Retrieved June 12, 2010, from http://reusability.org/read/chapters/wiley.doc

KEY TERMS AND DEFINITIONS

Cognitive Load: Refers to the load place on working memory during instruction. Cognitive load theory refers to the premise that the focus of an instruction should be the instruction itself rather than the efforts to obtain it.

Digital Immigrant: Coined by Marc Prensky (2001), describes a person who grew up without digital technology and adopted it later.

Digital Native: Coined by Marc Prensky (2001), describes a person for whom digital technologies already existed when they were born, and hence has grown up with digital technology

Learning Objects: Digital entities that can be used, reused, and distributed by any number of people for the for the purpose of instruction.

Long-Term Memory: The large, central location in which information is stored, managed and retrieved for later use.

Self-Regulated Learning: Learning that occurs from one's own influence as a result of self-generated thoughts, feelings, strategies, and behaviors.

Working Memory: The conscious locus of short-term information processing.

Chapter 5
Situation-Based and Activity-Based Learning Strategies for Pervasive Learning Systems at Workplace

Amel Bouzeghoub
Telecom SudParis, France

Serge Garlatti
Telecom-Bretagne, France

Kien Ngoc Do
Telecom SudParis, France

Cuong Pham-Nguyen
Telecom-Bretagne, France

ABSTRACT

The main topic of this chapter is about a pervasive learning at workplace, namely, work-based learning. The proposition is based on two complementary learning strategies: situation-based learning and activity-based learning to fulfill seamless learning across contexts and worked-based learning requirements. In situation-based learning, relevant activities and/or resources are recommended to the user. In activity-based learning, the user has to search and to select his/her activities and the corresponding resources. These strategies correspond to two different information dissemination approaches that can be distinguished, namely push and pull (Cheverst, Mitchell, & Davies, 1998). The authors of this chapter propose a pervasive learning environment where learners may follow different learning strategies. They may switch from one strategy to the other one according to their needs and/or the context change. These facilities are possible thanks to a set of models and adaptation processes developed for the P-LearNet project (Pervasive Learning Network). To illustrate this proposal, an example (or a use case) from this project is used.

DOI: 10.4018/978-1-60960-511-7.ch005

The chapter is organized as follows: the authors introduce some issues of technology-enhanced learning systems and define mobile, pervasive and ubiquitous learning and some closely related features: context, adaptation, situated learning, working and learning activities. Secondly, work-based learning features are described. Thirdly, situation-based and activity-based learning strategies are presented. Finally, the P-LearNet project is used to illustrate the proposal, and the conclusion summarizes the chapter and shows how and at which level this framework can be reused.

INTRODUCTION

Technology-Enhanced Learning (TEL) systems must have the capability to reuse learning resources and web services from large repositories, to take into account the context and to allow dynamic adaptation to different learners based on substantial advances in pedagogical theories and knowledge models (Balacheff, 2006). The computer-based reuse of learning resources and web services requires a precise information retrieval process by means of a search engine. A search engine based on a keyword approach has two main drawbacks: keywords are polysemic and the results are too numerous. Thus, it is impossible to select automatically the relevant resources and web services according to user's needs. A semantic web approach is able to prevent polysemy and to provide interoperability at semantic level. A semantic web search engine is based on common vocabularies or ontologies which are used to define a unique meaning for a given term or concept. Thus, it has the ability to select relevant resources and web services. Moreover, knowledge models and pedagogical theories can be fully represented by means of ontologies in a semantic web framework. The context-awareness and adaptation process is a refinement of the information retrieval process and it requires a very precise search engine provided by a semantic web. Context-awareness and adaptation are crucial issues in pervasive computing and learning.

In the "mobile learning" area, several expressions are used: mobile, pervasive and ubiquitous learning systems (Brodersen, Christensen, Gronboek, Dindler, & Sundararajah, 2005; Hundebol & Helms, 2006; Sharples, 2005; Siobhan, 2007). In computer science, mobile computing is mainly about increasing our capability to physically move computing tools and services with us. The computer becomes an ever-present device that expands our capabilities - by reducing the device size and/or by providing access to computing capacity over the network (Lyytinen & Yoo, 2002). In mobile computing, an important limitation is that the computing model does not change while we move. This is because the device cannot obtain information about the context in which the computing takes place and adjust it accordingly. In pervasive computing, the computer has the capability to inquire, detect and explore its environment to obtain information and to dynamically build environment models. This process is reciprocal as the environment also does it and becomes "intelligent". In ubiquitous computing, the main goal is to integrate large-scale mobility with pervasive computing functionalities.

Mobile learning is not just about learning at anytime, at any place and in any form using lightweight devices, but learning in context and seamless learning across different contexts (Balacheff, 2006; Sharples, 2006; Vavoula & Sharples, 2008). It is best viewed as mediating tools in the learning process (Sharples, 2006). In mobile learning, TEL systems do not have the capability to inquire, detect and explore their environments. In other words, the context is implicit. On the contrary, pervasive and ubiquitous learning systems are context-aware. Thus, we consider that mobile, pervasive and ubiquitous learning systems have the properties of mobile, pervasive and ubiquitous computing systems, respectively. Many definitions of pervasive learning are given in the literature (Bomsdorf, 2005; Hundebol & Helms, 2006; Jones & Jo, 2004; Siobhan, 2007). We can cite the following one "Pervasive learning environment is a context (or state) for mediating learning in a physical environment enriched with additional site-specific and situation dependent elements – be it plain data, graphics, information -, knowledge -, and learning objects, or, ultimately, audio-visually enhanced virtual layers" (Hundebol & Helms, 2006). One could consider pervasive learning as an extension to mobile learning where the roles of the intelligent environment and of the context are emphasized. In pervasive learning, computers can obtain information about the context of learning from the learning environment where small devices, sensors, pads, badges, large LCD screens, people, are embedded and communicate mutually. The physical environment is directly related to learning goals and activities. The learning system is dynamically adapted to the learning context.

Dey (2001) defines the context as "any information that can be used to characterize the situation of an entity. An entity is a person, place, or object that is considered relevant to the interaction between a user and an application, including the user and applications themselves". The main idea in this definition-centered application is that the context is the information, which describes the situation of an entity. Hence, a situation is a temporal state in the context. In context-aware applications, situations are external semantic interpretations of low level context used to high-level specification of human behavior and services. Situations inject meaning in the application and are stable and easier to define and maintain than basic contextual data. Adaptations in context-aware applications are then triggered for each change of situation (i.e. a change in a context value triggers the adaptation if the context update modifies the situation). The design and the execution of the applications become much easier with situations since the designer/programmer can operate at a high level of abstraction (situation) rather than all the contextual data, which create the situation. For example, (Loke, 2006) describes six different manners to specify the situation "at-meeting" based on contextual data: (1) co-localization of people and daily agenda information; (2) co-localization in a room of full cups of coffee; (3) weight sensors on the floor; (4) devices in the room (light, video-projector); (5) noise sensors; (6) video camera in the meeting room. The context consists of a set of structured information. It evolves continuously and is used as interpretation. The nature of information, as well as its interpretation, depends on the finality (Rey, 2006). The first issue is to determine the finality and from there, to define the necessary information to serve this finality. Brézillon states that the context is better specified by the current activity of the user: thus, the activity becomes a central notion of the context, it determines the relevant entities: intention, information, knowledge, objects of the environment, etc., those which are necessary to the achievement of the activity: "the context guides the focus of attention, i.e. the subset of common ground that is pertinent to the current task" (Brézillon, 2005). The activity itself is not separated from the context; it belongs to the context as it "guides" the situation of the user (Dourish, 2004). The knowledge or the recognition of the activity allows determining in a more precise way the relevant entities of the context. Activities, contexts and learning at workplace are closely related.

At Workplace, learning is centered on work activities in specific situations. Situated learning is able to provide the right learning support at the right time according to the situation parameters and to the goals in the working context. Situated learning increases the quality of learning and is attractive for learning at the workplace and for work-learning integration (Oppermann & Specht, 2006). As a consequence, the situation determines relevant activities and the learning needs linked to this activity. Consequently, pervasive learning can be viewed as a type of situated learning in which the location, the time, the environment, the tasks are taken into account. It provides the right learning support in the right time according to the situation parameters and to the goals of application at workplace. For pervasive learning at workplace, we can particularly mention the following issues: (1) Provide relevant entities (information, services, activities, communities, etc.) for employees according to the current situation (2) Manage seamless learning at workplace across contexts: After a situation change, the current activity of an employee could either be suspended or aborted or continued in the new situation. It is an important issue to manage activities seamlessly, that is to say to ensure the continuation of activities (suspended or not) in future situations by means the situation history (storing learning and working activities according to employee needs. For example, Anna is making an invoice and also learning how to make it on her laptop through the wireless network. As an incident occurs in the wireless network, the system can recommend her to use the 3G+ network or can switch the network to 3G+ to ensure the continuation of her working and learning activity. Nevertheless, the system could provide new entities to Anna to deal with the new situation (3G+ network, a smaller bandwidth). These entities could have the same semantic features - indexed by the same concepts and belonging to the same concept - and could be relevant in the new situation. These entities could be "smaller" according to the 3G+ bandwidth.

After a brief introduction on mobile, pervasive and ubiquitous learning, we pointed out the main roles of context and adaptation in pervasive learning and then we demonstrated that human activities are central issues to determine that context. We also showed that learning at workplace is focused on activities and that situated learning is relevant for learning at workplace. As pervasive learning is a kind of situated learning, we explicated the main issues in pervasive learning according to human activities at workplace.

As activities and situations are central issues for learning at workplace, we present two learning strategies (situation-based and activity-based) having the capability to link these issues and to manage pervasive learning issues (provide relevant entities and manage seamless learning across contexts). At workplace, activities can be explicit or implicit, predefined or not in a learning system, but in all these cases they can be realized only in situations or classes of typical situations. Situations and activities are closely linked. Two strategies of learning can be proposed to ensure this connection between situation and activities: one situation-based and the other activity-based. The situation-based learning strategy is used when the intention of the user is not known. It provides on the fly recommendations to an employee according to the current situation. These recommendations consist of relevant entities according to the current situation, i.e. information, documents, services, collaborators, equipment, etc. The activity-based learning strategy is used when an employee explicitly choices a known and predefined activity. We therefore make the assumption that we know the intention of the user that is described by the explicit representation of the activity in the system. This strategy provides to the employees a set of activities, coordinated and organized to support and integrate the learning and working processes. This strategy proposes relevant activities and the good way to carry them out and/or to continue them according to the current situation.

The main contribution of this chapter is to propose a mixed learning strategy at workplace that is based on the combination of a situation-based strategy and an activity-based strategy. The main goal of this strategy is to permit learning at workplace in the greatest number of possible situations. Indeed, each of these strategies treats only one part of the problem and only considers some situation classes. This mixed strategy authorizes the consideration of the two sets of situation classes but also their coordination. In other words, it is possible to switch from one strategy to the other but also to ensure continuity and consistency between the two strategies from the history of the situations and the learning and working activities. Thus it ensures seamless learning, i.e. the continuity and the coordination of activities, services and the access to resources.

LEARNING AND WORKING

At workplace, learning can occur in purposeful situations in which there is an explicit goal to learn as well as in incidental situations in which there is no explicit learning goal or interest. Working involves an activity or related set of activities that require effort and are aimed at achieving one or more objectives. Learning emphasizes on what a learner knows or is able to do while working is related to performance improvement (Michael-Spector & Wang, 2002). In other words, when performing at work, it often happens that learning occurs. The work's performance and quality are also enhanced after learning experiences. Working activities are mainly about solving tasks and particularly in knowledge-intensive organizations, which implies continuous learning. Solving the particular working tasks is prior - learning is just a means (Farmer, Lindstaedt, Droschl, & Luttenberger, 2004). The distinction between learning and working activity is blurring, working as a way of learning and vice versa. It is generally called work-based learning.

One of the principal characteristics of learning at workplace is to associate theory with practice and knowledge with experiment. On the contrary, the traditional model of learning in classrooms separates the theory from the practice and is not always very attractive for learners (Raelin, 2008). In general, the learning in classroom stresses on what a learner knows or is able to do while the learning at workplace is related to the improvement of the performance (Michael-Spector & Wang, 2002). The needs and the objectives of learning are devoted to competitiveness objectives, productivity and quality, which depend directly on the activities in the company (Drucker, 1999). The realization of working activity becomes priority; the learning is only one means to achieve and/or to improve the working activity (Farmer, et al., 2004). The learning can be done in situations where there are explicit objectives of learning, but also in accidental situations where there is no explicit objective of learning. It is thus sometimes difficult to differentiate the working activities from learning activities.

Among important obstacles for worked-based learning, there is a cognitive and structural disconnection between work, knowledge, and learning (Farmer, et al., 2004). The typical workplace of an employee and its structure consists of three separate spaces: a workspace, a knowledge space, and a learning space: (i) the workspace represents the user's devices (PC, PDA, etc.), the shared documents and tools and storage devices or a document management system. The workspace is mainly structured according to business processes; (ii) the knowledge space represents unconscious learning, past experiences applied to new situations, information retrieval and past used cases. It corresponds to the organizational memory; (iii) the learning space stands for conscious learning situations: seminars, e-learning courses, etc. These three spaces are cognitively disconnected. Each space has an inherent structure, which mirrors the mental model of the people who are using it. There is also a structural separation of the three spaces: spaces are implemented on different information systems, which are not interlinked. Thus, it is necessary that a worked-based learning system is able to integrate the three spaces and the corresponding information systems.

A. Simon, from "Yellow Edge Company" (http://www.yellowedge.com.au/), mentioned that traditional methodologies such as formal classroom and even Internet based, content oriented courses and programs have their place at worksite. Nevertheless, these methodologies are generally inflexible to the demands of contextualised, learner centred, and performance related challenges. Thus, learning processes need to be embedded in organisations, so that learning becomes pervasive and a natural part of work. In other words, learning and working are blurring. A. Simon gave the following example: "individual learning contracts with learning objectives precisely linked to an individual's performance targets within a performance management system can engender superior performance in an organizational context and also be used to guide learning for personal enrichment activities in a social context". Thus, learning methodology in the context of the organization can be viewed as a way of working. It is necessary to redefine learning as a work activity and to provide the infrastructure for seamless work-learning integration. The organizational system such as work flow, organizational design, organizational communication, etc. has to be explicitly defined so that learning processes can be integrated into these systems in appropriate ways. A learning program can be constructed around a performance and working problem to become performance driven and relevant for learning and working. This integration of working and learning into business processes could be viewed as a way to integrate the three previous spaces (knowledge space, working space and learning space).

In work-based learning, two general methods of learning are often used (Pimmer & Gröhbiel, 2008): the "just-in-case" and "just-in-time" learning. The "just-in-time" learning brings an immediate value to the working process. It includes the acquisition of knowledge and know-how during the work activity because of its promptness and relevancy. The "just-in-case" learning creates "potential values" for work. It is relevant to use it before or afterwards working activities. It seems interesting to use this kind of learning for a reflection on the working activities and the modification of the working processes. This learning method consists of a reflection on the passed experiments, the generation and the share of new knowledge and the learning with the others. Among the learning activities at workplace, we can mention the following categories: acquisition and revision of knowledge, online helps search, local or remote search. According to the type of the learning method "just-in-time" or "just-in-case" learning, the nature of these activities categories will be different, as they are not submitted to the same constraints.

This need for contextualization and setting in situation of the working and learning activities also can be taken into account by pervasive data processing (Naismith, Lonsdale, Vavoula, & Sharples, 2004). Indeed, pervasive computing objectives is to detect and explore the environment, build dynamically models to characterize the current situation and to use these models to adapt the suggested activities, services and contents. It is then possible to associate formal and practical learning, to reach relevant knowledge according to the working activity in which the learning is performed (Pimmer & Gröhbiel, 2008).

In conclusion, the learning at workplace is mainly characterized by: (i) improvement of the performance and the quality of the working activities and the company; (ii) the acquisition of new knowledge and their uses, but also meta-knowledge by thinking about the problem solving processes in the company; (iii) business processes and work-learning activities has to be tightly integrated and dedicated to performance. In other words, the integration of knowledge space, workspace and learning space must be done; and (iv) its close relationship with the working activities and their social contexts in the company. A pervasive learning system should propose to employees' only relevant information, services or activities, according to the situation. Indeed, it is the context and/or the current activity, which determines the requirements in terms for learning in a given community of practices. At workplace, activities and situations, tightly coupled to business processes, are the key issues to design and process work-learning activities. Consequently, situation-based and activity-based learning strategies fulfilling these requirements are relevant in our framework (the P-LearNet Project).

P-LEARNET PROJECT

The 3-year P-LearNet project (p-LearNet, 2006) is an exploratory study on adaptive services and usages for human learning in the context of pervasive communication. One goal of the project is to provide integrated learning and working support to employees during their working activities. This project claims to think about concepts and methodologies to facilitate such kind of learning. In such a framework, the main issues of the P-LearNet project are: work-integrated learning and customer learning support, continuous professional learning at the workplace, professional learning whatever the place, the time, the organisational and technological contexts of the individual or collective learning and working processes, context-as-construct and seamless learning. Learning focuses mainly on how to support individual and group learning processes through pedagogical guidance and how to enhance the learner's knowledge and know-how.

In this project, one of our corporate partners is an international retail company having chains of shops and hypermarkets wondering about seller learning at the workplace. Corporate partners identify the problems and requirements about quality and efficiency of information and services to increase market share and the corresponding learning goals. Several innovative scenarios have been set up according to two main learning and/or working situations for a seller and a customer as learners: (i) Seller or customer, outside the shop counters: seller in the back office or storage areas, customer at home or elsewhere; (ii) Seller in his department, alone or with a customer having resources from the Smart Spaces (large LCD screen, printers, RFID, etc.) surrounding them (Derycke, Chevrin, & Vantroys, 2007). In the e-retail framework, sometimes learning and working are interwoven in a pervasive environment. A substantial part of learning does not happen during training but during working activities. Just-in-time or just-in-case learning and professional activity support must therefore be integrated. A learning system must overcome three main obstacles: time pressure, inadequate learning support in the working context and cognitive and structural disconnection between work, knowledge, and learning (Farmer, et al., 2004). Some learning and working examples are as follows: (i) just-in-case learning: a seller equipped with a portable device, for example a PDA or a UMPC, close to shelves (without a customer) can revise their knowledge about products and selling techniques or can continue their previous learning activities to improve their knowledge; (ii) Just-in-time learning: during the selling process, the seller can use his/her mobile device as a coach to help him; and (iii) Just-in-time learning: a seller can communicate with customers while revising his/her knowledge, checking the inventory or contacting the supplier about products. Such type of learning and working support could be done by means of situation-based or activity-based learning strategies and/or the two strategies according to the context changes.

SITUATION-BASED AND ACTIVITY-BASED LEARNING STRATEGIES

Learning and working are mainly dedicated to situated working activities, which we want to improve performances. We can break down all employees' activities into three categories: not structured activities, semi-structured activities and structured activities. These last ones are activities, which we could describe as "procedural". "Procedural" activities have a quite stable structure, can guide the user step by step and determine precise situations of interaction. A structured activity is an activity that can be broken down into sub-activities that are precisely scheduled. The not structured activities cannot be composed of sub-activities. They are predefined activities that are not well known or are impossible to define as "procedures". Such type of activities mainly provides tools and resources to achieve them without guidance. The semi-structured activities consist of structured activities and not structured activities. The system can propose a set of predefined activities to the users (structured, not structured and semi-structured activities). This set is limited and does not take into account all employees' needs because some of activities are not known in advance or the user wants to achieve his activity on its own way. Employees must get entities according to their needs (information, services, activities, collaborators, etc.) to carry out predefined or not predefined activities. A pervasive learning system should only provide relevant entities suitable to the current situation. Indeed, it is the situation and/or the current activity that determines the requirements in terms of learning and working in a given community of practices. The activities can be explicit or implicit, predefined or not, but in all cases they can only be achieved in a particular situation or in typical situation classes. The situation and the activities are closely dependent.

Two learning strategies can be proposed to ensure this connection between situation and activities: situation-based strategy and activity-based strategy. The situation-based strategy provides recommendations to the employee according to the current situation. It proposes relevant entities according to the current situation (i.e., information, services, predefined activities, collaborators, equipment, etc.) in push or pull mode (Cheverst, Mitchell, & Davies, 1998). In the push mode, the system is proactive that means it decides when the employee is notified according to the situation changes. In other words, the system proposes automatically recommendations depending on the situation changes without any human intervention. Thus, an employee can select or not one of the given recommendations. On the contrary, in the pull mode, an employee searches himself for information. Thus, the employees "write queries" to express their specific needs to obtain the relevant entities according to the current situation.

The activity-based strategy is used as soon as an employee selects a predefined activity. The activity-based strategy provides a set of predefined activities integrating the learning and working processes to the employees. When an employee explicitly chooses an activity, the activity-based approach determines the good ways to execute the activity according to the current situation. Moreover it also ensures the orchestration of the sub-activities for the structured or semi-structured activities. Otherwise, it only provides relevant resources and tools to achieve activity

These two approaches (situation-based and activity-based) are now presented by specifying their characteristics as well as their advantages and drawbacks. Then we present the mixed approach that combines these two approaches.

Situation-Based Learning

The situation-based learning is an approach in which the situation of the employee is in the core of the learning. The situation serves either to detect learning needs by analyzing the current situation, or to represent the situated knowledge during work and to react to each change of situation of the employees.

The situations management in reactive systems needs to define a situation. To be able to detect situations, key information must be included in the definition of the situation:

- The events which can take part in the detection of situation;
- The context in which the detection of situation is suitable; and
- The semantic conditions which must be satisfactory in order to detect a situation;

In our case, we distinguish two functioning modes (see Figure 1): the push and pull modes. In the push mode, the strategy of learning based on the situations is activated with the arrival of a new event that modifies the current context

Figure 1. The situation-based learning strategy

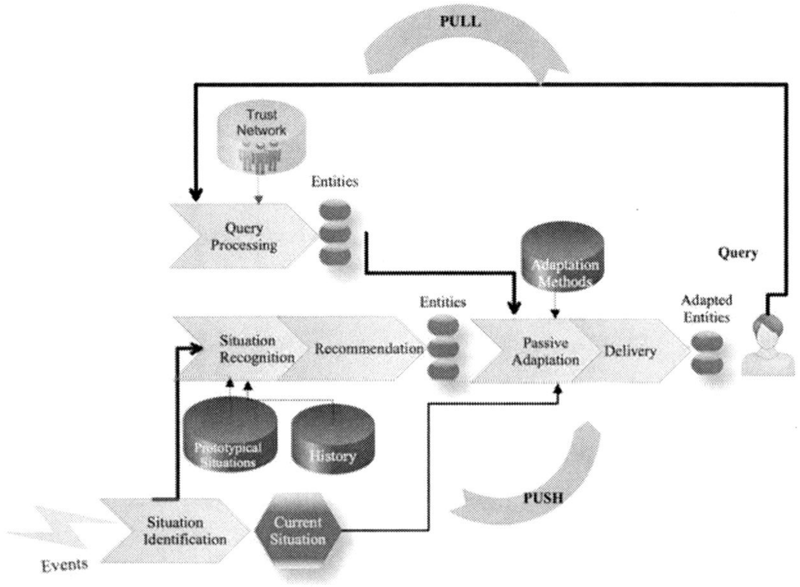

of learning (Bouzeghoub, Defude, Duitama, & Lecocq, 2006; Bouzeghoub & Do, 2008, 2009, 2010; Bouzeghoub, Do, & Lecocq, 2008). It is also called system-oriented approach because the system controls the recommendation of entities. A situation is a subset of properties accessible from the context at a given moment. The environment must adapt the process of learning and working to this new situation. The context adaptation mechanism of the learning processes is divided into two parts: (i) the current situation is recognized and identified among the set of the predefined situations and the corresponding adaptation rules are applied to propose recommendations. For example, the employee encounters difficulties and needs assistance and an author/designer has predefined the relevant actions for this help: to provide a resource and/or a simpler and more detailed activity or to propose the assistance of an expert (instructor) or a colleague which are connected; (ii) the current situation is not recognized. A mechanism of case-based reasoning is used to find a prototypic situation similar from which similar recommendations will be proposed. According to the number of resources and/or activities suggested, two processes of filtering can be used to reduce the number of recommendations according to knowledge and know-how and preferences to the employee.

In the pull mode, the user starts the process with a query. The result of the query is then adapted to the situation of the user and then filtered according to his knowledge and his preferences.

The seamless learning is managed at two levels: at learning object level and at domain concept level. After a change of situation, the system can propose to an employee to continue his "activity" in the new situation with the same learning objects if it is possible or with an equivalent one according to this new situation. The system may also propose the same entities or new entities to the employee allowing him to continue a knowledge acquisition on the same concepts.

Advantages and Drawbacks of the Situation-Based Strategy

This strategy offers the following advantages: (i) it proposes two interaction modes, the push mode in which the system takes the control and guides the employee in his training according to his current situation and the pull mode where the employee controls his learning path by choosing himself the learning entities that he wants to use; (ii) A recommendation system is started for each arrival of a new event. It proposes on the fly learning resources, equipments or collaborators according to the new situation; and (iii) Finally, this strategy generates links between employee's activities and social context by proposing collaborators in the geographical neighbourhood, having the same profile, doing the same activity or having already seen the same problem.

Among the drawbacks, we can mention: (i) the lack of scenario as support for learning make difficult the check of consistency. Indeed, the fact that the employee follows a free learning by choosing himself his path does not allow consistency checking; (ii) the intention of the employee is not known a priori.

The next section presents the activity-based strategy, which is complementary to situation-based strategy. The situation-based strategy is used when the employee's activity is not known or when the employee wants to achieve it in its own way (without guidance) or when the activity-based strategy is unable to provide relevant predefined activities according to the current situation. When an employee is willing to achieve a predefined activity and the activity-based strategy is able to provide a relevant one, the activity-based strategy can be used.

Activity-Based Learning

Activity-based learning strategy provides a set of activities to learners where activities and sub-activities are explicitly represented, structured and coordinated to support business and learning integrated processes (Pham-Nguyen & Garlatti, 2008a, 2008b; Pham-Nguyen, Garlatti, Lau, Barbry, & Vantroys, 2009a, 2009b; Pham-Nguyen, Lau, Barbry, Vantroys, & Garlatti, 2008). When the learner explicitly chooses an activity, several ways for achieving it are proposed according to the current context. In our framework, an activity is represented by task by means of a hierarchical task model, having the task/method paradigm. There are two types of tasks: abstract tasks and atomic tasks. An abstract task can be composed of atomic and abstract sub-tasks. An atomic task cannot be composed of sub-tasks. It consists of a simple procedure or function. Several methods can be associated to a task. A method represents a way to perform a task in a class of situations. It determines how to break down a task into sub-tasks and how to define their coordination. A method, associated to an abstract task, defines a control structure, which determines the task coordination, that is to say the recursive decomposition of tasks into sub-tasks (by means of operators) and the sub-task order at runtime.

A structured or semi-structured activity is represented by an abstract task, which is broken down into abstract sub-tasks and atomic sub-tasks. A not structured activity is represented by an atomic task providing tools and environments necessary to achieve the corresponding activity. In such a case, the activity is not controlled or guided by the system (hierarchical task model). In Activity-based Learning, the goal of generic scenarios is to describe the learning and working integrated activities to acquire some knowledge domain and know-how to solve a particular problem or to support working activities. An author/designer can manage a global activity consistency in a generic scenario. A generic scenario consists of an abstract task broken down into sub-tasks through methods. This abstract task is composed of a set of pairs (task, method) that represent the greatest number of learning and working situations for each abstract task. The context-aware and adaptive mechanism can be viewed as the selection of the relevant pair (task, methods) for a given task according to the current working and/or learning situation. Thus, at runtime, the adaptive mechanism chooses at least

Figure 2. The activity-based learning strategy

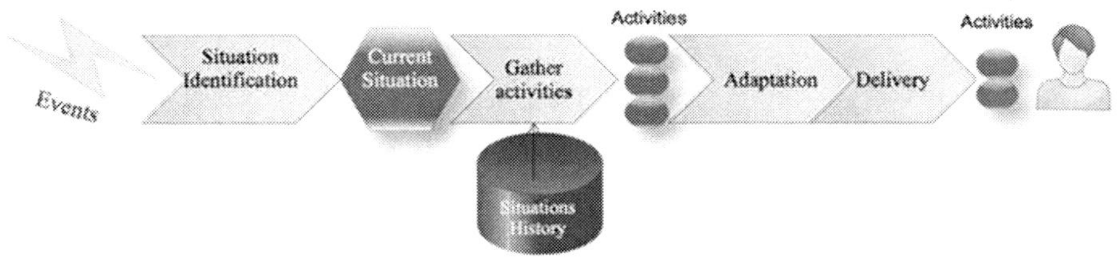

one pair (task, method) for a given task according to the current situation. The seamless learning and working strategy is based on generic scenarios.

Contrary to the situation-based strategy, there is only one mode available for the activity-based strategy, that is to say the push mode (see Figure 2). The adaptive mechanism selects the relevant pairs (task, method) according to the current situation. This mechanism is based on a matching between the contextual description of methods and the features of the current situation.

For each new event (push mode), the new situation is identified and the system gathers potential activities - pairs (task/method) - which are filtered by the adaptive mechanism according to the new situation. The adaptation mechanism is based on two integrated and ordered strategies: (1) by reflexivity (2) by contract (Chaari, 2007). The idea is to use metadata associated to context properties (reflexivity) and rules (by contract) to deal with the dynamic feature of context (in a given situation, you cannot ensure that all context properties are available). For properties managed by reflexivity, the first adaptation mechanism filters out the non relevant pairs (task, method). For others properties (managed by contract), rules are applied to classify the pairs (task, method) in equivalence classes. This second adaptation mechanism is strongly constrained in comparison with the situation-based strategy mechanism. The corresponding matching process is based on necessary and sufficient conditions to belong to an equivalence class defined by a conjunction of properties (at least two equivalence classes, "good" and "bad", mutually exclusive). Thus, sometimes all properties involved in the matching process are not available. Sometimes, it could be impossible to select a relevant activity. The adaptation mechanism provides to an employee all relevant activities and how to achieve them - pair (task, method). It can propose to continue the current activity, to take up suspended activities or to begin new activities. The employee is in charge to choose an activity - select a pair (task, method), to change the current activity and to finish an activity.

The seamless learning and working is managed as follows: (i) a generic scenario enable an author/designer to define a global and consistent organization of an activity set across contexts and their coordination; (ii) the adaptation mechanism is able to managed suspended activities after situation changes. The states associated to pair (task, method) are used to take up suspended activities and/or to continue the following activity according to the coordination of the sub-tasks of a particular pair (task, method); and (iii) A competence model, based on knowledge and know-how, is used to ensure the continuity of the learning process across contexts. It is necessary to be able to evaluate the employee levels of knowledge and know-how based on their performance and working quality. In other words, the adaptation mechanism has to provide the relevant pairs (task, method) according to the current levels of knowledge and know-how of an employee to ensure gradual learning and working.

Figure 3. The mixed learning strategy

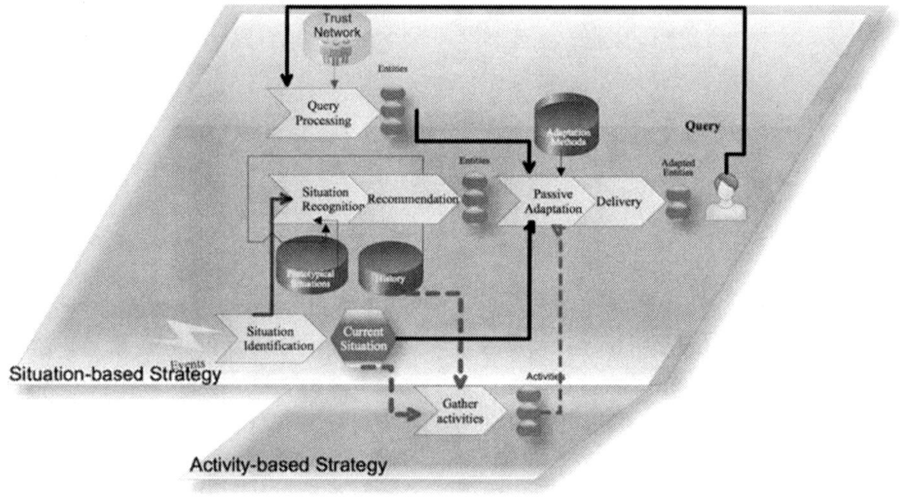

Advantages and Drawbacks of the Activity-Based Strategy

The main advantages of the activity-based learning strategies are as follows: (i) generic scenarios enable us to ensure global consistency of activities across contexts and provide a relevant integration of learning and working activities; (ii) as structured activities are well known, one can ensure a continuous learning process across contexts by means of the states associated to the pairs (task, method) and the competence model.

The main drawbacks are as follows: (i) all activities and situations of employees cannot be represented because some of them are not known or employees do not want to use a modelled activity; (ii) as the adaptation mechanism is constrained, sometimes the system is unable to provide relevant activities. The system can be unable to support the employee working activities; and (iii) collaborative working and learning activities are not explicitly represented. These activities are important for learning at workplace.

The mixed strategy, proposed in this chapter, aims at eliminating the drawbacks of the situation-based and activity-based strategies. The main idea is to deal with the greatest number of learning and working situations for employees.

MIXED STRATEGY

The mixed strategy combines the two learning strategies: situation-based and activity-based learning strategies. In this paragraph we will show how and when the two strategies are activated and how they are complementary. In other words, we will on the one hand define the conditions of transition from one strategy to another for the two modes push and pull and on the other hand list the contributions of the mixed strategy in term of seamless learning.

The mixed strategy life cycle modifies and integrates the two previous strategies life cycles (see Figure 3). We will describe on the one hand the two strategies in push and pull mode and on the other hand the conditions of transition from each strategy to the other. For each strategy, new functionalities are proposed: (i) the pull mode is now available for the activity-based strategy; (ii) the situation-based strategy can also propose activities in the two modes.

The situation-based strategy is activated by default in pull or push mode because it proposes more entities whatever the mode. When an employee writes a query on entities, the situation-based strategy is automatically activated in pull mode. The suggested entities can be contents, Web services, collaborators, predefined activities, etc. The system provides as a result a list of potential entities, which are then filtered by the adaptation mechanism (of the situation-based strategy) to keep only those, which are relevant in the current situation. Then, they are delivered to the employee.

The employee can also choose to be in the push mode. In this case, the system proposes him relevant entities (including activities) according to its current situation and starts again this process of recommendation at each time an event changes the current situation and generates a new situation.

The transition from this strategy to the activity-based strategy is done as soon as an employee selects an activity, in push or pull mode. In pull mode, the user may continue to achieve the selected activity as long as the current situation is suitable. If it is not the case, he can choose either to write a new query in pull mode or to pass in push mode to obtain automatically suggested entities. In pull mode, the system does not move itself to the situation-based strategy. It can only be the result of the employee intervention. In push mode, the system proposes to him relevant activities according to its current situation and starts again this process at each time an event changes the current situation and generates a new situation. If the employee chooses an activity, the process is the same as previously described for the activity-based strategy. The system can thus propose to continue interrupted activities or to begin new activities.

The system switches from activity-based strategy to situation-based strategy if the employee does not wish to use the suggested activities or if the activity-based strategy cannot propose activities any more. Indeed, the situation-based strategy is less constrained in terms of filtering and will be able to propose more activities and/or entities "semantically" related to the interrupted activity or entities already reached in similar situations. In this case, the system is never in a blocked situation and may always suggest an entity. The system changes its mode to deal with the drawbacks of the activity-based strategy.

USE CASES

Among the scenarios studied in the project, we are interested here in the postman scenario to illustrate the problems of the learning at work and to need to mix a situation-base strategy and activity-based strategy.

In his daily mission, each postman has a mobile terminal which enables him to interact with a knowledge base (e.g., query mode via a Web interface) and with a group of colleagues (e.g., standard Web interface). With each access, the terminal transmits data of geo-localization. The terminal is also equipped with an integrated camera. Confronted with a difficulty in its mission, the postman can:

1. **Ask the knowledge base:** This action is done in query mode (or pull mode) with a Web form. The postman profits from all the descriptions done by his/her colleagues as well as official indications. A postman starting a new round can thus see whether particular cases were previously announced. It can also consult a procedure or search an activity. We recognize in this example a "just-in-case" learning that is supported by the situation-based strategy. If the postman chooses an activity, the system switches in the activity-based strategy.
2. **Be informed in real time:** This application implies that the postman has activated the option "free path". This option allows a permanent follow-up of the postman trajec-

tory and can be useful for safety reasons (in countryside, transported funds, ...) This option notifies in real time (e.g. SMS, …), that the place where he will pass was the subject of an important remark classified by one of his/her colleagues previously in charge of this round. If the postman encounters a difficulty in the task which it is achieving, the application indicates to him (i) one or more colleague(s) having already made this round or a correspondent of a support service who are currently connected (ii) a reference to a course (iii) a linked activity allowing him to review the procedures to be followed. This example illustrates a "just-in-time" learning supported by the situation-based learning with push mode.

3. **Work during his free time or when he is not in round:** The application also enables postmen to work out of work hours to look further into a concept. This is another example of "just-in-case" learning supported by the situation-based learning with pull mode.

These examples show various types of learning at workplace in real contexts. They highlight the need of several types of learning for the same learner and the importance of switching from one strategy to another to guarantee seamless features.

CONCLUSION

We proposed a framework managing and mixing situation-based and activity-based learning strategies. This approach enables us to fulfill seamless learning across contexts at workplace and has been tested in the P-LearNet project.

To summarize, the main issue of the chapter is to show how two learning strategies work together in order to achieve a common goal: provide to learner the most relevant resources (activity, course, document, etc.) at the right moment depending on his/her context. In particular, we will show how learning processes, performed by switching from one learning strategy to another, ensure seamless learning across contexts.

The mixed learning strategy takes advantage of the two learning and working strategies. The switch from one strategy to the other is done (i) either explicitly by the user with choosing an activity or following a free course guided by the system. In this last case, recommendations of entities are proposed when necessary according to his current situation; (ii) or automatically after an interruption of the execution of the activity in progress; in this case the system analyzes the current situation of the user and recommends an entity in the continuity of the interrupted activity.

REFERENCES

Balacheff, N. (2006). 10 issues to think about the future of research on TEL. *Les Cahiers Leibniz, Kaleidoscope Research Report, 147.*

Bomsdorf, B. (2005). *Adaptation of learning spaces: Supporting ubiquitous learning in higher distance education.* Paper presented at the Mobile Computing and Ambient Intelligence: The Challenge of Multimedia, Dagstuhl Seminar Proceedings.

Bouzeghoub, A., Defude, B., Duitama, F., & Lecocq, C. (2006). A knowledge-based approach to describe and adapt learning objects. *International Journal on E-Learning*, 95–102.

Bouzeghoub, A., & Do, K. (2008). *A situation based metadata for describing pervasive learning objects.* Paper presented at the mLearn 2008: 1st International Conference on Mobile Learning, Wolverhampton, Ironbridge, UK.

Bouzeghoub, A., & Do, K. (2009). *Situation-aware adaptive recommendation to assist mobile users in a campus environment.* Paper presented at the AINA '09: The IEEE 23rd International Conference on Advanced Information Networking and Applications, Bradford, United Kingdom, 2009.

Bouzeghoub, A., & Do, K. (2010). *Active sharing of contextual learning experiences among users in personal learning environments using a peer-to-peer network*. Paper presented at the The 10th IEEE International Conference on Advanced Learning Technologies, ICALT, Sousse, Tunisia (à paraître).

Bouzeghoub, A., Do, K., & Lecocq, C. (2008). *A situation-based delivery of learning resources in pervasive learning*. Paper presented at the Creating new learning experiences on a global scale: Second European Conference on Technology Enhanced Learning, EC-TEL 2007, Crete, Greece.

Brézillon, P. (2005). *Task-realization models in contextual graphs*. Paper presented at the International and Interdisciplinary Conference CONTEXT 2005.

Brodersen, C., Christensen, B. G., Gronboek, K., Dindler, C., & Sundararajah, B. (2005). *eBag: A ubiquitous Web infrastructure for nomadic learning*. Paper presented at the Proceedings of the 14th international conference on World Wide Web.

Chaari, T. (2007). *Adaptation d'applications pervasives dans des environnements multi-contextes*. Institut national des science appliquées de Lyon, Lyon, France.

Cheverst, K., Mitchell, K., & Davies, N. (1998). Design of an object model for a context sensitive tourist guide. *Computers & Graphics*, *23*, 883–891. doi:10.1016/S0097-8493(99)00119-3

Derycke, A., Chevrin, V., & Vantroys, T. (2007). *P-learning and e-retail: A case and a flexible software architecture*. Paper presented at the Pervasive Learning 2007, an International Workshop on Pervasive Learning, Toronto, Ontario, Canada.

Dey, A. K. (2001). Understanding and using context. *Personal and Ubiquitous Computing*, *5*(1), 4–7. doi:10.1007/s007790170019

Dourish, P. (2004). What we talk about when we talk about context. *Personal and Ubiquitous Computing*, *8*(1), 19–30. doi:10.1007/s00779-003-0253-8

Drucker, P. F. (1999). Knowledge-worker productivity: The biggest challenge. *California Management Review*, *41*(2), 41–49.

Farmer, J., Lindstaedt, S. N., Droschl, G., & Luttenberger, P. (2004). *Ad-hoc-work-integrated technology-supported teaching and learning*. Paper presented at the 5th conference on Organisational Knowledge, Learning and Capabilities, Innsbruck, Austria.

Hundebol, J., & Helms, N. H. (2006). *Pervasive e-learning-in situ learning in changing contexts*.

Jones, V., & Jo, J. H. (2004). *Ubiquitous learning environment: An adaptive teaching system using ubiquitous technology*. Paper presented at the ASCILITE, Perth, Australia.

Loke, S. (2006). *On representing situations for context-aware pervasive computing: Six ways to tell if you are in a meeting*. Paper presented at the Proceedings of PerCom. Workshops 2006.

Lyytinen, K., & Yoo, Y. (2002). Issues and challenges in ubiquitous computing. *Issues and Challenges in Ubiquitous Computing, 45*(12).

Michael-Spector, J., & Wang, X. (2002). Integrating technology into learning and working: Promising opportunities and problematic issues. *Journal of Educational Technology & Society*, *5*(1).

Naismith, L., Lonsdale, P., Vavoula, G.-N., & Sharples, M. (2004). *Literature review in mobile technologies and learning*. Bristol, UK: NESTA FutureLab.

Oppermann, R., & Specht, M. (2006). Situated learning in the process of work. In S. Netherlands (Ed.), *Engaged learning with emerging technologies* (pp. 69-89).

p-LearNet. (2006). *p-LearNet project*. Retrieved July 2, 2010, from http://p-learnet.univ- lille1.fr/web/guest/home

Pham-Nguyen, C., & Garlatti, S. (2008a). *Context-aware scenarios for pervasive life-long learning*. Paper presented at the IEEE/WIC/ACM International Conference on Web Intelligence and Intelligent Agent Technology. WI-AT'08, Sydney, Australia.

Pham-Nguyen, C., & Garlatti, S. (2008b). *Pervasive technology-enhanced learning system integrating working and learning situations*. Paper presented at the 3rd International Conference on Information and Communication Technologies: From Theory to Applications. ICTTA 2008, Syria.

Pham-Nguyen, C., Garlatti, S., Lau, S., Barbry, B., & Vantroys, T. (2009a). An adaptive and context-aware scenario model based on a Web service architecture for pervasive learning systems. *International Journal of Mobile and Blended Learning*, *1*(3), 41–69. doi:10.4018/jmbl.2009092203

Pham-Nguyen, C., Garlatti, S., Lau, S., Barbry, B., & Vantroys, T. (2009b). Pervasive learning system based on a scenario model integrating Web service retrieval and orchestration. *International Journal of Interactive Mobile Technologies*, *3*(2), 25–32.

Pham-Nguyen, C., Lau, B.-Y.-S., Barbry, B., Vantroys, T., & Garlatti, S. (2008). *Pervasive learning system based on a scenario model integrating Web service retrieval and orchestration*. Paper presented at the mLearn 2008, the International Conference on Mobile Learning, University of Wolverhampton, School of Computing and IT, UK.

Pimmer, C., & Gröhbiel, U. (2008). *Mobile learning in corporate settings: Results from an expert survey*. Paper presented at the International conference on mobile learning. mLearn 2008.

Raelin, A. (Ed.). (2008). *Work-based learning: Bridging knowledge and action in the workplace.*

Rey, G. (2006). *Contexte en interaction homme-machine: Le contexteur*. Grenoble, France: L'université Joseph-Fourier-Grenoble I.

Sharples, M. (2005). Learning as conversation: Transforming education in the mobile age. *Proceedings of Conference on Seeing, Understanding, Learning in the Mobile Age* (pp. 147-152). Budapest, Hungary.

Sharples, M. (2006). *Big issues in mobile learning: Report of a workshop by the Kaleidoscope Network of Excellence Mobile Learning Initiative*. LSRI, University of Nottingham.

Siobhan, T. (2007). *Pervasive scale: A model of pervasive, ubiquitous, and ambient learning*. Paper presented at the An International Workshop on Pervasive Learning, in conjunction with Pervasive 2007, Toronto, Ontario, Canada.

Vavoula, G.-N., & Sharples, M. (2008). *Challenges in evaluating mobile learning*. Paper presented at the mLearn 2008, the international conference on mobile learning, University of Wolverhampton, School of Computing and IT, UK.

Section 2
Practicing Mobile Learning

Chapter 6
Mobile Learning in Action: Three Case Studies with the Net Generation

Andrew Kitchenham
University of Northern British Columbia, Canada

ABSTRACT

This chapter presents three m-learning case studies. The first is a school district in Northern Canada that uses one-to-one computing from intermediate to high school. The second case study is an auto-ethnographic approach to using m-learning in post-secondary education. The last deals with a United States school district that is planning to implement m-learning technologies across three schools in an attempt to address the needs of the Net Generation. The chapter concludes with a synopsis of the findings across the three case studies.

INTRODUCTION

As chapter authors in this book have demonstrated, mobile learning (hereafter, m-learning) is ubiquitous from kindergarten to graduate school from Northern Canada to Southern Queensland. The students in our schools come ready to learn in a fashion that accommodates their innate needs to discover answers to questions, to pose questions to be discovered, to use technology and not have it perceived as an add-on. They are the Net Generation (Tapscott, 2009) and the Digital Natives (Prensky, 2010) who are being taught by many teachers, instructors, trainers, and professors who are also members of that generation of learners. For those who are taught by the Digital Immigrants (Prensky, 2001) and the Digital Pilgrims (Kitchenham, 2009a), learning can be difficult and frustrating for them as their pedagogues and androgogues catch up and attempt to master the

DOI: 10.4018/978-1-60960-511-7.ch006

skills needed to address their needs. Of course, an average university classroom would be composed of Digital Natives, Digital Immigrants, and Digital Piligrims or Net Geners, XGeners, and baby boomers.

As Tapscott (2009) demonstrated when he and his colleagues interviewed young people from Canada and the United States (n=1750), Net Geners (*n*=5935) from Canada, the United States, the United Kingdom, Germany, France, Spain, Mexico, Brazil, Russia, China, Japan, and India (*n*=5935) and baby boomers and Gen Xers (*n*=800) from Canada and the United States, the needs of learners have fundamentally changed since the advent of the Internet, blended learning, and most recently, m-learning. They expect teachers to know the technology and to know that they learn differently than their teachers. These students "are bringing their demographic muscle, media smarts, purchasing power, new models of collaborating and parenting, entrepreneurship, and political power into the world" (p. 3). According to Tapscott, the Net Generation can be described typically by eight characteristics, or norms, that make them distinct from baby boomers, their parents: (1) they want freedom and freedom of choice to make their own decisions; (2) they desire customization so that they can make learning their own; (3) they collaborate naturally, without needed coercion from their teachers and fully enjoy a conversation or discussion more than a formal lecture; (4) they are apt to scruitinize the teacher and the school (and any other authority leader or organization); (5) they really honour and demand integrity for others; (6) they thrive on fun at school, work, and play; (7) they want speed and see it as a natural way to learn; and (8) they embrace innovation and change as natural parts of their lives.To be sure, the Net Generation is a force to be reckoned with and it will change the face of education for decades to come.

Similarly, Prensky (2010) argued that school teachers should consider the concept of partnering. That is, "letting students focus on the part of the learning process that they can do best, and letting teachers focus on the part of the learning process that they can do best" (p. 13). Students would accept responsibility for discovering and pursuing topics about which they are passionate, utilizing available technologies, tracking down information on the topics, answering questions and presenting their own ideas and further questions, practising assigned and chosen tasks, and presenting their findings through writing and multimedia. The teacher's responsibilities would include creating and presenting appropriate questions for discussion and research, providing guidance to the students, creating a context for the information, making time to explain ideas on a one-on-one basis, instilling rigour in the students, and ensuring that quality is paramount. He also makes the argument that critical to partnering through technology and m-learning is the distinction between verbs and nouns. Verbs are the skills that we want our students to acquire, practise, and master; they are constant and do not change or change very little over time. We strive for proficiency in such skills such as *thinking critically*, *presenting arguments*, *convincing others*, and *being open minded.* Verbs are, by their very nature, grounded in the subject areas and include general skills for achieving success. Nouns are the tools that the students use to master the verbs; they are ever-changing by necessity as they are task specific. Tools can include a textbook or a research essay but, with increasing frequency, are software and hardware. Such tools include interactive whiteboards and laptops, wikis and blogs, smartphones, iPhones, iPods, and iPads, and Facebook and MySpace. It is impossible to separate verbs and nouns, of course, but m-learning has revolutionized how the verbs can be addressed by the nouns.

When we consider Tapscott's and Prensky's sound arguments for addressing the needs of our students, it becomes apparent that we need to examine research that demonstrates m-learning in action so that we can better understand how the teaching and learning process can be changed to create defensible learning environments. This chapter outlines three case studies (Mills, Durepos, & Wiebe, 2009) that address m-learning with the Net Generation of learners. The first case study took place in a small school district in Northern British Columbia, Canada that has embraced one-to-one computing. The second is an autoethnographic study on the use of m-learning in my university teaching and learning at the University of Northern British Columbia, Canada. The last case study discusses the proposed plan for m-learning in the a school district in Wyoming, United States.

CASE STUDY ONE: NISGA'A SCHOOL DISTRICT

One-to-one computing involves a high degree of m-learning when the students have access to the laptops in and out of school (Kitchenham, 2009a; Livingston, 2009). I define 1:1 computing classrooms as places where every child in the class has a laptop computer with wireless Internet and printer capabilities for at least fifty percent of the day. The literature on 1:1 computing is full of proof that it changes the role and function of the teacher (Kitchenham, 2009a), increases student achievement in writing, analysis, and research skills (Bebell, 2005; Fadel, & Lemke, 2006; Jeroski, 2005, 2006; Livingstone, 2006; Russell, Bebell, & Higgins, 2004), increases standardized test scores (Stevenson, 1999), increases student and teacher technology use (Bebell & O'Dwyer, 2010), ensures that students come to school and stay in school (Barrios, 2004), promotes self-directed learning in students (Livingstone, 2006) and engagement (Cromwell, 1999; MEPRI, 2003), markedly lowers the number of disciplinary problems in a school (Baldwin, 1999), and creates an increase in, and lays the foundation for a disposition towards constructivist learning (Bransford, Brown, & Cocking, 2000; Light, McDermott, & Honey, 2002).

School District 92 (Nisga'a) is a typical rural school district in that the schools are small, include split grades, and are geographically spread out over hundreds of kilometres; however, it is atypical in that all Grade 4 for 10 students use laptops in their daily learning and have 24/7 access to wireless networks in their schools and villages. That is, the students have unlimited or controlled access, depending on the teacher-assigned task, and unlimited access to the laptops and Internet when outside of school. There are 407 students in the school district taught by 39 teachers (BC Ministry of Education, 2010); 98% of them are First Nations, primary from the Nisga'a Nation and 60% of the school district personnel are Nisga'a or other First Nations and a further seven percent are married to a First Nations spouse. The average class sizes range from 8.0 to 15.7.

In 2005, the BC Ministry of Education (BC Ministry of Education, 2005) invested 2.5 million dollars to be distributed across 12 school districts who were interested in the benefits of 1:1 computing in the classroom. Due to the tight link between funding and results, the focus was on student achievement which had been highlighted by Jeroski's (2005) early research in the Peace River School District. Primarly centred on the schools in the town of Fort St. John, the school district and Jeroski (2005, 2006) had reported significant improvement in reading and writing achievement.

The Nisga'a School District was one of the 12 selected schools; however, they had pilotted 1:1 computing in four Grade Six classes in 2003 and had added an additional grade until 2006. In fact, the District Principal of Technology had observed the Fort St. John 1:1 computing and initiated a wireless writing program in the Nisga'a School District shortly after his return. Those first few years were district funded which a testament to how much the teachers, students, and administrators were willing to invest to achieve success. Since receiving Ministry of Education funding,

Table 1. Percentage of students who are meeting or exceeding writing standards on the FSA tests in two-year intervals (Grade 4 and 7)

Grade 4	2004	2006	2008	2010
	75	90	77	81
Grade 7	64	40	87	84

all Grade 4 to 10 classrooms have embraced 1:1 computing and the results appear to be promising.

The implementation of the wireless writing program was supported by a group of outside-of-district and within-district professionals who provided both technical and instructional support for teachers two to three times per year. The professional development workshops were planned and facilitated by an independent researcher, a representative from Apple Canada, an educational consultant in Northwestern British Columbia who serves as the Literacy Coordinator for the Nisga'a School District, and a District Principal for Technology in the Nisga'a School District 92 (who has since left the district).

This commitment also resulted in a district-wide committee, Nisga'a on Wireless (NOW), whose mandate is to assist teachers and students in achieving their goals for literacy. In particular, the committee focussed on three main areas of concern that were in line with the district's accountability contract (a document filed with the Ministry of Education). First, they wanted to increase the percentage of intermediate students who met or exceeded the writing standards set out by the Assessment Branch of the Ministry of Education. Second, the committee wanted to increase the technological literact of all student involved in the project. Lastly, the committee planned to increase the capacity of teachers to teach writing and to integrate technology into their respective classrooms.

Writing Achievement

The Ministry of Education administers standardized tests, Foundational Skills Assessment (FSA), each year to obtain data on the performance of Grade Four and Seven students in the curricular areas of reading, writing, and numeracy. Table 1 portrays the district results after one year, three years, five years, and seven years of the 1:1 computing writing program. As can be seen, the students have gradually increased in their writing performance on the FSAs after each year of instruction with an overall seven-year average of 80.75% for Grade Four students and 68.75% for Grade Seven students. When compared with the provincial averages of 86.5% and 86.5%, respectively, the Nisga'a School District is performing admirably; however, since implementing the Nisga'a on Wireless (NOW) program, the writing performance of the students has improved dramatically as it used to be one of the lowest-performing school districts in the province. Additionally, the 1:1 computing program was not fully implemented in the Grade Seven classes until 2008 when you compare the 2010 school district result (84%) with the provincial average (81%), the Nisga'a wireless writing program is an obvious suceess (BC Ministry of Education (2010).

Additionally, the Nisga'a School Districts administers school-wide tests each year. The students compose writing samples and the samples are marked by the teachers across the school district. That is, the samples are graded by teachers who do not instruct the students in most cases. As can be seen from Table 2, there has been a gradual increase in the percentage of students

Table 2. Percentage of students who are meeting or exceeding writing standards on the school-wide writes in two-year intervals (Grades 1 - 9)

Grade 1	**2004**	**2006**	**2008**	**2010**
	40	43	73	72
Grade 2	53	29	70	73
Grade 3	37	44	94	91
Grade 4	**31**	29	68	79
Grade 5	41	45	79	88
Grade 6	46	**25**	61	71
Grade 7	47	29	81	46
Grade 8	26	31	**71**	82
Grade 9	24	15	40	59

Initial Grade Four student results appear in bold font.

who are meeting or exceeding expectations in their writing. Most notably, the Grade Four class that began with the laptops in the 2003/2004 school year increased from 31% to 71% in Grade Eight which represents a 40% increase over the six years of using laptops. Additionally, as the other students became more comfortable with using the laptops and m-learning initiatives, the percentage increased.

All in all, the writing achievement of the students involved in the NOW program have increased in their writing abilities as demonstrated by the scores on the provincial- and district-based assessments. They write better, using more words, and understand the conventions of writing much better now that they have had opportunities to use the laptops and accompanying m-learning software and hardware. It would also appear that being able to use the laptops at home or in any village has given the students more confidence.

Student Technological Literacy

As part of my earlier research (Kitchenham, 2009a), I not only interviewed the 1:1 computing teachers but also recorded information from the students as I visited the classrooms in three of the four schools. This information became a valuable addition to my research and augmented the questionnaires and interviews that were conducted during that research on teacher transformation.

Given that most students have been using laptops and m-learning strategies for one to seven years, their technological literacy is at a high level. The students use the laptops, IPods, iPhones, and smartphones for a greater part of their waking hours and are quite adept at using them. Each teacher can control the amount of access to the wireless networks so, over multiple encounters, the students have learned to respect their access as any violation affects the rest of the group. In this way, the students have built up a system of respectfully and responsibly using the m-learning software and hardware.

When I asked students the role that m-learning played in their lives, the responses were varied in discussing the uses but united in their belief that their learning was enhanced by both the 1:1 computing and wireless writing project and the use of mobile technologies. A sample explanation follows:

When teacher says that we need to find information on a topic, I can just Google or go to Wikipedia and look up stuff. ... Sometimes I use the laptop and sometimes we can borrow the [interactive] whiteboard or maybe a Ipod thing. ... It really makes learning fun and makes me want to look up stuff [but] ...before it was not so fun. (Grade 7 student)

Another discussed the use of iPods, when she said,

I used the laptop for a coupla years now but now I can use my iPod or my phone cuz I can just type in the request or look up YouTube videos to help me or my friends to find out information. I even helped [the teacher] check on some facts in the Science class when he was not sure. (Grade 9 student)

A younger student mentioned that they used their laptops at school to learn the Nisga'a language through YouTube videos and that she "could just pick up where I left off when I got home" (Grade 5 student). The Nisga'a language classes are supplemented with YouTube videos for language practice in which the students learn by watching a brief puppet show with an elder presenting the language to the puppet.

I also witnessed the facility with which the students mastered the technology and easily they applied the technology to their daily learning. For example, in one class the teacher was struggling to add bubbles to an Inspiration semantic map. One student volunteered assistance and said, "why don't you just press that lightning symbol and see if that fixes the problem?" Within seconds, the problem was solved and when I asked the teacher if that particular student was familiar with Inspiration, she stressed that this was the second time that she had used it in class. Clearly, that student was adept at technological literacy and really understand the applications of the technology.

Teacher Writing Instruction and Technology Integration

As previously mentioned, the teachers in this study had attended workshops on using 1:1 computing in the classroom. The time in attending workshops ranged from three to twenty hours so many teachers had acquired a great deal of information on how to use the laptops in the classroom from a competent researcher and a representative from Apple Canada with added "in-house" support from an educational consultant and a district principal. What they lacked was in-class application of how to properly teach writing and how to integrate technology through the use of mobile technologies.

When I spoke with the teachers, most had acquired a much larger repertoire of writing instruction strategies than m-learning strategies. The majority of elementary teachers used Inspiration to demonstrate how to pre-plan a story; some used interactive whiteboards so that the children could manipulate the texts but most used their laptops and a data projector. One teacher lamented, "I wish that I had a SMARTboard but we don't all have them in the classrooms yet [but] I know that the boards have a lot of potential for student and teacher interactivity" (Grade 5 teacher). Other strategies included having students post their assignments in a dropbox and then having the teacher distribute them to other students for peer evaluation; emailing assignments to a friend for comments on how to improve the writing; having the teacher put an example of a grammatical or stylistic point on the screen and asking students to discuss how to improve the structure; and having students or the teacher use track changes in a document so that the author could both understand the suggestion by reading an inserted comment or accept/reject the stylistic change.

In terms of integrating technology through mobile learning technologies, most teachers were in their infancy. Some teachers either used hardware such as iPhones and iPods or encouraged students to utilize the hardware and associated applications. One teacher reported that he "had students look up information on their laptops or iPods and then email it to their email accounts for use in research papers" (Grade 9 teacher) while enough used his "smartphone to connect to the data projector to show YouTube videos on dissection" (Grade 11 teacher).

It did become apparent that these teachers were the exception and that the vast majority of teachers did not use m-learning in their teaching or used it much more in their personal lives. One teacher said that he "tended to use his iPhone or laptop more and more frequently when he is at coffee or at Board meetings [so that] he can get work done" while attending other functions. Another teacher mentioned the convenience of using m-learning technologies since she could "get a lot of information for teaching and for [her] personal needs without having to run home to look it up or lug around her computer". In short, there is some progress being made but the teachers are in the experimentation stage at this point in their m-learning progress.

CASE STUDY TWO: UNIVERSITY OF NORTHERN BRITISH COLUMBIA

Autoethnography is a respected qualitative research method that involves self-observation and critical reflection and is situated within the ethnography field and can be analytic, subjectivist experiential, or post-structuralist/post-modernist in nature (Ellis, 2006). My study was analytic as I was personally engaged in a social group, setting, or culture as a full member and active participant but retain[ed] a distinct and highly visible identity as a self-aware scholar and social actor within the ethnographic text" (Maréchal, 2009, p. 43). In particular, I kept a journal, written and digital, for six months that described my m-learning experiences as a member of the post-secondary culture (Kitchenham, 2009b).

I am an Associate Professor in the School of Education who teaches undergraduate courses in educational technology and e-portfolios and graduate courses in technology, Special Education, and leadership so technology is part of my job description. I have mentored hundreds of students in their use, integration, and teaching of technology and continue to mentor my fellow professors so that they can become comfortable with using, integrating, or teaching with technology. I have also developed an on-line Master of Education in Special Education that will utilize all available technologies at this university. Technology has always been part of my life as I was an early adopter as a classroom teacher, an educational consultant in technology, and a frequent user in all five of my graduate programs. I have received an Excellence in Teaching Award and an Excellence in Research Award and some of that recognition can be attributed to my use of and research in m-learning. Having said that, I am not a member of the Net Generation but was probably borne into the wrong generation since I use technology with a facility and familiarity of someone who is much younger than I.

My m-learning companions for those months included: an HP TouchSmart tx2 tablet netbook, a Canon portable printer, and an iPhone. I would argue that any two of these systems could serve the purpose of a post-secondary instructor in the 21st century; however, I carried the netbook and the iPhone with me for the majority of each day.

Although I have had three netbooks, when I first received the netbook, I used it for six or seven hours without a break so that I could learn its features as they pertained to my teaching. My first journal entry was:

Because the screen is only 12.1" WXGA, it's easily portable without losing viewing quality which means that I can use it in any environment. After class today, I took the netbook and a jumpdrive to [a coffee shop] and marked assignments while enjoying my favourite beverage. I know that I could have the same drink in my office but this atmosphere does not have colleagues and students dropping in or emails and voicemails coming in while I mark. As I tell my students, I have a sickness that bars me from not reading and answering emails or returning voicemails. I have my iPhone turned to mute so I do not know whether someone is emailing, texting, or phoning. This netbook allows me to concentrate on the work while listening to [the coffee shop's] music or my own and I can use their Internet connection when I need to check on-line information or get around to the email.

As we have seen from others in this book, this idea of convenience and portability is common in the professional literature.

A few days later, I met with one of my 15 graduate students and we discussed the final stages of her thesis in her office. Rather than type notes or record the meeting, I used the tablet feature of the netbook to take notes and other related responsibilities. After the meeting, I used the voice memos feature of my iPhone to record a stream-of-consciousness entry for my journal:

Today I met with [a graduate student] to outline the next steps of the thesis but I didn't type notes. I used the Journal feature of the tablet and wrote notes. It was slick since I could write the notes, of course, [but ...] I could also sketch out a timeline and some basic diagrams to give her an idea of how to set up the figures in the piece [and] how to properly use APA headings. ... At one point, I was able to close the Journal feature and open her Word document to show her some sections to be developed.... I used the highlight function in Word but instead of the pen or switching to the mouse, I just used my finger so that I could control the highlighting with my finger. Cool. ... Later, I sent the notes to [her] so that she would have a record of our meeting. She emailed 10 minutes after the meeting and said how good it was to have those notes to look upon.

To be sure, all of the activities described in the above audio entry could be accomplished by using available technology but I was able to conduct this meeting with one technology that could fit in my briefcase and weighed less than five pounds. It was seamless and did not requirc opening myriad programs or rely on my adequate typing skills.

When I was conducting field research a few months ago, I utilized the netbook and iPhone again to record an interview. As many researchers know, recording an interview and taking notes can be distracting and can cause some anxiety on the part of the participants. As I demonstrate in this entry, the novelty of the experience was motivating for the participant.

When the first interviewee arrived, I reminded him that I would be recording the interview and would be taking notes. I pulled out my iPhone and showed him how I was using the recording feature to record the interview and I showed him the netbook and how I would take notes while he was chatting and that he could have the notes emailed to him right after the interview. ... He wanted to see a test of the recording on the iPhone and the he pulled out his iPhone and said that he would record the interview as well. ... After the interview, he said that he really appreciated learning the feature and would use it for his parent-teacher interviews and later, in his Masters project. I told him that I would import the interview into my computer then I could transcribe the interview on my computer which would be sent to him. A few minutes later, I emailed him my notes from the Journal scribblings.

This entry demonstrates how the convenience of the technology can be contagious and how a simple demonstration can turn on a person to using m-learning technology in his daily life.

As well, I have used both the netbook and the iPhone for an important part of my research, video-recording interviews of focus groups and one-on-one sessions. This entry expands on how I have used both of them in tandem.

I was interviewing a focus group this morning and decided to use the iPhone camera and a small flexible tripod on a table. I tested a three-minute segment to ensure that the positioning would be acceptable and then started the formal discussion. It went famously! The quality of video and sound was excellent and the file size was much smaller than the HD video cameras. In the afternoon, I used the same procedure for a one-on-one interview and sound/video quality was even better. Afterwards, I loaded the raw footage onto the netbook and did some quick edits. I was then able to send the edited version back to the participants for comment and authenticating. The response time was fantastic as all but one responded within 48 hours. As I am off to a conference tonight, I can view the footage on the iPhone and make notes for later analysis and interpretation.

Once again, the synchronicity between the iPhone and the netbook make the combination a very powerful form of m-learning for field research.

In one of courses, I ran a remedial math clinic that necessitated the graduate students working with one elementary-aged child while I observed and recorded notes. I used the netbook to take notes and after the meeting, used the Canon printer to print off the notes for the students so that they could place them in their portfolios. When I asked the students how they liked that system, their comments were effusive and I recorded them as they chatted with me. Entries included, "the students described the process as 'cool' and 'efficient'", "it was great to get the feedback so fast so that I could come to you after and ask for clarification or further guidance rather than wait for it to be emailed to me", "so many other professors say that we will get feedback but it never comes let alone within minutes of the end of the meeting", and "I would love to use that system when writing reports so that I could get work done while my daughter is at dance lessons and I am just sitting in the waiting room". Once again, that idea is apparent that m-learning affords opportunities for tearing down boundaries.

Like many people, I use my iPhone for pleasure but given the focus of this case study, I will discuss how it is used for research, teaching, and service purposes. When I am travelling, my iPhone is used a great deal for dealing with emails and attachments. For instance,

Today I was on my way to a conference in Washington state and a few critical emails came through for minor changes to an article. Rather than haul out the netbook, it was much easier to download the document in the airport (using their WiFi) and then edit it on the plane without experiencing the wrath of the air hostess. No real research on whether the wifi interferes with the instrumentation as far as I have found. ... Using [the] DocToGo [iPhone application], I could open the document and edit the few sentences that needed to be addressed. I was also able to open an Excel document for a book that I am editing so that I could update email addresses and send emails when I landed in the SeaTac airport.

Similarly, when I need to check on or order a book, I often use the iPhone to do so when the netbook is not an option such as when I am running around between schools and cannot always access a WiFi connection. Additionally, I use many apps for my teaching such as QuickDSM so that I can look up information on mental disorders in my learning disabilities and Special Education classes or the iSource apps for checking on APA, MLA, and Chicago referencing styles, depending on the course I am in or my Kindle to read professional literature or to look up obscure APA references. Once again, I could have used the netbook for the same functions but the iPhone was capable of performing the same operations while still attaching to my hip.

I have also found the iPhone to be useful when attending conferences. There are many uses for the apps in situations where I am interacting with colleagues over a short period of time. I wrote in my journal,

At this AACE conference, I get to see so many of my colleagues whom I visit with via email and video chats most of the year but it is not the same as that f2f contact. One of the major benefits is that we get to share resources and update contact information for each other and for colleagues who have moved on. I use the contacts app on the iPhone to do that which means all my contacts are loaded on the Exchange server as well. I also use the Notes app to take notes on resources or use the camera to take pics of the books at the book fair rather than take copious referencing notes! The Notes app was also great when I attended workshops today since I could jot down small details without having to worry that the battery would die like I do when using the netbook.

Additionally, I discovered that I could use Skype on my iPhone which was a major benefit for when I wanted to contact my family, graduate students, or colleagues back in my home city without incurring long distance charges.

All in all, these three technologies have served me well in the last six months. I have been able to increase my research productivity as I lose so much less time when I can take care of business very quickly and very efficiently. I also point out that these technologies were all purchased for less than two thousand dollars which is well under my allowable professional development funds, and, I would guess, within most professors' professional development monies.

CASE STUDY THREE: SWEET WATER SCHOOL DISTRICT

The Sweetwater County School District #1 is located in the southwestern corner of Wyoming and serves the communities of Rock Springs, Farson-Eden and Wamsutter. There are six K-Grade 4 elementary schools, one Grade 5-6 school, one Grade 7-8 school, one Grade 9-12 secondary, one Grade 9-12 alternate secondary school, one K-Grade 8 elementary/middle school, and one K-Grade 12 school. Across the school district, they enrol approximately 4,900 students.

This coming school year, 2010-2011, the district will be integrating iPod Touch and iPad units in the schools through the librarians (media specialists) in a project entitled, "M & M: Media and Mobile Learning". All teachers will be able to use the units through a standardized checkout system, using the Alexandria library system, which is in place now for the return of teachers in late-August. The teachers intend to use the iPod Touchs and iPads for fluency (recordings), podcasting, e-books, specific apps for instruction, interventions (Tier 1), and enrichment, along with the traditional uses like word processing and Internet searches. They will also utilize various multimedia programs such as iTunes, GarageBand, and iMovie, Microsoft Office software, and services such as EducationCity.com, BrainPOP, and United Streaming. On occasion, students will make use of Promethean boards, document cameras, laptop computers, and printers. The media teachers will use these mobile learning devices to teach students 21st century skills such as technology, problem solving, thinking strategies, collaboration, work ethic, communication, and research (C. Magagna McBee, personal communication, July 20, 2010).

In August 2010, Dr. Cristy Magagna McBee, an instructional coach and coordinator for the project, will deliver professional development workshops with the media specialists, paraprofessionals, and instructional facilitators to train them and set up the program. Additionally, the media specialists and Dr. McBee will ensure that the professional development is set up for teachers, that they will have a system of measurement of success in place to track engagement, usage, and effectiveness, and that they will continue to locate funding sources to add additional units.

At present, they have created a a standard template that each school will use to record the iPod and iPad apps, e-books and other resources available which will be unique to each school so that teachers know what they can use. They will also use this system to track for which functions the teachers used the units.

As mentioned, the training will be conducted by Dr. McBee and will include troubleshooting, using the devices, syncing the devices, managing and using iTunes, checking out the devices through the Alexandria system, the paperwork, ordering of apps and other resources, and planning to provide information to each of the buildings. Using a trickle-down process, these media specialists and the instructional facilitators will then meet with building staff to explain the program and the process for checkout and use. The media specialists will model the units during the first week of school with the students during media class time so that they can ensure that they reach all students. Teachers can observe then, attend a session before or after school, or have a coaching session from an instructional facilitator to gain their initial professional development to be ready to use the devices in the classroom.

The program will begin with $200 in iTunes money, but the team will be going to local businesses to have them adopt a school (purchasing additional iTune cards). PTO, civic groups, and local grants will also be explored. The media specialists will keep an up-to-date website and a blog, allowing them to present their apps lists, provide a troubleshooting section, link and explain resources and apps, and describe and post lessons of the week. The website will open to public access and will be updated on a regular basis.

Specifically, the M & M project will have three goals: (1) each student or group will complete two media-based research projects (podcast, PowerPoint, iMovie, etc.) on their iPod or iPad during the 2010-2011 school year; (2) each student or group will keep a digital journal, on the iPod or iPad, reflecting on experiences during instructional time; and (3) all students will use iPod or iPad apps and/or e-books on to gain new knowledge (enrichment) or to review skills (intervention). When the school year concludes, students will have met these three goals. They will be able to create two different kinds of media presentations using research skills, reading strategies, thinking strategies, writing skills, cooperative learning, technology standards, and work ethics. The projects will be completed using the iPod Touch or iPad technology along with such programs as GarageBand, e-books from iMovie, PowerPoint, and so forth Additionally, students will improve writing and reflection skills through the digital journal that they will be keeping on their iPod or iPad units. These entries could be based on books red; pluses and deltas for work accomplished on projects; and anything else that students work on in media and Learning Assistance courses. Finally, students will make use of e-books and apps to review skills and to gain additional knowledge and experience with various topics meeting the district's curriculum maps in the areas being used (C. Magagna McBee, personal communication, June 14, 2010).

In short, this school district has a solid foundation for improving literacy in its schools by firmly embracing m-learning technologies. They had spent many hours researching the promising practices for m-learning teaching and will be successful in their endeavours.

M-LEARNING ACROSS THE CASES

These three case studies have clear commonalities. In particular, they demonstrate that m-learning with the Net Generation is a fait accompli and that we have to assist students in achieving 21st century skills (Brooks-Young, 2010). The commonalities across the three cases include flexibility, transportability, convenience, and addressing needs.

Flexibility

The whole notion of flexibility in m-learning is apparent in each of the three case studies. Many of the students and teachers in the Nisga'a school district reported how they found the freedom of choice inherent in the use of laptops and iPods to be a major advantage. As I indicated in my journal entries, I found that I could use my iPhone or netbook for myriad tasks whenever I needed to in my teaching, research, and service. The Sweetwater County School District is also providing evidence that flexibility is an important element to their decision in the Media and Mobile Learning project as they want the students to be able to use the iPods and iPads when they see the need during media and Learning Assistance classes.

Transportability

An obvious advantage of m-learning for the Net Generation is the idea of transportability. The students and teachers in the Nisga'a school district have demonstrated that the ease with which they can access their laptops within the school and then take them to their villages after school is a major asset for a school district that has limited text-based resources. By downloading assignment at school or accessing books online to take home, the students and teachers can transport volumes of information. Like many academics, I used to carry pounds of books and assignment to and from my office to my home to a conference to my office to my home; however, as my journal entries have shown, I now have all that information contained within my briefcase which contains my netbook, a portable printer, two jumpdrives, and an iPhone which, on any given week, would encapsulate 10,000 pages of reading material. Transportability is a major concern for the Sweetwater County School District because they want their students to be able to transport a great deal of information at any given time in support of the districts literacy initiative.

Convenience

The Nisga'a students and teachers argued that having the freedom to just grab an iPod or laptop was a major advantage in the Nisga'a on Wireless (NOW) program. The students were able to complete many assignments or tasks by using their iPods for searching for information in and out of the school. As well, I found that having the power of my iPhone on my hip allowed me much more freedom to complete the plethora of tasks in my daily work. The added ability to Skype or phone colleagues and family members was also extremely convenient and cost saving. Additionally, the Sweetwater County School District has clearly planned carefully to utilize convenience by having a standard signout and tracking system so that, on the one hand, teachers can access the iPods and iPads when needed but, on the other hand, the media specialists and the project coordinator can ascertain how frequently and for what purpose the conveniently-available units are being used. Like the Nisga'a School District, the students and teachers will find the facility with which information can be accessed from such tiny but powerful hardware and software a major advantage for these m-learning technologies.

Addressing Needs

Lastly, the m-learning opportunities afforded the participants in these case studies clearly addressed the needs of the Net Generation. The Nisga'a students and teachers found that students learned better, attended to tasks more closely, improved their writing abilities, and were engaged more readily whenever they used laptops and iPods.

As previously mentioned, I am not a member of the Net Generation but I engage with technology with the same adeptness and comfortableness as my students who are 20 years my junior. I found that the netbook and iPhone especially met my needs as there were no tasks that could not be performed by one or the other. I could easily find information, analyze information, present information, or write about information without having to switch hardware for each task. Additionally, the Media and Mobile Learning project planned by the Sweetwater County School District is focussed on meeting the needs of the Net Generation and they are planning activities that allow for creativity, collaboration, partnering, and deconstructing.

These students are the Net Generation and they are ready to revolutionize education throughout the next decade. They expect information to be sought out, analyzed, discussed, and synthesized rather than handed out. They prefer to uncover information rather than have their instructors cover the information. And, they expect m-learning technologies to be a major part of that learning process.

REFERENCES

Baldwin, A. (1999). Taking the laptop home. *Appalachia, 32*(1), 10–15.

Barrios, (2004). *Laptops for learning: Final report and recommendations of the Laptops for Learning Task Force.* Retrieved November 18, 2006, from www.etc.usf.edu/L4L/Cover.pdf

BC Ministry of Education. (2005). *$2.1 million for computer pilot projects in schools.* Retrieved March 12, 2006, from http://www.bced.gov.bc.ca/onetoone/

BC Ministry of Education. (2010). *Student statistics-2009/10 92-Nisga'a.* Retrieved July 12, 2010, from http://www.bced.gov.bc.ca/reports / pdfs/student_stats/092.pdf

Bebell, D. (2005). *Technology promoting student excellence: An investigation of the first year of 1:1 computing in New Hampshire Middle Schools.* Boston, MA: Boston College, Technology and Assessment Study Collaborative.

Bebell, D., & O'Dwyer, L. M. (2010). Educational outcomes and research from 1:1 computing settings. *Journal of Technology, Learning, and Assessment, 9*(1). Retrieved June 26, 2010, from http://www.jtla.org

Bransford, J., Brown, A., & Cocking, R. (Eds.). (2000). *How people learn: Brain, mind, experience, and school.* Washington, DC: National Academy Press.

Brooks-Young, S. (2010). *Teaching with the tools kids really use: Learning with Web and mobile technologies.* Thousand Oaks, CA: Corwin.

Ellis, C. (2006). *The ethnographic I: A methodological novel about autoethnography.* Lanham, MD: Rowan & Littefield.

Fadel, C., & Lemke, C. (2006). *Technology in schools: What the research says.* Culver City, CA: Metiri Groups's Technology Solutions that Work/Cisco Systems.

Jeroski, S. (2005). *Research report: The Wireless Writing Program 2004-2005.* Vancouver, BC: Horizon Research & Development Inc.

Jeroski, S. (2006). *Research report: The Wireless Writing Program 2004-2006.* Vancouver, BC: Horizon Research & Development Inc.

Kitchenham, A. D. (2009a). Laptops and teacher transformation. In Rogers, P., Berg, G., Boettcher, J., Howard, C., Justice, L., & Schenk, K. (Eds.), *Encyclopedia of distance learning* (pp. 1313–1318). Hershey, PA: IGI Global.

Kitchenham, A. D. (2009b). Diaries and journals. In Mills, A., Durepos, G., & Wiebe, E. (Eds.), *Encyclopedia of case study research* (pp. 299–301). Thousand Oaks, CA: Sage.

Light, D., McDermott, M., & Honey, M. (2002). *Project Hiller: The impact of ubiquitous portable technology on an urban school*. Retrieved December 13, 2006, from www2.edc.org/CCT/publications_report_ summary.asp?numPubId=129

Livingstone, P. (2009). *1:1 learning: Laptop programs that work* (2nd ed.). Washington, DC: ISTE Publications.

Maine Education Policy Research Institute. (2003). *The Maine Learning Technology Initiative: Teacher, student, and school perspectives mid-year evaluation*. Retrieved January 2, 2007, from www.cepare.usm.maine.edu

Maréchal, G. (2009). Autoethnography. In Mills, A., Durepos, G., & Wiebe, E. (Eds.), *Encyclopedia of case study research* (pp. 43–45). Thousand Oaks, CA: Sage.

Mills, A., Durepos, G., & Wiebe, E. (Eds.). (2009). *Encyclopedia of case study research*. Thousand Oaks, CA: Sage.

Prensky, M. (2001). Digital natives, digital immigrants. *On the Horizon*, 9(5).October 2001. MCB University Press. Retrieved November 2, 2009, from http://www.marcprensky.com/writing/ Prensky%20%20Digital%20Natives,%20 Digital%20Immigrants%20-%20Part1.pdf

Prensky, M. (2010). *Teaching digital natives: Partnering for real learning*. Thousand Oaks, CA: Corwin Press.

Russell, M., Bebell, D., & Higgins, J. (2004). *Laptop learning: A comparison of teaching and learning in upper elementary classrooms equipped with shared carts of laptops and permanent laptops*. Boston, MA: Boston College, Technology and Assessment Study Collaborative.

Stevenson, K. R. (1999). *Evaluation report–year 3, Middle School Laptop Program*. Retrieved December 14, 2006, from www.beaufort.k12.sc.us/ district/evalreport3.htm

Tapscott, D. (2009). *Grown up digital: How the net generation is changing your world*. New York, NY: McGraw-Hill.

ADDITIONAL READING

Adhikari, R. (2009, November 12). Teaching mature markets new m-commerce tricks. *E-Commerce Times*. Retrieved April 4, 2010, from http:// www.ecommercetimes.com/ story/68626.html

Ardito, C., Buono, P., Costabile, M. F., Lanzilotti, R., & Pederson, T. (2007). Re-experiencing history in archaeological parks by playing a mobile augmented reality game. In Proceedings of the OTM MONET'07 Workshop, Springer LNCS 4805, pp. 357-366. Retrieved May 5, 2008, from http://ftp.cs.umu.se/~top/pubs/MONET2007%20 Arditoetal_w_confinfo.pdf

Babcock, C. (2010). *Management strategies for the cloud revolution*. Toronto: McGraw-Hill.

Baron, N. S. (2008). *Always on: Language in an online and mobile world*. New York: Oxford University Press.

Brooks-Young, S. (2010). *Teaching with the tools kids really use: Learning with web and mobile technologies*. Thousand Oaks, CA: Corwin.

Cochrane, T., & Bateman, R. (2009). Transforming pedagogy using mobile web 2.0. *International Journal of Mobile and Blended Learning, 1*(4), 56–83. doi:10.4018/jmbl.2009090804

Collins, A., & Halverson, R. (2006). *Rethinking education in the age of technology: The digital revolution and schooling in America*. New York: Teachers College.

Kaptelinin, V., & Nardi, B. (1997). Activity theory: Basic concepts and applications. In *CHI 97 Electronic Publications*: Tutorials. Retrieved February 5, 2010, from http://www.sigchi.org/ chi97/proceedings/ tutorial/bn.htm

Kitchenham, A. D. (2006). Teachers and technology: A transformative journey. *Journal of Transformative Education*, *4*(3), 202–225. doi:10.1177/1541344606290947

Kitchenham, A. D. (2008). One-to-one computing and teacher transformation. In Hansson, T. (Ed.), *Handbook of research on digital information technologies: Innovations, methods, and ethical issues* (pp. 82–101). Hershey, PA: Information Science Reference.

Kitchenham, A. D. (2009a). Laptops and teacher transformation. In Rogers, P., Berg, G., Boettcher, J., Howard, C., Justice, L., & Schenk, K. (Eds.), *Encyclopedia of Distance Learning* (pp. 1313–1318). Hershey, PA: IGI Global.

Kitchenham, A. D. (2009b). E-professional development and rural teachers: Finding the blend. *International Journal of Mobile and Blended Learning*, *1*(3), 70–85. doi:10.4018/jmbl.2009092204

Kittl, C., Edegger, F., & Petrovic, O. (2009). Learning by Pervasive Gaming: An Empirical Study. In Ryu, H., & Parsons, D. (Eds.), *Innovative Mobile Learning: Techniques and Technologies* (pp. 60–82). Hershey, PA: Information Science Reference.

Kukulska-Hulme, A., & Traxler, J. (Eds.). (2007). *Mobile learning: A handbook for educators and trainers*. New York: Taylor & Francis.

Macdonald, J., & Creanor, L. (2010). *Learning with online and mobile technologies: A student survival guide*. Burlington, VT: Gower Publishing.

Montgomery, K. C. (2007). *Generation digital: Politics, commerce, and childhood in the age of the Internet*. Cambridge, MS: MIT Press.

Palfrey, J., & Gasser, U. (2008). *Born digital: Understanding the first generation of digital natives. New York*. Basic Books.

Prensky, M. (2010). *Teaching digital natives: Partnering for real learning*. Thousand Oaks, CA: Corwin Press.

Ryu, H., & Parsons, D. (Eds.). (2009). *Innovative mobile learning: Techniques and technologies*. Hershey, PA: IGI Global.

Scornavacca, E., Huff, S., & Marshall, S. (2009). Understanding the value of interactive SMS for large classes. In Ryu, H., & Parsons, D. (Eds.), *Innovative Mobile Learning: Techniques and Technologies* (pp. 48–59). Hershey, PA: Information Science Reference.

Sharma, S., & Kitchens, F. (2004). Web services architecture for m-learning. *Electronic Journal of e-Learning, 2,* Article 2. Retrieved March 4, 2008, from http://www.ejel.org/volume- 2/vol2-issue1/issue1-art2.htm

Small, G., & Vorgan, G. (2008). *iBrain: Surviving the technological alteration of the modern mind*. Toronto, ON: Harper.

Spikol, D., Kurti, A., & Milrad, M. (2009). Collaboration in Context as a Framework for Designing Innovative Mobile Learning Activities. In Ryu, H., & Parsons, D. (Eds.), *Innovative Mobile Learning: Techniques and Technologies* (pp. 60–82). Hershey, PA: Information Science Reference.

Tapscott, D. (1998). *Growing up digital: The rise of the Net Generation*. Toronto: McGraw-Hill.

Thomas, M. (2006). Book Review-mLearning: Mobile Learning and Performance in the Palm of your Hand. Retrieved January 30, 2008 from http://kt.flexiblelearning.net.au/tkt2006/ edition-9/book-reviewmlearning-mobile-learning- and-performance-in-the-palm-of-your-hand

Tomei, L. (2009). (Ed.). *Information Communication Technologies for enhanced education and learning: Advanced applications and developments*. Hershey, PA: IGI Global.

United Nations Department of Economic and Social Affairs. (2008, August). *UN Millennial Goals Report*. Retrieved January 20, 2010 from http://mdgs.un.org/unsd/mdg/Resources/ Static/Products/Progress2008/MDG_Report_2008_En.pdf.

Wall, S. (2008).Easier said than done: Writing an autoethnography. *International journal of qualitative methods, 7*(1), 38-53.

Webster, J., & Watson, R. T. (2002). Analyzing the past to prepare for the future: Writing a literature review [Electronic version]. *Management Information Systems Quarterly*, *26*(2), xiii–xxiii.

KEY TERMS AND DEFINITIONS

1:1 Computing Classrooms: Places where every child in the class has a laptop computer with wireless Internet and printer capabilities for at least fifty percent of the day.

Digital Immigrants: Learners who are akin to immigrants to a new world who stumble with the language and must rely on translations and references to understand the new technologies.

Digital Natives: Learners who are native to the culture of technology and m-learning and naturally acquire the skills without reliance on outside sources and who learn through practice and experimentation.

Digital Pilgrims: Learners who are searching for some mecca or messiah to take them out of the darkness of technology and m-learning into the light of knowledge.

M-Learning: The use of mobile devices such as cell phones, PDAs, and iPods as educational tools. Also known as mLearning, and mobile learning.

Media and Mobile Learning (M&M) Project: A research project proposed by the Sweetwater County School District in which they plan to give teachers and students access to iPods and iPads as well as to multiple forms of m-learning technologies.

Net Generation: The generation of learners between the ages of 13 and 33.

Nisga'a on Wireless (NOW) Project: A research project conducted in the Nisga'a School District in which students from Grade 4 to 10 use laptops and iPods to support their improvement in writing achievement.

Chapter 7
Investigating Undergraduate Student Mobile Device Use in Context

Yanjie Song
The University of Hong Kong, China

ABSTRACT

This chapter reports on an in-depth one-year empirical research into examining five undergraduate student mobile device uses in context. Data collection methods include: student reflective e-journals, student artifacts, observations, interviews, field notes, and memos. Three complementary streams were involved in the data analysis. Seven interacting factors in context that could either facilitate or inhibit mobile device use were identified and discussed.

INTRODUCTION

Mobile devices have been increasingly used in education. Mobile devices are considered by a variety of researchers to have unique characteristics of portability, connectivity, convenience, expediency, immediacy, accessibility, individuality and interactivity. These characteristics make educational uses of mobile devices in relation to functions distinct from other technologies such as desktop or laptop computers. Many researchers claim that this technology use is highly dependent on the context (Sharples, Taylor, & Vavoula, 2007; Wali, Winters, & Oliver, 2008). In practice, numerous studies have been designed to exploit the capabilities and constraints of mobile learning systems in specific settings rather than build upon sound educational underpinnings in a broader context (Patten, Arnedillo-Sanchez, & Tangney, 2006). Only a few empirical studies on mobile technology educational uses have addressed the concept of context in higher education (Jones & Issroff, 2007; Sharples et al., 2007). What is in context that influences the mobile device use has not been fully explored. This research endeavors to

DOI: 10.4018/978-1-60960-511-7.ch007

work towards the direction of examining factors in context that influence student mobile device uses.

LITERATURE

In mobile technology educational practices, Seale (2008) posits that understanding the relationship between educational uses and technology is "all about understanding context" (p. 2). Mobile device uses have to be understood in "multiple virtual and physical contexts" (Seale, 2008, p. 2). In Wali *et al.*'s (2008) words, we need to make sense of the use in "context crossing and social setting - the intersection between context as change in location and context as change in social settings" (p. 55). More specifically, Gay and Hembrooke (2004) propose that context needs to be redefined as a "multidimensional construct that has overlapping and interacting layers" (p. 75). Thus, context includes "the external physical context, the context that the individual brings to the situation, the context of the tool or device, the information context, and the context that is created by the activity itself" (p. 75). Some studies have proposed conceptual frameworks for designing mobile learning environments from a socio-historical perspective (e.g., Uden, 2007; Sharples *et al.*, 2007, Wali *et al.*, 2008, Zurita & Nussbaum, 2007). These views on context in mobile device educational uses enrich the literature regarding the concept of context in theory. However, they also impose challenges on how to apply these theoretical views of context in mobile device educational uses. In practice, the majority of mobile device educational research has presupposed a specific setting where students have used mobile technologies to perform teacher/researcher-led tasks (e.g., Lan, Sung, & Chang, 2007; Motiwalla, 2007). Only a few empirical studies on mobile device educational uses have been carried out in a social context (e.g., Jones & Issroff, 2007; Sharples, Taylor, & Vavoula, 2007; Waycott, 2005), but they have not addressed the concept of context in student mobile device uses in higher education.

Study practice is embedded in social contexts (Cole, 1996; Lave, 1988). Lave (1988) makes a theoretical distinction between two aspects of a social context. What she calls the arena that constitutes the "objective" social context, to be described in physical, economic, sociological, cultural or political terms; setting, on the other hand, is used to refer to the context as experienced by a participant or set of participants with reference to a social context. Cole (1996) distinguishes between context as "that-which-surrounds-the-object" which is often represented as a set of concentric circles representing different levels of context (p. 135), and context as "that which weaves together" (p. 135). According to Cole, instead of considering context as a container of objects and behaviors, an "act in its context" (p. 136) understood in regard to the weaving-together metaphor requires a relational interpretation of an individual person, the tool and the context. He further states that to understand the human action in context, we need to analyze not only the individual person and the tool, but also his/her purposes and the environment in which the action is embedded. Humans do not act directly on the world; rather, actions are mediated by tools (Vygotsky, 1978). Because the individual interacts with the environment through the tool, the individual mind must be seen as distributed in the tool which is woven together, which also weaves together individual human actions in concert with and as a part of the changing events of life. These contexts emerge from the weaving together of various levels of contexts to form a particular pattern of learning activities mediated by the tool. The combination of goals, tools, and social context (setting and arena) constitutes simultaneously the context of a behavior and ways in which mind is related to that context. The context-as-weaving metaphor suggests that "[t]he relevant order of the levels of context will depend crucially upon the tools through which one interacts with the world, and these in turn depend upon one's goals and other constraints on action" (Cole, 1996, p. 137).

This conviction of context shows that to study human action in context requires a relational interpretation of the individual, the tool and the context. It implies that context has to be understood in multiple layers or dimensions, including goals, tools, and other elements in social context (setting and arena), which are all woven together in human practice. It also implies that context elements and the order of elements that are introduced into context are not fixed, but dynamic. Therefore, it will be meaningless to distinguish which layers or dimensions the elements belong to because these elements are "mixed together" in the context in which the action takes place. To put it simply, human action in context needs to be interpreted in the framework of relationships between the individual, the tool and the context, namely, the relational elements of goals, tools, and other factors to be explored. Furthermore, Jarvelä, Lehtinen, and Salonen (2000) posit that goals cannot be automatically built into the context without an individual's interpretations of the context in which human action takes place. The individual's interpretations are embedded and influenced by constraints and possibilities of the context.

The above convictions of human action in context shed light on my understanding of student mobile device use under the framework of relationships between the individual student, the mobile device and the context that I termed "the framework of mobile device use in context". To be specific, to examine mobile device use in context is to examine the relationships of the interacting elements in terms of goals, the mobile device, individual interpretations and other factors to be explored. Currently, few studies have attempted to investigate mobile device use within such a framework.

METHODS

This research sought to investigate educational uses of the mobile device from a student perspective under the "framework of mobile device use in context". The question investigated is: How did undergraduate students use the mobile device to support their studies in context?

The mobile device used in this study was the PDA with phone functionality and a 1 G mini SD card were provided to each of the participating students for one year use, free of charge. It was a wireless enabled Dopod 818 Pro with phone functionalities and ran the Windows Mobile operating system, which was compatible with computers. In addition, a one-year mobile service package was granted to each student to encourage them to maximize their use of the mobile device.

This research adopted a qualitative research methodology based on a descriptive multiple-case study approach for a one year period to gain a deeper understanding of the student use of the mobile device to support their studies in natural settings at a university. The multiple-case study approach should add a great deal of confidence to the research findings (Merriam, 1998; Yin, 2003). Five individual first-year undergraduate students from different disciplines at a university in Hong Kong participated in this research, and are identified using the following pseudonyms: Curt, Amy, Betty, Eric and Fred. Although all five of them were working towards the ultimate goal of obtaining a Bachelor degree after three years of study, they had different backgrounds and goals. Table 1 illustrates the profiles of the students.

To understand students' interpretation and use of the mobile device as they engaged in their studies in context, data collection instruments employed in this study included: student reflective electronic journals (e-journals), student artifacts that were a collection of mobile device screenshots that showed what the students did using the mobile device to support their studies, retrospective interviews (face-to-face, emails, phone calls or

Table 1. Participant profile

Participant*	Age at the time of the study	Major	Gender	Year of study	Nationality at the time of the study
Curt	20	Mechanical Engineering	M	1	Hong Kong
Amy	19	Journalism	F	1	Mainland China
Betty	20	Nursing and Biochemistry	F	1	Hong Kong
Eric	20	Bio-technology and Social sciences	M	1	Hong Kong
Fred	19	Translation	M	1	Hong Kong

*Pseudonyms are used to protect the identities of the students.

SMS) based on the questions raised from reading students' e-journals and artifacts, in-depth face-to-face interviews, observations, follow-up interviews, field notes and memos. Table 2 shows a summary of the collected data in this research. Member check with the data sources was conducted after the data was transcribed and the case report was written. The multiple sources of data provided the opportunity for me to get a holistic understanding of students' mobile device uses grounded in this research.

The data analysis process was an ongoing, recursive and iterative process, in tandem with data collection. Three complementary streams of data analysis were involved: (a) "a preliminary exploratory analysis" was used to obtain an understanding of the data (Creswell, 2008, p. 250); (b) categorizing strategies were used to code categories of factors that influenced student mobile device uses (Maxwell, 1996); and (c) contextualizing strategies were employed to understand better the factors in context that contributed to the student mobile device use (Maxwell, 1996). The data was analyzed with the assistance of the computer-based qualitative analysis software – NVivo 7.

Table 2. Data summary

Data	Method of collection	Language
Demographic information	Demographic information was collected at the end of the orientation seminar for participant selection.	English
Focus group discussions	Five focus group discussions were conducted in order to select the participants.	English
Student e-journals	Participants were expected to submit e-journals once a week over one-year period.	English
Student artifacts	Participants were expected to submit artifacts once a week over one year from.	Images
Retrospective interviews	The interviews were conducted whenever I had questions or problems in reading student submitted e-journals or artifacts in forms of face-to-face, email, SMS or MSN communication.	Chinese
In-depth interviews	In-depth individual face-to-face interviews were conducted three times.	Chinese
Observations	One-week observations were conducted for each participant.	
Follow-up interviews	Follow-up interviews were conducted six times during the one-week observations for each participant.	Chinese
Field notes	Field notes were taken by the researcher during the observations.	English
Memos	Memos were made by the researcher throughout the one-year research to keep alert and responsive to the data collection and analysis processes	English

RESULTS AND DISCUSSIONS

Based on data sources from the five participating students, this research examined the interacting factors of goals, tools, and other factors in context that contributed to or inhibited the student mobile device use. The findings reveal that student mobile device use was mainly affected by the interacting factors of *goals, tasks, learning resources, mobile device capabilities and constraints, time and place, social factors,* and *individual interpretations*.

The Mobile Device Use towards Learning Goals

In this study, each of the participating students had their own educational goals. The most basic and tangible goal for all students was always to pass their exams and assignments and obtain a Bachelor's degree. These goals were achieved through performing a series of tasks. In the course of performing tasks to achieve certain goals, new goals might emerge, which diverted the student to perform tasks that were favorable to the new goals. For example, in order to get good grades in his exams and assignments, Curt took advantage of the PDA to help study the course material. When he was studying, he encountered new academic terms that hindered his ability to get through the material. Therefore, he used the mobile device as a tool to access a downloaded dictionary to solve his vocabulary problems. He mentioned, "Exam is coming… many English words I did not know. Hopefully, I installed a dictionary which helped me to solve the problem". Just as Oliver (2005) posits that, if the tool cannot be used to undertake an action to achieve our goal, "we can find an alternative or make a new tool that does" (p. 412). In this case, Curt's new study goal of overcoming his vocabulary problems was shaped by the use of the tool to consult the downloaded dictionary. At the same time, using downloaded the dictionary on the mobile device was Curt's solution to overcoming the limitation posed by the lack of built-in dictionary on the device.

Except for the goals of completing exams and assignments, the students had other goals such as learning language, broadening knowledge, preparing for future careers, and socialization. These learning goals, together with different interpretations of how the students perceive their immediate situations, resulted in different mobile device use strategies. These strategies made them either negatively anticipate learning to take place or support learning using the device. For example, Amy wanted to be a journalist in Hong Kong after her graduation. Therefore, she tried her very best to learn to write traditional Chinese characters that she did not learn when she was in Mainland China. She only knew to write and type simplified Chinese characters. After she received the mobile device, she installed the Chinese software CE Star, allowing her to type traditional Chinese characters in Notes. After that, she referred to the words in Notes to practice writing the characters. In this case, Amy utilized the mobile device to learn how to write traditional Chinese characters, which was important for her future career.

Jones and Issroff (2007) posit that technologies can support students in defining their own goals and ways of accessing and utilizing resources. In this study, the mobile device provided opportunities for the five participating students to develop new goals by allowing them to adopt the device as a tool that was more beneficial for them to pass exams. On the other hand, different goals drove the students to make use of the mobile device as a tool for different purposes. Therefore, tools shape the goals of students who use the tools (Kaptelinin, 1996). Tools are also shaped by students who appropriate the tools for their use to achieve their goals (Cole, 1996).

The Mobile Device Use to Perform Tasks

The findings of this research show that learning tasks of the students varied. They were: tasks defined by the teacher (e.g., writing project reports), tasks defined by the student (e.g., course review), and tasks emerging in context (e.g., recording lectures in class). Compared to the tasks assigned by the teachers, the other two kinds of tasks were more dynamic. Very often emerging tasks arose from the assigned tasks by the teacher (e.g., photographing lab results for writing lab reports), or self-defined tasks (e.g., resource sharing and discussion in order to learn new academic terms). Take Betty's use of the mobile device for example. Betty mentioned that when she was in the Department of Nursing Studies, she used the device in order to access learning materials more frequently before exams. She reported:

Before having the device, I had to remember and review them [learning materials] in my dormitory. It took time. I always worried whether I could remember them by heart. This made me stay up late. After using the PDA, I didn't have to worry about it. I saved the files on the PDA, and could review the useful material on bus or on campus at least two or three times. I didn't have to finish reviewing it in my dormitory. (30 May 2006, retrospective interview)

In this example, Betty utilized the mobile device as a tool for her course revision which changed the way she studied. Before, she used to do all her studying in her dormitory, and often stayed up late. By using the PDA tools, she could study anywhere, anytime, which made her days longer. According to Murphy-Berman and Berman (2003), the tasks and requirements of the institution and the related requirements of career involve a relatively high degree of choice, planning and decision making. Certainly, studying course material could be considered a task defined, planned and performed by the students themselves in order to pass exams required by the university. The above mentioned task of using the device to help study for exams, was directed towards the general goal of obtaining a Bachelor's Degree upon graduation. Bedney and Harris (2008) point out that general study goals usually consist of a series of task-goals which are achieved by doing various tasks. However, not all tasks have clear goals. "For many everyday tasks, goals and intentions are not well specified: they are opportunistic rather than planned" (Norman, 1998, p. 48). Doing opportunistic tasks concerns the behavior that takes advantage of the situations rather than engaging in extensive planning and analysis. The findings in this research confirm this argument.

In many instances, the students used the mobile device as different tools to perform emerging tasks to achieve new or emerging goals that arose from other assigned or self-defined tasks. Barnard, Yi, Jacko, and Sears (2007) posit that often there are multiple tasks taking place in mobile educational uses, but the tasks to be performed using mobile devices are often secondary, which is why the context of use must be taken into account. In this research, the students chose and used the mobile device as different tools triggered by immediate context for emerging tasks to deal with emerging goals. Take Curt's model-making project for instance. In the assigned task of making a mold to produce a toy during his summer training course, Curt connected the mobile device to the internet via GPRS, and used the device as a tool to search for useful information and/or pictures about how to make a mold. This explorative task was intentionally and opportunistically triggered by the original model-making task, which resulted in the strategic use of the device. Although the explorative task was not the primary task in the model-making process, it played a key role by providing Curt the necessary information to successfully perform the design. Curt's mobile device use was inseparable from the institutional requirements of doing and submitting the project work for the training course. It was also necessitated by the fact that the facility where the training took place had no internet connections. At the same time, the way Curt approached the model-making task was altered by his use of the mobile device as a tool. Therefore, the tool use shaped the way that the task was performed, and the task fashioned the way that the tool was used in the complex social context.

Students achieve their goals through their performance on a variety of study tasks (Bedney & Harris, 2008). These tasks were performed through, or partly through, the use of mobile device as a tool. In this sense, tool use altered the ways that the tasks were performed. In the meantime, students appropriated the tool to cater for the task performance. Therefore, the tasks to be performed also shaped the tool use.

The Mobile Device Use in Relation to Learning Resources

To achieve educational goals through task performances, the students had to make use of learning resources that included (a) learning material such as hardcopy and softcopy lecture handouts, and other course related materials provided by the professors; and (b) learning material explored, collected and created by students. The resources related to various courses provided by the professors influenced how the student used the mobile device. For example, Amy used the mobile device for visualizing purposes to view video clips provided by the teacher for course revision or doing assignments. Brock and Smith (2007) state that video clips delivered via the mobile device is a convenient and potentially powerful way to visualize delivered messages to support student studies. However, such resources were rarely provided by the professors.

In many cases, the students used the mobile device as a tool to review downloaded course materials, which were usually in Microsoft Word, PDF (six slides on one page) and/or PowerPoint formats, before, during or after class. However, except for Amy, few students continued using this tool to access the learning resources due to the constraints of the technology, particularly the small screen size for reading course materials with detailed word description. In general, the students believed that reviewing course related material provided by the course professors was not appropriate for mobile device use, which prevented them from making good use of the material on the device to prepare for exams. Some studies on mobile device educational uses also reported such issues (e.g., Colevins, Sond, & Clark, 2006). The literature indicates that mobile device compatible and friendly (e.g., short passages) course material should be developed and provided for students.

Because the resources provided by the professor were limited, or just not suitable, to be accessed using the mobile device, the students found ways to obtain appropriate resources to support their studies. Ito (2005) states that digital kids consume multimedia resources that are created by others and created by themselves, engaging in "two-way literacies" (p. 1514) in the cultural production of knowledge. It was the same with the undergraduate students. In this study, although the course related materials available in softcopy were inadequate, the students themselves, created, explored and downloaded additional resources for a variety of uses. Some of the resources were searched and downloaded using a computer and saved onto the mobile device. Some of the resources were created or searched and accessed using the device directly. For example, all students searched the internet and downloaded a dictionary on the device, and used the device as a tool to overcome their vocabulary problems in their academic studies. This might be partly due to the fact that all of the students believed they had the need to improve their English because they were studying at an English medium university.

Eric and Fred downloaded several kinds of newspapers or accessed the news online to keep informed of the world news whenever they had spare time to read them. Amy downloaded BBC news podcasts and New Concept English to improve her English. She also accessed BBC news online and Chinese news sources to expose herself to a variety of writing styles. Students used the mobile device as a tool to take notes and photographs, and record lectures in class, lab experiments or other places "just-in-time" for later use to support their studies. The students reported that these uses benefited their studies. Indeed, just as Kress and Van Leeuwen (2001, p. 2) point out, individuals will "shape and re-shape the resources" they have in order to ensure that their uses match their needs/intentions.

Learning resources both supported and constrained student mobile device use to perform tasks to achieve their goals. On the other hand, the tool use also shaped the learning resources accessed, used or created by students. Edward (2005) posits that acting within context "requires us to read the situation and draw on the most effective resources available in it to support our actions" (p. 60). The way in which the student allocated resources when s/he was using the mobile device as a tool was very important (Barnard *et al.*, 2007). Study practices could not happen without making use of learning resources through the tool. The learning resources might be used in various ways, or not used at all, depending on how the students perceived that they fit into the possibilities and constraints of a particular context.

The Mobile Device Use and the Mobile Device Capabilities and Constraints

Wertsch (2002) asserts that new forms of technology can be enabling as well as constraining. It is the same with the mobile device. Given that both capabilities and constraints of the mobile device existed, in some particular contexts, the mobile device was appropriated to support student studies. For example, in her academic reading, Betty used the mobile device as a tool to take a picture out of her anatomy textbook with names of the body parts. As there was not a softcopy version of that particular diagram, she wanted that specific page out of the textbook to refer to it whenever she had time because the professor had told the students to remember both the anatomical picture and the names of the body parts for the course exam. However, the Camera on the device could only capture either the diagram itself or the names of the body parts clearly, but not both. She finally made the decision to take a clear picture of the diagram with blurred names to help her recall and reflect on the names of the body parts on the image. In this case, Betty's mobile device uses were shaped by the enabling property of having the ability to take pictures with the mobile device, as well as the constraint in the form of poor image quality of the camera on the device in the context of preparing the anatomy exam.

However, some of the mobile device uses were tried but given up because of the constraints of the device in some situations. For example, after having used the mobile device, Fred intended to review the handouts on the mobile device to make use of the commuting time and save himself the printing costs. He mentioned that during study period before exams, he saved a lecture handout in PDF format on the device to review it when he was commuting. However, he found that PDF files were not easy to read due to the small screen size. He had to drag the mouse up and down, left and right while reading. In this way, the object of reviewing the lecture note using the device as was overtaken by operating the device interface by adjusting the screen up and down, and left and right. "I didn't like it… I felt dizzy reading the small screen on the bus" (in-depth interview 1, Fred). In this case, the intended goal of studying the course material could not be achieved due to the limited screen size of the device and the discomfort caused by the limitations of the device. As a result, Fred chose not to appropriate this use.

All in all, the mobile device use depended on how the students interpreted, and then interacted with the capabilities and/or constraints of the PDA tools, and must be examined in that context, because the capability and constraints might be restricted in one context and be enabled in another.

The Mobile Device Use and Time and Place

In this research, the students reported that they used the mobile device on campus, outside physical buildings, in lecture rooms, on public transport vehicles, at home, in the street, in the library when the computers were occupied or not convenient to access, and so on. However, it is too simplistic to understand student mobile device use only in terms of different time and places. In practice, many of the study episodes using PDA tools were a continuation of previous episodes or connected with other study episodes, either accompanied with or without using other technologies, and across contexts to accomplish a variety of activities (Sharples *et al.* 2007). For example, on the way to a press conference, Amy, as a trainee journalist, met an important figure who was heading for the conference. She interviewed him and used the mobile device as a tool to take notes. When she arrived at the press conference, she used the mobile device as a tool to draft her news story. After the conference, when she was back in her office, she synchronized the mobile device with the desktop computer to do the final revisions and sent it out straight away. In this example, first, there was a rapid series of changes in the mobile device tool Amy used (from note-taking to drafting news story), and in conjunction with the desktop computer use to deal with the fast changing tasks to achieve the goal of submitting the news report in time. Second, she used the mobile device as a tool to cope with the task "outside the space" of the mobile device to take notes in the interview and then, moved "inside the space" of the mobile device when she was drafting the news report using the notes previously taken on the device. Third, to submit her news report on time, she used the device before the press conference while moving, in the conference while sitting, and after the conference while working in the office. The news report writing task was accomplished across different contexts in terms of different time and places, different device uses, different resource uses, and social factors such as the importance of the interviewee, and the desire for quick publication of the news story.

In sum, time and place must be taken account into mobile device use as they also played a part in shaping its use. On the other hand, the mobile device could also be used to take advantage of time and place. Sharples *et al.* (2007), in addressing mobile devices and studies, argue that studies supported by mobile devices should be distinguished from other forms of studies by showing that: (a) students' studies are supported across space as they take ideas and learning resources gained in one location and apply or develop them in another; (b) students' studies are supported across time by revisiting knowledge gained earlier in a different context for lifelong learning; (c) students move from topic to topic by managing a range of personal study tasks; (d) students move in and out of engagement with technology. This calls for new thinking about mobile device uses to support studies as these uses are not triggered by constant contextual proprieties such as fixed settings and scheduled time, but by a mixture of dynamic and diverse contextual components, including time and place.

The Mobile Device Use and Social Factors

Student mobile device use was shaped by social factors with respect to institutional, community, and physical conditions, and vice versa. These factors played a part in enabling or restricting mobile device uses. In terms of institutional factors, the provision of SMS services by the library helped to inform students of their book status effectively. In addition, the capability of using the mobile device to access the library website provided students a convenient means to look up books online anytime, anywhere.

With respect to institutional requirements, in order to obtain a bachelor degree, student mobile device use seemed to be driven by exams and assignments required by the university. The most common practice among the students was to use the device to review course materials. However, due to the lack of course material available in softcopy, the inability to access e-books from the library and the course material posted on WebCT using the mobile device which restricted students to making use of the resources for "just-in-time" needs, the ability of the students to use the device to help them study their course material was definitely restricted. On the other hand, despite these restrictions, the students themselves used the device as different tools to create and access resources for their study, such as the use of the device to create resources, and use the of the device in conjunction with computers to search for and download useful resources (e.g., dictionaries, e-books, audio language material), and access them for study purposes. Hence, institutional factors shaped how the students made use of the mobile device to support their study.

Regarding the community, the students reported that, in general, they only used the mobile device as a tool to interact with their friends or peers, and considered it impolite or inappropriate to interact with their professors using the device. Harley, Winn, Pemberton, and Wilcox (2007) assert that text message dialogues among students provide emotional and social peer support, and facilitate an informal way of interdependent study. Indeed, in this research, the students communicated with peers primarily for socializing, informative or coordination purposes. Harley *et al.* further state that if communications from the university staff are introduced into the dialogue among students, they can enhance the existing peer support system and help students to integrate into university life. The students reported that few peers had mobile devices that were compatible with the device they used. This prevented them from sharing files that could facilitate group projects with their classmates. This is similar to Laru and Järvelä's (2008) study on identifying social patterns in mobile technology that supported collaboration among distributed members of the professional distance education community. Their findings show that the members made very little use of mobile devices to support collaboration among participants in the offline community. The reasons lie in the fact that because the members were sufficiently separated, they tended to do their own things and did not have a need for mobile technology to support collaboration.

Further, some student mobile device uses were influenced by a culture of intense competition, raising background and the like. For example, Curt came up with many creative uses of the mobile device, most of which were geared towards getting high marks in exams and assignments in order to get a first-class honors degree. In addition, the culture that a student was raised in could affect how he/she used the mobile device. Take Amy's PDA use for an example. Amy was raised in Mainland China, she did not learn to write traditional Chinese characters, and did not know how to speak Cantonese. When she studied in Hong Kong, she perceived the necessity of learning to write traditional Chinese characters and speak Cantonese, especially for her internships and future career development. Therefore, she used the mobile device as a tool to improve her Cantonese and learn to write Chinese characters. As the other four students were raised in Hong Kong, none of them put the device to this type of use. Take the students' phone card use for another example. Normally, the participating students from Hong Kong carried two phone cards because they preferred to make phone calls than other means of communication using the device. However, Amy who was raised in Mainland China, only kept one phone card. She made phone calls when she was in Hong Kong. And when she went back to Mainland China, she chose to send more SMS than phone calls. On the one hand, it was cheaper to send SMS messages than to make phone calls in Mainland China. A large number of people in Mainland China seem to prefer sending SMS messages to making phone calls to communicate. Thorne (2003) posit that individuals from two communities carry out their interactions using communication technology according to their own rules or norms, "often resulting in a clash in expectations and, ultimately, misperceptions about the other group" (p. 45). She termed such technology use as "cultures-of-use" (p. 38).

Regarding physical conditions, the students could only use the mobile device to access the internet using Wi-Fi free of charge on campus, including the lecture rooms. They reported that they would have used the device more to support their studies if they could use the device to access internet across Hong Kong free of charge.

To sum up, student mobile device use was shaped by social factors. The social factors were usually more constant, therefore could not be altered overnight. However, they were also gradually constructed by an increasing number of mobile device users. Thus, student mobile device use was shaped by the institutional factors, and vice versa.

The Mobile Device Use and Individual Interpretations

In this study, all of the participating students had the goal of getting a bachelor's degree. Their mobile device use was oriented towards such a goal, either directly or indirectly. In addition to this main goal, Curt aimed at a higher goal of achieving a first-class honors degree, and Amy endeavored to become a journalist in Hong Kong after graduation. In the five participating students, Curt and Amy were the most active students in taking advantage of the mobile device to help perform tasks to achieve their goals. Both Curt and Amy set up very clear goals, hence actively performed many tasks using the device to achieve their goals. Although they faced different challenges in using the device, they made every effort to adapt themselves to various constraints and opportunities of the context to use the PDA to support their studies.

For example, Amy also took great advantage of using the device to support her studies, especially in the areas of improving her language abilities to get a bachelor degree and become a journalist in the future. As her study tasks were concerned more with listening, speaking, reading and writing, she primarily used the device for accessing and collecting learning resources. She downloaded a number of softcopy resources prepared by the professor. In addition, she wanted to improve her English because she believed that her English was not good enough and it was important for her to become proficient for both her course work and her future career development. Some studies on Chinese language learners reveal that millions of Chinese students consider English not only as a core academic subject, but also as a crucial quality in deciding how far they can move forward in the future (e.g., Gao, 2007). Amy demonstrated a great deal of eagerness and enthusiasm towards making her English more "idiomatic" or "standard". She downloaded classical English passages onto the device and recited them whenever she got time; she read online BBC news whenever she had a few spare moments; she read downloaded e-books chapter by chapter; and she also formed the habit of writing down useful expressions using Notes on the device during her academic readings, or attending lectures. Her episodes of mobile device use to support her studies historically focused more on activities related to language learning, journalism and internships as a journalist trainee. Van Lier (1996) posits that if performing a task "realizes a free choice and is an expression of what a person genuinely feels and believes", then the person is "intrinsically motivated" (p. 13).

As for Betty, she did not have ambitious goals in the first half year of her study as she was not interested in nursing. She always sought ways to transfer to another department. At that time, her goal was to pass her exams and assignments in order to avoid doing make-up exams or assignments once she failed to get her transfer. She primarily used the mobile device as a tool to study for exams or do group projects; to take notes for course exams; and to organize student activities. After she successfully transferred to biochemistry, she reported that due to the requirements of the professors in the department, she often had to take her laptop along with her. Because she wanted to be a "good student" in the new department, she tried to record what the teacher talked about in class using the mobile device. However, due to the poor quality of the recordings, she abandoned this particular tool use. She did not find ways in dealing with the recording constraints of the device. She reported that she perceived very few needs for the device to support her studies in the new context. Gradually, she even gave up her habit of note-taking using the device that she adopted in the first half year of the study period.

Tools cannot impose on the users to use them. They are useful only when users perceive their potential and use them in context (Oliver, 2005). Different users interpret the context in which the tools are embedded differently. This is true for the students in this research. The "subjective interpretations" of the context made by the students, can either make students negatively anticipate learning to happen or support spontaneous involvement in a learning task (Jarvelä *et al.,* 2000, p.304). The interpretations of the context are closely related to the goals and motivations of the student in question. Students perform best if they are actively involved in tasks and integrate new information with their prior knowledge to achieve their goals (Lajoie & Azevedo, 2006).

FUTURE RESEARCH DIRECTIONS

The findings of the research show that student mobile device uses are highly dependent on context. Student mobile device use in context requires interactions between the student interpretation of the situation and other factors regarding individual goals, tasks, learning resources, the mobile device, time and place and social factors. These interactions form a connected whole contributing to affordances and constraints in which mobile technology educational practice takes place. This is in line with the two intertwined focuses suggested by Edwards (2005) when examining student learning: (a) how learners interpret and act on their worlds, and (b) the opportunities afford them for those interpretations and actions. In educational technology use, Jonassen, Hernandez-Serrano and Choi (2000) express a similar view. The authors posit that from a constructivist perspective, if learning technologies are tools for mediating the practice of learning, and if we examine the potential of learning technologies from a student perspective, then "the affordances of any [learning] technology are the properties of that environment that enable the affectivities of the technology, the abilities of the learner to take learning actions" (p.113). The technology, therefore, mediates learning depending on: (a) whether the context allows the technology to be enacted; and (b) whether the student has the capabilities (and also willingness) to use the technology to mediate the learning activity in the context. Although it is demanding to lay two focuses simultaneously (Edwards, 2005) on examining study actions mediated by mobile devices, it is advisable for educators, practitioners and designers to maximize the possibilities that the context provides for mobile device use to support student studies.

Mobile technology is distinctive from other educational technology in relation to its striking characteristics of portability/mobility, immediate accessibility, and connectivity. Hence, the uses and the impact of these uses are also largely different from other technology use. As is shown from the findings of this research, mobile devices, in fact, have changed the way student learn. Thus, in order to maximize the possibilities of mobile devices to mediate studies, future research should shift from emphasizing technical aspects of developing and designing mobile learning systems to pedagogical practices and social context, especially in terms of pedagogic designs, resources development and provision, pedagogically sound mobile technology tool development, and institutional support.

CONCLUSION

This research, adopting a qualitative multiple-case study approach, examined the student mobile device use in context. Various data collection methods were employed to triangulate and strengthen the research findings. Grounded from the data collected, the research pinpointed seven relational and interwoven factors in context that either contributed to or inhibited student mobile device use. They are: goals, tasks, learning resources, time and place, social factors, the mobile device and individual interpretation. Although these factors have been previously cited in other studies (Jarvelä *et al.*, 2000; Jones & Issroff, 2007; Sharples *et al.*, 2007) respectively, they have not been addressed as a unity that interact simultaneously to create a context that influences mobile device use. The findings in this research enrich the literature on mobile technology educational research by seriously analyzing the individual student, the mobile device and the context.

REFERENCES

Barnard, L., Yi, J. S., Jacko, J. A., & Sears, A. (2007). Capturing the effects of context on human performance in mobile computing systems. *Personal and Ubiquitous Computing*, *11*(2), 81–96. doi:10.1007/s00779-006-0063-x

Bedny, G. Z., & Harris, S. R. (2008). Working sphere/engagement and the concept of task in activity theory. *Interacting with Computers*, *20*(2), 251–255. doi:10.1016/j.intcom.2007.07.005

Brock, T. P., & Smith, S. R. (2007). Using digital videos displayed on personal digital assistants (PDAs) to enhance patient education in clinical settings. *International Journal of Medical Informatics*, *76*(11/12), 829–835. doi:10.1016/j.ijmedinf.2006.09.024

Cole, M. (1996). *Cultural psychology: A once and future discipline*. Cambridge, MA/London, UK: Belknap Press of Harvard University Press.

Colevins, H., Bond, D., & Clark, K. (2006). Nurses get a hand from handhelds. *Computers in Libraries*, *26*(4), 6–48.

Creswell, J. W. (2008). *Educational research: Planning, conducting, and evaluating quantitative and qualitative research* (3rd ed.). Upper Saddle River, NJ: Pearson/Merrill Prentice Hall.

Edwards, A. (2005). Let's get beyond community and practice: The many meanings of learning by participating. *Curriculum Journal*, *16*(1), 49–65. doi:10.1080/0958517042000336809

Gao, X. (2007). Language learning experiences and learning strategy research: Voices of a mainland Chinese student in Hong Kong. *Innovations in Language Learning and Teaching*, *1*(2), 193–207. doi:10.2167/illt011.0

Gay, G., & Hembrooke, H. (2004). *Activity-centered design: An ecological approach to designing smart tools and usable systems*. Cambridge, MA/London, UK: MIT Press.

Harley, D., Winn, S., Pemberton, S., & Wilcox, P. (2007). Using texting to support students' transition to university. *Innovations in Education and Teaching International*, *44*(3), 229–241. doi:10.1080/14703290701486506

Ito, M. (2005). Technologies of the childhood imagination: Yu-gi-oh, media mixes, and everyday cultural production. In J. Karaganis & N. Jeremijenko (Eds.), *Structures of participation in digital culture*. Durham, NC: Duke University Press. Retrieved January 16, 2010, from http://www.itofisher.com/mito/archives/ technoimagination.pdf

Jarvelä, S., Lehtinen, E., & Salonen, P. (2000). Socio-emotional orientation as a mediating variable in the teaching-learning interaction: Implications for instructional design. *Scandinavian Journal of Educational Research*, *44*(3), 293–306. doi:10.1080/713696677

Jonassen, D. H., & Carr, C. S. (2000). Mindtools: Affording multiple knowledge representations for learning. In Lajoie, S. (Ed.), *Computers as cognitive tools* (2nd ed., pp. 165–196). Mahwah, NJ/ London, UK: L. Erlbaum.

Jones, A., & Issroff, K. (2007). Motivation and mobile devices: Exploring the role of appropriation and coping strategies. *ALT-J*, *15*(3), 247–258. doi:10.1080/09687760701673675

Kaptelinin, V. (1996). Computer-mediated activity: Functional organs in social and developmental contexts. In Nardi, B. A. (Ed.), *Context and consciousness: Activity theory and human computer interaction* (pp. 45–68).

Kress, G. R., & Van Leeuwen, T. (2001). *Multimodal discourse: The modes and media of contemporary communication*. London, UK/ New York, NY: Arnold; Oxford University Press.

Lajoie, S. P., & Azevedo, R. (2006). Teaching and learning in technology-rich envrionments. In Alexander, P. A., & Winne, P. H. (Eds.), *Handbook of educational psychology* (2nd ed., pp. 803–821). Mahwah, NJ: Lawrence Erlbaum.

Lan, Y.-J., Sung, Y.-T., & Chang, K.-E. (2007). A mobile-device-supported peer-assisted learning system for collaborative early EFL reading. *Language Learning & Technology*, *11*(3), 130–151.

Laru, J., & Järvelä, S. (2008). Social patterns in mobile technology mediated collaboration among members of the professional distance education community. *Educational Media International*, *45*(1), 17–32. doi:10.1080/09523980701847131

Lave, J. (1988). *Cognition in practice: Mind, mathematics, and culture in everyday life*. Cambridge, UK: Cambridge University Press. doi:10.1017/CBO9780511609268

Maxwell, J. A. (1996). *Qualitative research design: An interactive approach*. Thousand Oaks, CA: Sage Publications.

Motiwalla, L. F. (2007). Mobile learning: A framework and evaluation. *Computers & Education*, *49*(3), 581–596. doi:10.1016/j.compedu.2005.10.011

Murphy-Berman, V., & Berman, J. J. (2003). *Cross-cultural differences in perspectives on the self*. Lincoln, NE/ London, UK: University of Nebraska Press.

Norman, D. A. (1998). *The design of everyday things* (1st MIT Press ed.). London, UK: MIT Press.

Oliver, M. (2005). The problem with affordance. *e-learning*, *2*(4), 402-413.

Patten, B., Arnedillo-Sanchez, I., & Tangney, B. (2006). Designing collaborative, constructionist and contextual applications for handheld devices. *Computers & Education*, *46*(3), 294–308. doi:10.1016/j.compedu.2005.11.011

Seale, J. K. (2008). Developing e-learning experiences and practices: The importance of context. *ALT-J. Research in Learning Technology*, *16*(1), 1–3.

Sharples, M., Taylor, J., & Vavoula, G. (2007). A theory of learning for the mobile age. In Andrews, R., & Haythornthwaite, C. (Eds.), *The Sage handbook of e-learning research* (pp. 221–247). London, UK: Sage Publications.

Trouche, L. (2003). *Managing the complexity of human/machine interaction in a computer based learning environment (CBLE): Guiding students' process command through instrumental orchestrations.* Paper presented at the Third CAME Symposium: Learning in a CAS Environment: Mind-Machine Interaction, Curriculum & Assessment, Reims, France.

Uden, L. (2007). Activity theory for designing mobile learning. *International Journal of Mobile Learning and Organisation, 1*(1), 81–102. doi:10.1504/IJMLO.2007.011190

van Lier, L. (1996). *Interaction in the language curriculum: Awareness, autonomy, and authenticity.* London, UK: Longman.

Vygotsky, L. S. (1978). *Mind in society: The development of higher psychological process.* Cambridge, MA: Harvard University Press.

Wali, E., Winters, N., & Oliver, M. (2008). Maintaining, changing and crossing contexts: An activity theoretic reinterpretation of mobile learning. *ALT-J. Research in Learning Technology, 16*(1), 41–57.

Waycott, J. (2005). Appropriating tools and shaping activities: The use of PDAs in the workplace. In Hamill, L., & Amparo, L. (Eds.), *Mobile world: Past, present and future* (pp. 119–139). London, UK/ New York, NY: Springer.

Wertsch, J. V. (2002). Computer mediation, PBL, and dialogicality. *Distance Education, 23*(1), 105–108. doi:10.1080/01587910220124008

Zurita, G., & Nussbaum, M. (2007). A conceptual framework based on activity theory for mobile CSCL. *British Journal of Educational Technology, 38*(2), 211–235. doi:10.1111/j.1467-8535.2006.00580.x

ADDITIONAL READING

Chen, C. M., & Chung, C. J. (2008). Personalized mobile English vocabulary learning system based on item response theory and learning memory cycle. *Computers & Education, 50*(1), 77–90. doi:10.1016/j.compedu.2006.03.004

Chen, G. D., Chang, C. K., & Wang, C. Y. (2008). Ubiquitous learning website: Scaffold learners by mobile devices with information-aware techniques. *Computers & Education*, 77–90. doi:10.1016/j.compedu.2006.03.004

Chen, Y.-C., Chiu, H.-C., Tsai, M.-D., Chang, H., & Chong, C.-F. (2007). Development of a personal digital assistant-based wireless application in clinical practice. *Computer Methods and Programs in Biomedicine, 85*(2), 181–184. doi:10.1016/j.cmpb.2006.11.001

Churchill, D., & Churchill, N. (2008). Educational affordances of PDAs: A study of a teacher's exploration of this technology. *Computers & Education, 50*(4), 1439–1450. doi:10.1016/j.compedu.2007.01.002

Clough, G., Jones, A. C., McAndrew, P., & Scanlon, E. (2008). Informal learning with PDAs and smartphones. *Journal of Computer Assisted Learning, 24*(5), 359–371. doi:10.1111/j.1365-2729.2007.00268.x

Colevins, H., Bond, D., & Clark, K. (2006). Nurses get a hand from handhelds. *Computers in Libraries, 26*(4), 6–48.

Evans, C. (2008). The effectiveness of m-learning in the form of podcast revision lectures in higher education. *Computers & Education, 50*(2), 491–498. doi:10.1016/j.compedu.2007.09.016

Godwin-Jones, R. (2007). Emerging technologies: E-texts, mobile browsing, and rich internet applications. *Language Learning & Technology, 11*(3), 8–13.

Henricksen, K., & Indulska, J. (2006). Developing contextaware pervasive computing applications: Models and approach. *Pervasive and Mobile Computing*, *2*(2), 37–64. doi:10.1016/j.pmcj.2005.07.003

Herrington, A. (2008). Using a smartphone to create digital teaching episodes as resources in adult education, in J. Herrington, A. Herrington, J. Mantei, I. Olney, & B. Ferry (eds.), *New technologies, new pedagogies: Mobile learning in higher education* (pp. 28-35). Faculty of Education, Australia: University of Wollongong, 2009.

Hilleson, M. (1996). "I want to talk with them, but I don t want them to hear": an introspective study of secondlanguage anxiety in an English-medium school. In Bailey, K. M., & Nunan, D. (Eds.), *Voices from the Language Classroom: Qualitative Research in Second Language Classroom.* Cambridge: Cambridge University Press.

Klopfer, E. (2008). *Augmented learning: research and design of mobile educational games.* Cambridge, Mass.: MIT Press.

Kukulska-Hulme, A., & Shield, L. (2008). An overview of mobile assisted language learning: From content delivery to supported collaboration and interaction. *ReCALL*, *20*(3), 271–289. doi:10.1017/S0958344008000335

Liu, C. C., & Hong, Y. C. (2007). Providing hearing-impaired students with learning care after classes through smart phones and the GPRS network. *British Journal of Educational Technology*, *38*(4), 727–741. doi:10.1111/j.1467-8535.2006.00656.x

Liu, C. C., & Kao, L. C. (2007). Do handheld devices facilitate face-to-face collaboration? Handheld devices with large shared display groupware to facilitate group interactions. *Journal of Computer Assisted Learning*, *23*(4), 285–299. doi:10.1111/j.1365-2729.2007.00234.x

Martinez-Torres, M. R., Toral, S. L., Barrero, F., & Gallardo, S. (2007). Improving learning performance in laboratory instruction by means of SMS messaging. *Innovations in Education and Teaching International*, *44*(4), 409–422. doi:10.1080/14703290701602854

Matsuura, K., Gotoda, N., Niki, K., Kanenishi, K., & Yano, Y. (2008). Supporting multi-step annotation to promote reflective learning: triggered by a cell-phone. *International Journal of Mobile Learning and Organisation*, *2*(2), 119–132. doi:10.1504/IJMLO.2008.019764

Murphy, E., & Manzanares, M. A. R. (2008). Instant Messaging in a context of virtual schooling: balancing the affordances and challenges. *Educational Media International*, *45*(1), 47–58. doi:10.1080/09523980701847180

Owen, M. (2009). From individual learning to collaborative learning - Location, fun, and games: Place, context, and identity in mobile learning. In Ryu, H., & Parsons, D. (Eds.), *Innovative mobile learning: techniques and technologies* (pp. 102–122). Hershey, PA: Information Science Reference.

Peng, H., Su, Y.-J., Chou, C., & Tsai, C.-C. (2009). Ubiquitous knowledge construction: mobile learning re-defined and a conceptual framework. *Innovations in Education and Teaching International*, *46*(2), 171–183. doi:10.1080/14703290902843828

Ryu, H., & Parsons, D. (2008). A learner-centred design of a location-aware learning reminder. *International Journal of Mobile Learning and Organisation*, *2*(2), 187–200. doi:10.1504/IJMLO.2008.019768

Selwyn, N. (2008). An investigation of differences in undergraduates' academic use of the internet. *Active Learning in Higher Education*, *9*(1), 11–22. doi:10.1177/1469787407086744

Song, Y. (2008). SMS enhanced vocabulary learning for mobile audiences. *International Journal of Mobile Learning and Organisation*, *2*(1), 81–98. doi:10.1504/IJMLO.2008.018719

Song, Y., & Fox, R. (2008). Using PDA for undergraduate student incidental vocabulary learning. *ReCALL*, *20*(3), 290–314. doi:10.1017/S0958344008000438

Stockwell, G. (2008). Investigating learner preparedness for and usage patterns of mobile learning. *ReCALL*, *20*(3), 253–270. doi:10.1017/S0958344008000232

Uzunboylu, H., Cavus, N., & Ercag, E. (2009). Using mobile learning to increase environmental awareness. *Computers & Education*, *52*(2), 381–389. doi:10.1016/j.compedu.2008.09.008

Wali, E., Winters, N., & Oliver, M. (2008). Maintaining, changing and crossing contexts: an activity theoretic reinterpretation of mobile learning. *ALT-J. Research in Learning Technology*, *16*(1), 41–57.

Zurita, G., & Nussbaum, M. (2007). A conceptual framework based on activity theory for mobile CSCL. *British Journal of Educational Technology*, *38*(2), 211–235. doi:10.1111/j.1467-8535.2006.00580.x

Chapter 8
Mobile Learning in Medicine

Kalyani Premkumar
University of Saskatchewan, Canada

ABSTRACT

Mobile learning (m-learning) is particularly important in medical education, and the major users of mobile devices are in the field of medicine. The contexts and environment in which learning occurs necessitates m-learning. Medical students are placed in hospital/clinical settings very early in training and require access to course information and to record and reflect on their experiences while on the move. The work of postgraduates and physicians involves a high degree of mobility between distributed sites and instant communications within work environments. Distributed sites where physicians work and in which students are placed are often in remote and rural areas. The technological advances can be capitalized to promote and facilitate situated learning.

This chapter describes the medical context and characteristics of medical students, residents, and medical professionals and implications for m-learning. Some technologies used and examples of usage, benefits, outcomes, and barriers at the undergraduate, postgraduate and continuing medical education are explored.

INTRODUCTION

The medical education system in any country plays a pivotal role in training physicians with the skills required to address the priority health concerns of the community, region, and/or nation they have a mandate to serve. Although the priority health concerns are to be identified jointly by governments, health care organizations, health professionals and the public (Boelen 2000), each country has its own national voice.

DOI: 10.4018/978-1-60960-511-7.ch008

In Canada, the Association of Faculties of Medicine of Canada (AFMC) is responsible for providing guidance on undergraduate, postgraduate and continuing medical education and for managing a rigorous system of accreditation. In USA, the Association of American Medical Colleges (AAMC) and in UK, the General Medical Council (GMC) are equivalents. The duration of basic medical training is typically 4 years where a student completes an acceptable program of medical education (undergraduate medical education: UGME) followed by 2-6 years of accredited residency training in a specialty (postgraduate medical education: PGME) such as family medicine, surgery, psychiatry, etc. Licensing examinations have to be passed to legally practice medicine. Even after beginning to practice, medical professionals are expected to be involved in continuing medical education (CME) to maintain their license.

The learning needs of UGME, PGME and CME, while overlapping in many ways are also quite distinct. In UGME, the learner focuses primarily on obtaining foundational knowledge, skills and attitudes and learning is generally driven by external factors such as curriculum and examination (Davis et al, 2007). In the early years, the student usually learns within a university setting or closely supervised hospital/clinic setting.

In PGME, the learner builds on this knowledge, with increasing opportunities to practice skills and to work under minimal supervision or independently in a hospital or clinic. Here, learning is influenced by self-motivation and relevance to clinical practice. In CME, the focus is more on reviewing knowledge and skills, maintaining competence and learning about new developments in the field. Thus, as in almost all professions, in order to make informed decisions in practice, medical personnel have to be provided with opportunities to participate in Life Long Learning.

In medical education, though the learning needs are different at the UG, PG and CME levels, because of the varying settings in which learning occurs, opportunities have to be provided in the context of the "anything, anywhere, anytime" imperative (von Jan, Ammann & Matthies, 2008), and mobile learning may be one of the solutions.

BACKGROUND

What is Mobile Learning?

Mobile learning or m-learning has been defined in very many ways (Attewell & Savill-Smith, 2004; Collis & Moonen, 2001; Keegan, 2002; Kukulska-Hulme & Traxler, 2005; Laouris & Eteokleous, 2005; Metcalf, 2006; Mitchell, 2003; O'Malley et al. 2003; Van Barneveld & Shaw, 2006). While eLearning is defined as learning supported by digital electronic tools and media, m-learning is defined as eLearning using mobile devices and wireless transmission (Milrad in Attewell & Savill-Smith, 2004). Others (Keegan) define m-learning as "the provision of education and training on PDAs/palmtops/handhelds, smart phones and mobile phones" (p. 6) number). O'Malley et al, with a focus on mobility, define m-learning as "any sort of learning that happens when the learner is not at a fixed, predetermined location, or learning that happens when the learner takes advantage of the learning opportunities offered by mobile technologies." While some (Milrad, Keegan) focus on the mobile use of technology, others (O'Malley) focus on the mobility of persons. M-learning is defined by yet others who emphasize learning as "the acquisition of any knowledge and skill through using mobile technology, anywhere, anytime, that results in an alteration of behavior" (Geddes, 2004, p. 1). In this chapter, the emphasis will be on learning.

The goal of m-learning should be to develop content for mobile applications in order to accommodate the needs often described as just-in-time (JIT), location based learning (LBL), just-in-location learning (JILL), learning-on-demand (LOD), and what-I-need-when-I-need-it (WINWINI) (Van Barneveld & Shaw, 2006).

Attributes of M-Learning

The attributes of m-learning are described as spontaneous, personal, informal, contextual (situated), portable, ubiquitous (available everywhere), ambient (surrounding us completely), unobtrusive or pervasive (integrated with daily activities that it is hardly noticed) (Kukulska-Hulme & Traxler, 2005; O'Malley et al. 2003). This means that the device used should be small, so that it can be easily carried in the pocket or hand, blending with the environment, and easily accessible to the person at all times. It should have reduced boot up time and be instantly ready as soon as you switch it on. It should be easy to exchange and synchronize data with other devices. Because of these requirements, the smallness of the device, and smallness of the screen (in those that have one), tiny or no input device, and increased vulnerability for breakage, loss or theft, may be seen as barriers for certain types of tasks.

The benefits of mobile devices are not limited to learning at a distance or individually. They have their benefits in large classes as well. For instance, it is much easier to accommodate many more mobile devices than desktops in a classroom. It is easier to carry (e)books stored in these devices than actual hardcopies. Students can take handwritten or typed notes and even draw directly into the device. It is more natural for some students to write with a stylus than type using a keyboard and mouse. Students can collaborate by transferring their projects and assignments using the infrared function or wireless network (Mahmoud, 2008). With the right software, handheld devices can also be used as student response systems for formative and summative evaluations and for active learning during lectures (Al-Ubaydli, 2006; Premkumar & Coupal, 2008).

Types of Mobile Technologies

Technologies that may be used for m-learning range from simple single-purpose devices such as audio players, to complex multipurpose handheld devices such as Personal Digital Assistants (PDAs) and smart phones that combine a number of features including audio and video. Other devices not considered as typical for m-learning but still useful are USB drives, and student response systems (clickers). Wearable devices such as biosensors are technologies unexplored, but with potential uses for m-learning.

Some of the functionality of mobile devices includes tools for

- **Organization:** organizer (address book, scheduling, calendar, memo/note pad)
- **Communication:** email, phone, SMS (Short Messaging Service), MMS (Multimedia Messaging Service), camera, video
- **Relaxation:** music, games, movie player, eBooks, camera
- **Information:** web browser, GPS compass, eBooks, dictionaries
- **Other applications:** Office (word processing, presentations, data management); databases; e-Book readers etc.

Most mobile devices are designed for personal information management or communication within the workplace, home and society. As such, current patterns of educational uses are adaptations of these features to enhance teaching and learning in educational contexts by innovative educators and researchers.

Some of the more frequently used devices are described in Table 1.

In this chapter, examples of the technologies used for m-learning at the UGME, PGME and CME levels will be described along with the benefits, outcome and barriers. Where available, models for implementation will be addressed. Every attempt has been made to focus on the learning as opposed to the technology used.

NEED FOR M-LEARNING IN MEDICINE

M-learning is particularly important in medical education. Unlike most undergraduate students, medical students are placed in hospital/clinical settings within the first two years of training. While on the move, they are required to record and reflect on their experiences as part of formal assessments and handheld devices come in handy. Postgraduates and physicians are no different. As they move from ward to ward and between clin-

Table 1. Description of specific mobile devices

Device	Features/function	Learning opportunities	Pros/Cons of device for m-learning
Single Purpose Device			
iPod (portable media player)	Downloads music, audio books, podcasts, photos, video, address book, mass-storage device	Download podcasts of audio and video lectures; audio books; exchange information files, collaborate on projects, review coursework, prepare for exams; microphone can be added to capture material for educational use	*Pros:* popular with students; compact, light *Cons:* not affordable for all students; one-way communication; no interactivity; small screen
MP3 Player	Plays music and audio files; may have voice recorder	Download podcasts of audio lectures; read audio books; review course material; prepare for exams; some devices can be used to record lectures	*Pros:* compact; light; increased battery life; excellent audio quality *Cons:* time-consuming to transfer data; one-way communication; no interactivity
E-Book Reader	Downloads large volume of text-based material; possible to magnify, highlight; search capabilities; bookmark	Store e-books and other instructional material; read resources on demand; conduct research; continue learning from where you stopped; highlighters to facilitate learning	*Pros:* large screen; backlighting facilitates reading in the dark; digital bookmarks *Cons:* single purpose
Mobile DVD players	Portable; plays movies	Play interactive video-based learning and simulations	*Pros:* large screen (compared to other devices) *Cons:* single purpose; no interactivity
Mobile Phone (cell phone)	Telephone; text messaging; may support additional services such as email, access to internet; gaming, Bluetooth, infrared camera with video recorder, MMS (multimedia messaging service), GPS (global positioning system)	Send announcements; contact students using text messages or calls; allows audio communication between students; collaboration via moblogs	*Pros:* almost all students/ instructors own and use one; familiarity with technology; low cost; light weight *Cons:* batteries have to be charged; limited purpose; cannot be used in all locations (e.g. restrictions within hospitals); limited storage capacity for data; only short messages (up to 160 characters)
Multi-purpose Devices			
Personal Digital Assistant (PDA)/ palmtops/handheld PCs/iPod Touch/iPad)	Computing capability; Internet access; networking features; calendar; notepad; address book; productivity tools; programmable; Bluetooth-enabled*; Wi-Fi- equipped; pen/stylus input interface; can be synchronized with desktop; touch-sensitive screen; handwriting recognition (not all devices)	Download podcasts, vodcasts, Flash movies; display and edit text; access to e-mail web content; supports text message; mass storage; take notes; supports interactive learning	*Pros:* large screen (for a portable device); data entry using screen keyboard, stylus or add-on input device; readily uploads/downloads information; no boot time; password protected *Cons:* comparatively bulky (some devices); not efficient for entering long texts; battery charging necessary; more expensive than other devices

continued on following page

Table 1. Continued

Smart Phone (a type of mobile phone)	Combination of telephone with PDA; camera; video; mass storage; MP3 player; internet access; networking features. Newer phones allow video calls (speakers can see each other); view news clips	Download podcasts, vodcasts, Flash movies; display and edit text; access to e-mail; web content; supports text message; mass storage; supports interactive learning	*Pros:* combination of communication and computing features; compact *Cons:* small screen; small keys inefficient for text entry; expensive
Ultra-Mobile PC (UMPC)	smaller-sized tablet PC; audio; video; gaming; internet browsing; Internet access; networking features; Bluetooth enabled; Wi-Fi-equipped; Ethernet enabled	Download podcasts & vodcasts of lectures; Flash movies of case scenarios; create and edit assignments; send e-mails; text messages; access course websites; allows collaboration; supports interactive learning	*Pros:* smaller than tablet PC; 7-inch touch sensitive screen great for Web surfing and viewing multimedia *Cons:* small keyboard inefficient for text entry; expensive (costs more than high-powered PC); password protected
Laptop/Tablet PC	audio; video; gaming; internet browsing; Internet access; networking features; Bluetooth enabled; Wi-Fi equipped; Ethernet enabled; handwriting recognition; voice-to-text conversion	Download podcasts; vodcasts; Flash movies; create and edit assignments; send e-mails; text messages; access course website; allows collaboration; supports interactive learning	*Pros:* most power and capabilities of all devices *Cons:* expensive; large size (not a handheld device); cannot be used while walking; some boot time
Smart Pen (Computer within a pen)	records writing/drawing; simultaneously records audio; replays audio; allows transfer of written notes to a computer via USB port; search capabilities	Excellent note-taking tool - take notes/draw and record lecture simultaneously; transfer written material to computer and share; search notes; review lectures	*Pros:* Pen-like device; unobtrusive; compact, portable *Cons:* requires special notebook; ink cartridges have to be replaced; batteries have to be charged; limited purpose compared to other portable devices
Others			
USB Drive	Mass storage; attaches to computers and other devices	Store coursework; audio/video files; easy to share files for projects; transfer work to and from computers; save and submit work to instructor	*Pros:* very small; portable; mass storage *Cons:* single purpose
Wearable devices (de Freitas & Levene, 2003) Wristwatch computer (WatchPad)	Worn on wrist; tells time; supports calendar scheduling; address book functionality; to-do-lists; send and receive short email messages; Bluetooth connectivity; wireless access to web services	Hold student schedules; provide location sensing; messaging allows for some interactivity	*Pros:* wearable; both hands free *Cons:* very small screen
Xybernaut Mobile Assistant	Wearable computer; desktop/laptop capabilities; wireless web connectivity; email; location sensing; hands-free voice recognition and activation; head mounted or flat panel display; touch screen activated; allows pen input; wrist strapped mini-keyboard	Calculate; word process; watch multimedia; access computer; play educational games; useful for students with special needs	*Pros:* wearable; both hands free; can be voice activated *Cons:* very small screen
iButton	Computer chip enclosed in stainless steel can; each device has unique identification address; base can have different functionality (memory, clock, security, temperature sensing etc.)	Allows registration of students - access to classrooms; web pages; computers	*Pros:* wearable; small; unobtrusive *Cons:* limited functionality; educational potential still unexplored

continued on following page

Table 1. Continued

MIThril (wearable vest)	Has wide range of sensors and different kinds of interfaces for user interaction; multiple untested applications	Can serve as a reminder delivery system – e.g. student user specific reminders	*Pros:* wearable *Cons:* limited functionality; educational potential still unexplored

*Bluetooth – wireless, short-range communication technology that can be used to transfer data between other Bluetooth-enabled devices

ics and hospitals they often refer to information in order to make evidence-based decisions. It is therefore not surprising that the major users of mobile handheld devices are in the field of medicine. This use of mobile devices and m-learning is only going to increase in medicine in years to come.

In Canada, approximately one-third of the population lives in rural communities of 10,000 people or less but only 10% of general practitioners and 4% of specialist physicians practice in rural areas (Lau & Bates, 2004). Similarly, there is a physician shortage in rural communities of USA and Australia. To combat this shortage, undergraduate and postgraduate medical students are placed in rural areas very early in training as research has shown that early exposure increases the likelihood of graduating physicians choosing to practice in rural and remote communities (Rourke, 2006).

Thus, virtually every medical school in Canada has some distributed medical education activities, i.e. educational activities beyond the walls of their institutions such as family physicians' offices, satellite campuses, other hospitals, rural and remote communities. In some schools, the whole program is given at a distant site. Others have a variety of rotations, electives and community experiences of shorter duration.

Many CME activities are distributed to practicing physicians via video and computer technologies in distributed sites and m-learning is an important solution to facilitate learning in such situations.

Another dimension relating to mobility that has implications for these physicians is the introduction of mobile devices in patient care. Emerging mobile technologies such as smart phones, wireless tablet PCs, wearable wireless biosensors, and monitoring devices are increasingly being used for monitoring patients remotely. As part of CME, these physicians have to learn about this form of patient care as well.

Search Strategies

In order to identify how m-learning is utilized in medicine, a number of databases such as ERIC (education), HMIC (health management), PUBMED (Medline), Web of Science (Social Science Citation Index), and AACE Digital Library were searched. The search strategy used the phrases 'm-learning', 'blended-learning', 'mobile learning', and 'e-learning', limited by the terms 'medicine' and 'medical students.' In addition, keywords such as PDAs, iPods, and mobile phones were used. The bibliographies of selected articles were also assessed for relevant items. A separate search of report literature and relevant websites was also done. Relevant articles were then categorized as mobile learning in UGME, PGME and CME.

Unfortunately, at the time of writing this chapter, similar to a systematic review of m-learning done by Masters (2008), most research is still on the technology rather than the process of m-learning in medicine. Examples of m-learning and usage are generally pilot studies and case studies where specific mobile devices have been given to selected medical students or health professionals and their usage monitored (Dearnley et al, 2009; Dearnley, Haigh & Fairhall, 2008; Dearnley & Walker, 2009; Fisher, Stewart, Mehta, Wax &

Lapinsky, 2003; Garett & Jackson, 2006; Walton, Childs & Blenkinsopp, 2005).

Learning Context and Characteristics of Medical Students, Residents and Medical Professionals and Implications for M-Learning

Many medical schools require students to have their own computers upon admission and often, the computers and other mobile technologies are supplied by the school (Lau & Bates, 2004; Jackson, Ganger, Bridge & Ginsburg, 2005). Medical students have been found to be the largest undergraduate users of PDAs within UK higher education (Savill-Smith & Kent, 2003). However, possession of a computer does not necessarily translate into being adept at the use of technology. A number of medical students have difficulty with e-learning and some are even opposed to it (Link & Marz, 2006). Gender differences have also been identified, with females being less inclined to use technology for learning. Medical students have also been shown to spend the least time at the computer (an average of 8 hours per week) as compared to students of other disciplines and to be less familiar with other program types (Link & Marz).

Early placement of medical students in distributed sites (e.g., rural, clinic, hospital), access to internet, year of training (1st, 2nd year, etc.), level of education (undergraduate, postgraduate, continuing), domain of learning (knowledge, skills or attitudes), emphasis on patient-centered learning, and moves towards inter-professional learning are some factors that have implications for teaching and learning in medicine using mobile technologies.

In the early years, undergraduate learners need continued access to content, course material and resources. In some situations, attendance is also tracked to promote professionalism. In distributed sites, they may require access to other learners within the community. They may need to gather data and complete assignments while in the workplace and require ample support with personal information management. This is mainly because undergraduate learning is generally driven by external factors such as curriculum and examinations (Davis et al, 2007) and students are learning foundational knowledge and getting accustomed to the new environment.

The learning context of medical residents itself is unique. With the rapid growth of medical knowledge, it is challenging for busy physicians and residents to fulfill their social obligation to stay updated on recent developments occurring in their specific field of practice. This calls for efficient and new methods and technology that can support them. In addition, their work involves a high degree of unpredictability and mobility, and is fraught with disturbances and interruptions (Bertelsen, Kanstrupand & Christiansen, 2007). These professionals generally cannot prepare, plan or gather knowledge ahead of time as new patients are unknown cases. Their work also involves a high degree of mobility – walking from ward to ward, often between physically separated buildings. Their work is constantly disturbed by interactions with other professionals and patients and by other activities such as writing notes, etc. Most often, residents work alone and are forced to make difficult decisions on their own. Residents do not have personal computers at work and have to share computers in the wards. All this implies that learning has to occur on the fly and in short spurts. To cope with the situation, residents carry notebooks, guides, references and other accessories in their numerous pockets – contents weighing as much as 2.6 kg (Bertelsen, Kanstrupand & Christiansen)! Carefully chosen mobile devices with multiple capabilities can reduce the weight (Burdette, Herchline & Oehler, 2008).

Residents learn at patients' bedsides as apprentices, gaining experience and reflecting on action under the supervision of expert physicians. By studying real practice, their clinical diagnostic reasoning is honed. By viewing patient records and comments made by other physicians, and by keeping records and notes, reflection-on-action is facilitated.

The environment of practicing physicians is similar to that of residents but these individuals are the medical experts. However they, too, need access to the latest medical research at the point of care. They also need to communicate with physicians and colleagues at a moment's notice. Although the trend is changing, in many hospitals pagers are the sole connection between physicians, staff, colleagues, and patients. As a result, physicians, like residents, use pagers, cell phones and PDAs simultaneously to communicate and to access medical information or calendar function. Patient tracking and patient billing are additional important tasks of physicians.

Use of Mobile Phones

Dearnley et al (2009) expand on five case studies that were part of the Mobile Technologies Pilot Project for the Assessment and Learning in Practice Settings (ALPS) undertaken across five UK higher educational institutions. Although students of other health professions were also involved, the lessons learned are applicable to medical students.

In one case, dietetic and physiotherapy students and tutors were issued with mobile phones and MediaBoard was used to set up web-based multimedia message boards. They (students) contributed by sending SMS (text) and MMS (text, picture, audio) messages from their devices. These messages enabled students to record and share experiences in the practice setting in multimedia format. Students also collected information about the learning experiences in a multimedia blog (weblog). The device was also used by tutors to contact groups of students using the text alert feature.

Students reported that it was advantageous to be able to record evidence in multimedia format. They were pleased with the ability to form social mobile networks and enjoyed keeping in touch with their peers via blogs and to read about what others were doing. Students used their phones to post reminders and found it convenient to have a work phone as a point of contact.

Some of the technical challenges reported were problems connecting to the university server, problems with some devices, loss of data, and loss and theft of devices. Some instructors found the new technologies daunting. Gaining ethical approval to use the devices in practice settings was another issue.

A relatively new technology – Moblogging, also known as mobile blogs, can be created using any mobile phone. Winksite (http://www.winksite.com) allows users to create pages that can be accessed via cell phone. This has potential for encouraging students to share and archive in a secure area what they have learned and seen in clinical practice.

Implications for Learning and Teaching

Mobile phones, though possessing limited capabilities, can be effectively used in undergraduate teaching by learners to easily keep in touch with fellow learners, thus dispelling feelings of isolation, especially when placed in a distributed environment. They can be used, as in this case, for capturing real life experiences. Because of the ability of mobile phones to capture text, audio and photos, learning can be facilitated by requiring students to capture, for example, photos of pathologies or short audio recordings of patient encounters. Later, these can be annotated and shared in a blog or wiki or used to create a personal learning e-portfolio.

Instructors can use mobile phones for contacting students to inform them about changes in schedules or to motivate by using text or voice messaging. This technology allows for groups of students or individual students to be contacted easily in a timely manner.

There is potential for instructors to use this technology in other innovative ways to deliver "bite-sized" content, introductions, and revision tips. Research in memory and second language vocabulary learning indicates that spaced repetition in vocabulary learning results in retention and is a better way to learn (Cuddy & Jacoby, 1982). Mobile phones are excellent tools for sending short messages. For example, students can be sent new medical terms, definitions, and context questions relating to anatomy, physiology or other subjects at appropriately spaced intervals. This technology allows for instructors to prepare messages that are carefully integrated to topics being dealt with or courses in the curriculum in advance and send them out at a given time.

Use of iPods

iPods are popular as they are small, portable, and can be dropped into a pocket to free both hands. They can be used to download audio books, podcasts, photos and videos. Podcasts and vodcasts are audio and video recordings that can be downloaded onto mobile devices. A special feature of such broadcasting is the fact that they can be automatically downloaded using software capable of reading new additions to a website (RSS feeds). The use of podcasts in medicine is gaining momentum partly because of the ease of use and the availability of many open and low cost software and hosting options for creation. Audio recordings of textbooks are being used for learning (Meng, 2005). Downloadable libraries of high-resolution heart and respiratory sounds are available to medical students to learn and review.

Podcasts are particularly useful in CME. For example, New York University offers CME programs via podcasts (CVMD) and many journals such as the New England Journal of Medicine podcast weekly audio summaries (NEJME). McGraw-Hill's AccessMedicine weekly podcasts of grand rounds and medical text updates (McGraw-Hill), Cleveland Clinic's recorded lectures (Cleveland Clinic) and links in libraries to health-related podcasts (Health Sciences Library) are just some examples of resources that can be easily downloaded to iPods and used on the go by busy medical personnel. Medical vodcasts (video casts) are also available. However, they cannot be used while driving or walking and require more focused attention.

Implications for Learning and Teaching

Though iPods allow only for one-way communications and are not interactive, they can be used in a number of situations. To make instruction more portable, instructors can convert their lectures to podcasts for easy download. These podcasts can then be used by students to review lectures and prepare for exams. Using RSS feeds, students can download information automatically from courses in which they are registered. Thus workplace learning can be connected to institutional learning and accessibility improved. However instructors need to realize that it is not practical for students to listen to a whole lecture in the form of podcasts. Instead, they should be of short duration and used only to summarize key concepts. This implies that additional time is needed for preparation of lectures into podcasts. One drawback of using this technology for teaching and learning is that it caters to auditory learners.

Production of podcasts can also be incorporated as student assignments, thus facilitating critical thinking and reflection. Its potential is yet to be explored in medical education.

Use of PDAs and Smartphones

Given that medical students are the largest undergraduate users of PDAs within UK higher secondary education (Savill-Smith & Kent, 2003) and a majority of doctors use mobile devices (Walton, Childs & Blekinsopp, 2005), it is of interest to examine how these devices are used in teaching and learning. PDAs are carried at all times and provide guaranteed access to a computer. They are password protected to prevent unauthorized access, and a backup is created every time they are synchronized with a PC. They are standalone products that do not depend on local information technology infrastructure. They allow for clinical activities to be recorded on the fly. Literature indicates that PDAs are mostly used by residents and physicians for care assistance, i.e. to look up drug information, clinical guidelines, decision aids, and reference books (Dearnley, Haigh & Fairhill, 2008, Dearnley & Walker, 2009). With the increasing wireless connectivity and the increasing availability of patient data in digital form, PDAs are also used for electronic prescribing, billing, real-time access to medical records and point-of-care literature searches (Rothschild et.al.2006). More recently, the use of smartphones is on the increase (Burdette, 2008).

Some disadvantages of these devices are the small screens that limit the amount of information that can be seen in one screen, and greater risk of damage, loss or theft. In medicine, because patient information can be stored in PDAs, loss of confidential information can have far-reaching implications. With the availability of multiple devices, each with its unique features and capabilities, it is important for physicians to compare various devices and to review user experiences before purchasing one that is suitable for their specific situation.

The KNOWMOBILE case study (Smordal & Gregory, 2002) at the University of Oslo expounds interesting issues in the use of PDAs by medical students. The aim of this project was to develop and evaluate net-based and mobile solutions for knowledge access in distributed training of medical students. In this project small groups of medical students were given PDAs and their usage monitored in three settings: while involved in problem-based learning when physically dispersed; living together in a distributed site; living at home and commuting to hospital.

The researchers found that use of websites was limited by time to download and inability of many websites to adapt to the screen size. Students did not use some of the e-books that had been downloaded as they did not support just-in-time interactions. Students also experienced problems in working across different applications and information resources on the PDA. SMS, though specifically designed for the group on the PDA, was not used as existing infrastructure such as cell phones and paging systems in hospitals proved to be more efficient. The researchers conclude that "the PDAs should no longer be regarded only or primarily as Personal Digital Assistants, but rather as potential gateways in complicated webs of interdependent technical and social networks."

Reflection is an important component of Kolb's experiential learning cycle (Kolb & Fry, 1975) and in medical education. PDAs can be used to reflect-in-action and reflect-on-action (Schon, 1983). In their study that examined PDAs as performance support to medical professionals, Bertelsen, Kanstrupand & Christiansen (2007) describe the learning context of residents and how PDAs can be effectively used to move them from novice to expert and promote their learning practice, i.e. apprenticeship learning, problem-based learning and reflection in- and on- action. PDAs allow for note taking (memory aid), accessing references quickly (situated learning) – even while walking, communicating with peers and others (cooperative learning). They are light weight and easy to carry, and operable with one or both hands – important and welcome features for these highly mobile individuals.

PDAs have also been used for ePortfolios (Dearnley et al., 2009) and blogs. Students were encouraged to maintain written reflections of their practice experiences that would be used for formative and summative assessments (Dearnley et al). Completed documents were transferred to the instructor's device using Bluetooth® wireless technology. Electronically signed, corrected documents were sent back to the students as well as to the central database maintained at the university. The researchers found that such usage facilitates learning by encouraging reflection and by providing timely, useful feedback. Students liked the fact that they could edit and extend their e-portfolios on an on-going basis and incorporate multimedia. They also appreciated the fact that their e-portfolios can be kept neat and tidy. In relation to blogs, PDAs are more useful for posting reminders and short notes on the go. Students can then expand while sitting in front of a PC.

In one study (Krippendorf, Simpson, Schiedermayer, 1999), teachers were given PDAs with a list of characteristics associated with effective clinical teaching and asked to record their perceptions of their teaching in lieu of these characteristics. Later, the teachers compared their perceptions with the evaluations given by learners. The teachers reported that this exercise prompted them to reflect on their effectiveness as educators.

Jackson et al (2005) of Wayne State University Medical School describe their experiences in the implementation of handheld devices in all four years of the undergraduate curriculum. Students were required to purchase PocketPCs with a prespecified configuration. Following orientation to the device and software, students and faculty used the devices in a variety of settings. Attendance was tracked in didactic presentations where attendance was mandatory by students tapping a check-in button. The unique IP address from which the student checked in was recorded to prevent check-in from other locations. Course evaluations were also completed using these devices. Since the evaluations are listed as tasks to be completed, returns were improved. The note-taking feature also allowed for comments.

The authors also describe the use of PDAs to enhance interactive learning by utilizing them as student response systems. Questions were prepared ahead of time or on the fly and sent to the devices during lectures and the responses projected to the class. Students also used the devices to post questions to lecturers during the class. These strategies enhanced interactivity. Students could also access recorded lectures using the devices and download the audio portion of the lecture to listen offline.

Students were given access to licensed medical decision-making software to promote diagnostic skills. The devices were used to monitor patient encounters using an application developed for the purpose. Since many of the remote placement sites lacked wireless capabilities, students logged their encounters offline. The data were transmitted to the centralized base when the devices were synced with a PC. Data could also be transmitted via infrared kiosks in specific locations. The medical school found that this strategy eased the administrative burden tremendously.

Some of the issues encountered during implementation were the lack of sufficient technological infrastructure; the need for coordination between multiple players (such as Biomedical communication, Office of Academic and Student Programs, and Medical Library Services); and students requiring hands-on assistance with the device; incidences of lost data caused by battery drain or accidental resetting of device, among others.

Implications for Teaching and Learning

Because of the extensive computing capabilities, PDAs can be used for teaching and learning in many ways. They can be used for accessing prepackaged information i.e. standalone learning resources that do not require connection to the internet or IT infrastructure. Such packages can be in the form of learning objects e.g. small video interactive files and virtual patient cases (Smith, 2007). Many such resources are available in peer-reviewed learning repositories such as MedEdPORTAL (http://services.aamc.org/30/mededportal/servlet/segment/mededportal/information/), HEAL (Health Education Assets Library) and MERLOT (Multimedia Education Resource for Learning and Online Teaching).

Since PDAs have synchronizing capabilities with desk tops and other devices, students can do assignments requiring minimal text input or collect data in the field using PDAs and upload them when internet connections are available. The availability of input devices such as compact, foldable keyboards that allow input of longer texts, increase the potential for m-learning.

PDAs are ideal for undergraduate students to access their calendar and schedules while on the go and for accessing reference information. The communicating capabilities can be used for interactions between peers – for organizing group activities and sharing information in multiple media. They can also be used by students to log patient encounters and procedures performed. One way of assessing competence in medicine is to keep track of the different types of student-patient encounters. Details of every encounter can be easily recorded in the field and later downloaded onto a PC. This information can then be used to improve the student experience and to identify areas and topics that require further exposure. Other perceived benefits include improved instructor and peer support, better access to information, and the ability to record and reflect clinical encounters in real time.

PDAs can also be used to transfer files from one device to another using infra-red communication, facilitating cooperative learning and sharing in large classes.

PDAs are not only useful for learners but also for preceptors as an important tool for CME. They can be used to document observations in the workplace – an important form of assessment in clinical practice. In medicine, checklists are used frequently to observe students and PDAs may be very useful tools for OSCEs (objective structured clinical examinations). Their large storage capacity allows physicians to access information and learn even during consultations. Studies (Rothschild et al 2002, Houston et al 2003) have shown that patients are positive about the use of PDAs during consultations. There is also evidence that medical errors can be decreased by their use (Fischer et al, 2003; Grasso, Genest, Yung & Arnold, 2002; Tooey & Mayo, 2003). Currently, there are many application packages for physicians but relatively few for students.

From existing studies it can be seen that the usability and acceptability of PDAs to students and others in a practice setting can be a challenge. Many students find the introduction of new technology in an already overloaded curriculum daunting. Support systems need to be in place to help students facing technical problems with the devices. Because of the smallness of the devices, many features taken for granted in PCs are not available. Therefore, programs have to be modified according to the device being used and its capabilities. Students also worry about their movements being monitored via these devices, introducing a trust issue.

Another challenge can be the difficulty of accessing university websites and other virtual sites while within a hospital where firewalls and higher security is in place. Since mobile devices have an inbuilt camera, consent forms are required from patients if photographs of patients are used. Not all devices are socially acceptable among students and this should be given some consideration before introducing these devices institution-wide.

Other Devices

Smart pens are newer devices that have not been explored as yet in medical education. However, there is potential for them to be used for note taking and recording in the busy clinic or at the bedside and then transferring the data to a PC. Since they are programmable, creative and innovative uses for them in medical education are likely to emerge in the near future.

Wearable devices, while being increasingly used in patient care, have not been researched in medical education. Some of the potential uses are in monitoring heart rate, temperature etc. as labs for physiology teaching.

LEARNING THAT OCCURS IN M-LEARNING

From the limited literature, the role of mobile devices seems to have a greater impact on the knowledge domain – 'drip, drip' learning, i.e., little and often; learning in context; and greater interaction with content (e.g. annotation, bookmarking). Their effects on the psychomotor and affective domain need to be explored further.

It can be concluded that the success of using mobile technology in learning has more to do with the pedagogical task than the technology (Kristiansen, 2001). In choosing the technology for m-learning, a number of issues have to be considered:

- *Learning*
  - Is the task at hand related to the knowledge, psychomotor or affective domain?
  - Will this technology promote surface or deep learning? Is the task at hand better learned in a self-directed manner or collaboratively?
- *Mobile device*
  - Can the learning objective be achieved without the use of the device? Is a mobile device the best way to achieve the task at hand? Given the desired learning outcome, which of the devices is most suitable?
- *The learner*
  - Is the learner amenable to using the technology? What individualized training is required to learn the technology? What is her/his learning style?
  - Is the learner an undergraduate? Postgraduate? Practicing physician? Given the level of training, is the technology feasible for use, based on lifestyle and other commitments?
- *Instructors/Preceptors/Facilitators*
  - Are they open to using technology? What training is required? Do they have the skills and time to develop or identify learning experiences using mobile technologies that are appropriate for the task at hand? What are the evidence-based, pedagogically sound principles that should be used to facilitate learning using this technology?
- *Location*
  - Where will learning occur? Remote, rural or urban site? Clinic? Ward?
- *Connectivity*
  - Is the technology infrastructure in place and available? Will the mobile technology interfere with current technologies? What issues of security and confidentiality have to be addressed? Does learning require synchronous or asynchronous connection or no internet connection?
- *Technological support*
  - Are resources available for technical support on an ongoing basis? Can support be given in a timely manner?
- *Communication and Coordination*
  - What kind of communications and coordination are required between clinics, hospitals, university, technical support, administrators and others to implement this mobile technology seamlessly?
- *Costs*
  - What is the cost to the institution? The learner? The instructor? Maintenance cost?

FUTURE DIRECTIONS

The changing trends in medical education from university-based to a distributed model and the push to increase the number of physicians in rural and remote areas reinforces the important role of m-learning in medical education, especially in countries such as Canada, USA and Australia. The availability of pilot projects and case studies in m-learning in medicine is evidence that we are in the early stages of Rogers' change cycle – innovators and early adopters (Rogers, 2003). In order to move forward, large-scale surveys and studies of m-learning are required. At the same time, emerging mobile technologies as well as features currently available (e.g. GPS) in mobile devices have to be explored to identify how they can facilitate learning. It has also to be kept in mind that, if ineptly managed, technology can cause more problems than it solves. Also, mobile technologies cannot be used in isolation to support learning. An appropriate mixture of learning experiences and technologies has to be chosen in any given circumstance. At present, the impact of mobile technologies in *learning* has been the focus. In medicine, the gold standard should be improvement in quality of patient care. Does m-learning have an impact on quality of patient care? This is yet to be demonstrated unequivocally.

REFERENCES

Al-Ubaydli, M. (2006). *The doctor's PDA and smartphone handbook: A guide to handheld healthcare*. London, UK: Royal Society of Medicine Press.

Association of Faculties of Medicine of Canada. (2006). *Results of a survey of distributed medical education activities at Canadian Faculties of Medicine*. Retrieved from http://afmc.ca/education-distributed -med-edu-e.php

Attewell, J., & Savill-Smith, C. (2004). Mobile learning and social inclusion: Focus on learners and learning, (pp. 3-11). *Proceedings of MLEARN 2003 Conference*. London, UK: LSDA.

Bertelsen, P., Kanstrupand, A. M., & Christiansen, E. (2007). Problem-based learning in medicine: Performance support with m-learning. [nd *International Conference on e-learning.* New York, USA.]. *Proceedings of the ICEL, 2007*, 2.

Boelen, C. (2000). *Towards unity for health: Challenges and opportunities for partnership in health development, a working paper*. World Health Organization. Retrieved March 21, 2010, from http://www.who.int/hrh/documents/ en/TUFH_challenges.pdf

Burdette, S. D., Herchline, T. E., & Oehler, R. (2008). Practicing medicine in a technological age: Using smartphones in clinical practice. *Clinical Infectious Diseases, 47*, 117–122. doi:10.1086/588788

Cleveland Clinic. (n.d.). *Videos, webcasts & podcasts*. Retrieved from http://my.clevelandclinic.org/health _edge/about_podcast.aspx

Collis, B., & Moonen, J. (2001). *Flexible learning in a digital world: Experiences and expectations* (pp. 154–155). London, UK: Kogan.

Cuddy, L. J., & Jacoby, L. L. (1982). When forgetting helps memory: An analysis of repetition effects. *Journal of Verbal Learning and Verbal Behavior, 21*(4), 451–467. doi:10.1016/S0022-5371(82)90727-7

CVMD. (2010). *The cardiovascular multimedia information network*. Retrieved March 21, 2010, from http://www.cvmd.org/

Davis, J., Chryssafidou, E., Zamora, J., Davies, D., Khan, K., & Coomarasamy, A. (2007). Computer-based teaching is as good as face to face lecture-based teaching of evidence based medicine: A randomized controlled trial. *BMC Medical Education, 7*(23). doi:.doi:10.1186/1472-6920-7-23

de Freitas, S., & Levene, M. (2003). Evaluating the development of wearable devices, personal data assistants and the use of other mobile devices in further and higher education. Developing the next generation of education. *Proceedings of the EDEN 10th Anniversary Conference*, Stockholm, June 10-13, 2001.

Dearnley, C. A., Haigh, J., & Fairhall, J. (2008). Using mobile technologies for assessment and learning in practice settings: A case study. *Nurse Education in Practice*, *8*(3), 197–204. doi:10.1016/j.nepr.2007.07.003

Dearnley, C. A., Taylor, J., Hennessy, S., Parks, M., Coates, C., & Haigh, J. (2009). Using mobile technologies for assessment and learning in practice settings: Outcomes of five case studies. *International Journal on E-Learning*, *8*(2), 193–207.

Dearnley, C. A., & Walker, S. A. (2009). Mobile enabled research. In Vavoula, G. (Ed.), *Researching mobile learning*. Oxford, UK: Peter Lang.

Fischer, S., Stewart, T. E., Mehta, S., Wax, R., & Lapinsky, S. E. (2003). Handheld computing in medicine. *Journal of the American Medical Informatics Association*, *10*(2), 139–149. doi:10.1197/jamia.M1180

Garrett, B. M., & Jackson, C. (2006). A mobile clinical e-portfolio for nursing and medical students, using wireless personal digital assistants (PDAs). *Nurse Education in Practice*, *6*(6), 339–346. doi:10.1016/j.nepr.2006.07.015

Geddes, S. (2004). *Mobile learning in the 21st century: Benefit for learners*. The knowledge tree. Retrieved from http://knowledgetree.flexible-learning.net .au/edition06/download/geddes.pdf

Grasso, B. C., Genest, R., Yung, R., & Arnold, C. (2002). Reducing errors in discharge medication lists by using personal digital assistants. *Psychiatric Services (Washington, D.C.)*, *53*, 1325–1326. doi:10.1176/appi.ps.53.10.1325

Health Sciences Library. (2010). *Health related podcasts*. University of Colorado Denver. Retrieved March 21, 2010, from http://hsclibrary.uchsc.edu/podcasts

Jackson, M., Ganger, A. C., Bridge, P. D., & Ginsburg, K. (2005). Wireless handheld computers in the undergraduate medical curriculum. *Medical Education Online, 10*, 5. Retrieved from http://www.med-ed-online.org

Keegan, D. (2002). *The future of learning: From e-learning to m-learning*. Forschungsbericht. FernUniversitat, Hagen. Retrieved from http://deposit.fernuni-hagen.de/1920/

Kolb, D. A., & Fry, R. (1975). Toward an applied theory of experiential learning. In Cooper, C. (Ed.), *Theories of group process*. London, UK: John Wiley.

Krippendorf, R. L., Simpson, D. E., & Schiedermayer, D. (1999). Promoting reflective teaching with personal digital assistants. *Academic Medicine*, *74*, 577. doi:10.1097/00001888-199905000-00042

Kristiansen, T. (2001). M-learning. Experiences from the use of WAP as a supplement in learning. In Szucs, A., Wagner, E., & Holmberg, C. (Eds.), *Learning without limits. Developing the next generation of education*. Oslo: Fornebu Knowation.

Kukulska-Hulme, A., & Traxler, J. (Eds.). (2005). *Mobile learning: A handbook for educators and trainers open & flexible learning series*. London, UK/ New York, NY: Taylor & Francis Routledge.

Laouris, Y., & Eteokleous, N. (2005). We need an educationally relevant definition of mobile learning. *Proceedings of the MLEARN 2005 Conference*. Melbourne, Australia. Retrieved from http://www.mlearn.org.za/CD/papers/ Laouris%20&%20Eteokleous.pdf

Lau, F., & Bates, J. (2004). A review of e-learning practices for undergraduate medical education. *Journal of Medical Systems*, *28*(1), 71–87. doi:10.1023/B:JOMS.0000021522.30587.ff

Link, T. M., & Marz, R. (2006). Computer literacy and attitudes towards e-learning among first year medical students. *BMC Medical Education*, *6*, 34. Retrieved from http://www.biomedcentral.com/1472-6920/6/34. doi:10.1186/1472-6920-6-34

Mahmoud, S. S. (2008). A proposed model for distributed e-courses content through mobile technology architectures. *World Academy of Science. Engineering and Technology*, *37*, 274–289.

Masters, K. (2008). M-learning: How much of what has been diffused? A systematic literature review. *Proceedings of World Conference on Educational Multimedia, Hypermedia and Telecommunications,* Chesapeake, VA. (pp. 5790-5795).

McGraw-Hill. (2010). *AccessMedicine weekly podcasts*. Retrieved May 21, 2010, from http://books.mcgraw-hill.com/podcast/acm/

Meng, P. (2005). *Podcasting & vodcasting: Definitions, discussions & implications*. A White paper by IAT Services at University of Missouri. Retrieved May 12, 2010, from http://edmarketing.apple.com/adcinstitute/wp- content/Missouri_Podcasting_White_Paper.pdf

Mitchell. J. (2003). *E-business and online learning: Connections and opportunities for vocational education and training*. Australian National Training Authority, Melbourne. ISBN-1-74096-143-9

New England Journal of Medicine. (2010). *About podcasts*. Retrieved Mach 21, 2010, from http://content.nejm.org/misc/podcast.dtl

O'Malley, C., Vavoula, G., Glew, J. P., Taylor, J., Sharples, M., & Lefrere, P. (2003). *MOBIlearn WP4-guidelines for learning/teaching/tutoring in a mobile environment.* Retrieved March 21, 2010, from http://www.mobilearn.org/download/results/guidelines.pdf

Premkumar, K., & Coupal, C. (2008). Rules of engagement-12 tips for successful use of clickers in the classroom. *Medical Teacher*, *30*(2), 146–149. doi:10.1080/01421590801965111

Rogers, E. M. (2003). *Diffusion of innovations* (5th ed.). New York, NY: Free Press.

Rothschild, J. M., Fang, E., Liu, V., Litvak, I., Yoon, C., & Bates, D. W. (2006). Use and perceived benefits of handheld computer-based clinical references. *Journal of the American Medical Informatics Association*, *13*(6), 619–626. doi:10.1197/jamia.M2103

Rourke, J. (2006). Social accountability in theory and practice. *Annals of Family Medicine*, *1*(4), S45–S48. doi:10.1370/afm.559

Savill-Smith, C., & Kent, P. (2003). *The use of palmtop computers for learning: A review of the literature*. London, UK: Learning and Skills Development Agency. Retrieved March 21, 2010, from http://www.m-learning.org/docs/ the_use_of_palmtop_computers_for_learning_sept03.pdf

Schon, D. (1983). *The reflective practitioner*. New York, NY: Basic Books.

Smith, S. R., Cookson, J., McKendree, J., & Harden, R. M. (2007). Patient-centred learning–back to the future. *Medical Teacher*, *29*(1), 33–37. doi:10.1080/01421590701213406

Smordal, O., & Gregory, J. (2002). *Mobile learning book – the KNOWMOBILE case study*. Retrieved March 21, 2010, from http://academic.research.microsoft.com/ Paper/4947211.aspx

Tooey, M. J., & Mayo, A. (2003). Handheld technologies in a clinical setting: State of the technology and resources. *AACN Clinical Issues*, *14*, 342–349. doi:10.1097/00044067-200308000-00009

van Barneveld, A., & Shaw, S. (2006). Mobile learning: A review of implementations. *Proceedings of World Conference on E-Learning in Corporate, Government, Healthcare, and Higher Education 2006,* Honolulu, Hawaii. (pp. 980-985).

von Jan, U., Ammann, A., & Matthies, H. K. (2008). Generating and presenting dynamic knowledge in medicine and dentistry. In C. Bonk, et al. (Eds.), *Proceedings of World conference on E-Learning in Corporate, Government, Healthcare, and Higher Education 2008* (pp. 209-214). Chesapeake, VA.

Walton, G., Childs, S., & Blenkinsopp, E. (2005). Using mobile technologies to give health students access to learning resources in the UK community setting. *Health Information and Libraries Journal*, *22*, 51–65. doi:10.1111/j.1470-3327.2005.00615.x

Chapter 9
Unleashing the Potential of Mobile Learning through SMS Text for Open and Distance Learners

Zoraini Wati Abas
Open University, Malaysia

Tina Lim
Open University, Malaysia

Ruzita Ramli
Open University, Malaysia

ABSTRACT

Malaysia has a population of about 28 million people but there are, incredibly, more than 30 million mobile phone subscriptions. Sixth in the world in terms of SMS (Short Message Service) volume, Malaysians appear to be addictive SMS texters. With over 98 percent of its students having mobile phones and 82 percent of the students ready for learning through mobile phones, Open University Malaysia initiated a project that first experimented with podcasts and SMS texts later. This chapter describes how the institution conceptualized, planned, and created a mobile learning environment using SMS to enhance its current blended learning model in general, and in particular, one of its courses with over 1,000 students enrolled. The chapter also describes the categories used for formulating the SMS content, use of Twitter and Facebook to support the SMS sent and discusses the feedback received on the initiative as well as the issues and challenges.

INTRODUCTION

Open University Malaysia (OUM) was established in August 2000 to provide opportunities to its working adult population who wish to obtain tertiary qualifications. As an institution, it seeks to become a leader in providing flexible education and believes that opportunities should be provided to all, regardless of time, location, or one's age or socioeconomic status. OUM enrolled its first batch

DOI: 10.4018/978-1-60960-511-7.ch009

of 753 students in 2001. Established in response to the government's call to democratise education and to provide opportunities to working adults in particular without having to leave the workforce or study away from the family, the institution aims to widen access to education by leveraging on technology, adopting flexible modes of learning, and at the same time provide a stimulating environment to engage learners during the learning process. Having enrolled almost 100,000 94,000 learners in 61 learning centres throughout the country since its first intake of students in 2001, OUM seeks to put learning in the hands of its learners through mobile learning. It is believed that mobile learning will increase the flexibility of learning support in terms of approaches used to engage learners during their learning process. It has previously introduced various learning technologies from the most basic print learning resource in the form of modules for all its courses to the more sophisticated form of learning material such as multimedia courseware, learning objects, video streamed tutorials and podcasts (MP3 and MP4 formats), depending on the suitability of the need for selected courses.

In 2008 OUM decided that it was time to seriously explore the use of mobile technologies to enhance its blended learning model. Following the decision, the Institute of Quality, Research and Innovation (IQRI) was entrusted to lead and explore how mobile learning can be implemented. IQRI formed a mobile learning team to research on various uses of mobile technologies and how these could be used to enhance its current blend of learning approaches comprising: self-managed learning, online learning and face-to-face interaction. To ensure acceptability, the team distributed a survey to 3,000 students in all its learning centres throughout the country. The objective was to determine students' receptiveness and readiness to mobile learning. Through the survey, it was found that 98 percent of OUM learners have at least one mobile phone each while close to 82 percent stated that they would be ready for mobile learning when implemented within six to twelve months (Abas, Chng & Mansor, 2009). Generally, the survey generated a positive response from the 2,837 students who responded. It was, to say the least, encouraging.

The objective of the chapter is to share OUM's mobile learning initiative, specifically, on how a very common device, the mobile phone, and how a tremendously popular form of communication, the SMS text; was successfully used to engage distance learners during their learning process and hence, enhance their learning. It is, potentially, a tool that may support ubiquitous learning.

BACKGROUND

If one were to look at data on mobile phone subscriptions world-wide, the landscape could be described as spiky with the tallest peaks located in Taiwan, Luxembourg, Hong Kong, Italy and Iceland (International Telecommunication Union as cited in Nationmaster.com (2010). In an article published online entitled "The World is Spiky", Richard Florida the guru of the Creative Class pointed out that when the world is portrayed through three-dimensional graphs for aspects such as population, light emissions, number of patents and scientific citation, it appears spiky with sharp, tall peaks, hills and valleys. As for mobile cellular subscriptions by region, there has been a phenomenal increase for all regions across the world. In 2008, for example, mobile penetration in Europe reached 118 percent. This was followed by the Commonwealth of Independent States (113%), Americas (82%), Arab States (63%), Asia Pacific (46%) and Africa (32%) as compared to the world's mobile penetration (for 2008) average rate of 60 percent (International Communication Union, 2010).

The penetration rate for mobile phones in Malaysia was of a similar trend. From 21.8 percent in 2000, the penetration rate had increased to 74.1 percent in 2005 and then on to an amazing 106.2

percent in the last quarter of 2009 (Malaysian Communications and Multimedia Commission, 2010). By the end of 2009, the total number of subscriptions (pre-paid and post-paid) was 30,379,000 for a population of approximately 28,310,000.

THE EMERGENCE OF MOBILE LEARNING

Herrington and Herrington (2007) observed that the use of mobile devices has become increasingly popular in our daily lives due to its "decreasing cost and increasing social currency." Kirkwood (1998) as cited in Motlik (2008, p. 5) however, warned that the educational community needs to be careful not to "let the tail of technology wag the pedagogical dog." In other words, adopting technology for the sake of technology is not acceptable. However, Keegan, Kismihok, Mileva and Rekkedal (2009) stated that distance education research strongly indicates that "it is not technologies with inherent pedagogical characteristics which succeed in distance education, but technologies that are generally available to citizens" (p. 5). They therefore believed that if that was true, mobile telephony is thus far most suited for distance education.

According to Hayes, Joyce and Pathak (2004), mobile learning refers to the use of mobile devices in teaching and learning. Price (2007) extended the definition to refer to "the use of handheld technologies enabling the learner to be 'on the move', providing anytime anywhere access for learning" (p.33). Nyiri (2002) explained that mobile learning is fundamentally delivered through mobile devices such as Personal Digital Assistants (PDA), smart phones and any other handheld devices and mobile learning is seen as the next generation of e-learning and important instrument for lifelong learning (Sharples, 2000). According to Traxler (2009) mobile learning is defined in terms of its technology and hardware such as PDA, smart phones or wireless laptop computers, which are portable. In terms of characteristics, mobile learning could be described as personal, spontaneous, opportunist, informal, pervasive, situated, private, context aware, bite-size, portable and ubiquitous learning.

Many consider mobile learning a natural extension of e-learning while others have labelled it a new frontier in e-learning and have claimed that it holds great promise (Caudill, 2007). Features inherent in mobile learning that seem to endear educators, practitioners, designers, and developers of instruction include the fact that learning is made more convenient and flexible as the devices are portable and easy to use. Further, these mobile devices are cost-effective and efficient.

Meanwhile Parsons and Ryu (2006) pointed out that it is not the quality of the software (or even the mobile device, for that matter) that solely determines the quality of the learning experience, but also" the conceptual basis upon which the learning experience is constructed" (p. 1). Ryu and Parsons (2009) were of the opinion that the most important feature in the mobile environment is that educational providers and the learners are able to be in contact with each other while "outside the reach of conventional communication spaces ... (and it) ... allows learners to access knowledge resources where and when they want them" contributing positively to students' autonomy (p. 9). This may be treated as a great asset of distance learning where the dimensions of time and distance need to be addressed. Hayes et al. (2004) concurred that m-learning affords students opportunities to make use of "dead time" or non-productive time such as when one is travelling on the bus or while waiting in queue to review course materials.

In designing mobile learning environments, Valentine (2004) emphasized that it needs to be considered within a blended learning mode in the same way as other learning delivery modes. Further, she noted that mobile learning is meant to enhance learners' learning experience, and ought not to be used as a primary method for delivering a course. She stressed that m-learning is a very powerful method for engaging learners especially for ODL learners whose face-to-face meeting is very limited. Mellow (2005) concurred and opined that the use of learning management systems to complement face-to-face interactions "may not be enough to fully engage our students or offer them the flexibility in their study life that is recognised as being important to provide an environment for deep learning and understanding to take place. .. (but) the integration of mobile devices will offer true flexibility for our students and fit in with their digital lifestyle" (p. 469).

Learning anytime anywhere, accessible to the masses, but tailored to the individual (Thomas, 2005) is considered a desirable environment for learning in the digital age. This is particularly crucial in ODL. The flexibility offered by m-learning can support self-managed learning and help engage learners in various learning activities. Ryu and Parsons (2009) believed that mobile learning is also able to augment or enhance learning experiences in that it can support collaborative activities in different ways and by "strengthening the organization of the learning material and information, supporting communication among group members, and helping the co-ordination between learning activities" (p. 10). Other advantages include transforming attitudes and feelings about a learning activity, maximizing engagement in the learning process, increasing motivation to learn and heightening satisfaction about the whole learning experience.

However, hand-held devices such as mobile phones and PDAs are not without their unique disadvantages. The constraints of using mobile devices need to be taken into consideration. Among the limitations as noted by Devinder and Zaitun (2006) are that the devices:

1. Have small screens and therefore not much information may be displayed at one go;
2. Possess limited storage capacities;
3. Are less robust than other non-mobile equipment like desktops; and
4. Do not have much power storage capacity.

Further, the effectiveness of these devices very much depends on the bandwidth available; they might not function as expected or required if the bandwidth available cannot cope with the number of users in the system.

According to Caudill (2007), mobile learning can be utilized on different scales and for different purposes such as administrative and academic support. The range of uses of mobile devices for learning that has been documented include the use of SMS, MMS, mobile games, podcasts, interactive resources, real time collaborative learning, and laboratory data analysis. This new mode of learning has been used for courses ranging from yoga and personal health to languages, mathematics, and science.

THE USE OF SMS FOR LEARNING

Like all other educational innovations, the issue of access remains a crucial point of consideration when it comes to mobile learning. Viteli (2000) as cited in Caudill (2007) very aptly pointed out that "The first demand for a successful application of m-Learning is one of scale; without a saturation of the technology in the target audience the system will fail" (p.5).

Of all the applications that have been developed for mobile phones, the most useful and most used is the Short Message Service (more commonly known as SMS), also known as text messaging. Markett, Sanchez, Weber and Tangney (2006) labeled the SMS as the "killer" application of mobile phones as its usage has surpassed all expectations. Four major reasons why educational institutions should seriously consider using SMS as part of their efforts to make learning flexible and ubiquitous are:

1. It is the lowest common denominator of all mobile technologies,
2. People are very familiar with receiving and sending SMSes,
3. It can be used in all types of mobile phones, and
4. SMSes are comparatively cheap and convenient modes of communication with the masses.

SMS has been used as a standalone application in mobile learning as well as in tandem with other applications. The most basic functions are for communication and information delivery and retrieval. Like other information and communication technologies such as the email, the push and pull metaphor commonly used in business, especially marketing may be applied to mobile learning via SMS. When an education provider sends out information to students, it may be considered a push mechanism. If it involves interactivity as when an education provider communicates to students about their courses through their mobile phones and the students reply with questions or provide details that have been requested via the mobile phone, a push and pull mechanism is present.

Notable achievements in the use of SMS for higher education include the immense impact of administrative support using SMS in the University of Pretoria, South Africa in motivating distance students for various aspects such as reminding them of contact session dates and registration deadlines. The initiative was hugely successful in reducing "perceptual distance between learners and the university" and in reducing the drop-out rate of at-risk students (Ericsson Global, n.d.). Similarly, the University of Ulster was able to reduce student drop-out by sending timely SMS to students who had not been attending classes and largely had the misperception that nobody cared (Keegan et al., 2009)

According to Keegan (2007), the use of SMS messaging for tutoring is well-established as a form of academic support that is successful in enabling students to focus on the academic content, to come better prepared for tutorials and to be more ready to participate actively in discussions. When learners receive SMSes ahead of their tutorials, there is an "expansion of time" in that students have more time to reflect and react to the information they receive. As for academic support, Markett et al. (2006) reported that students asked more questions, and more freely at that, in an in-class SMS system while Motiwalla (2007) noted that content delivery is more effective when a combination of push and pull mechanisms are used. Further, "nuggets" or small chunks of content sent via SMS are more easily absorbed and have been found to be effective in helping learners learn facts (Uday Bhaskar & Govindarajulu, 2008).

However, writing on authentic mobile learning in higher education, Herrington & Herrington (2007) reported that although m-learning has been described as "an emergent paradigm in a state of intense development, few universities have adopted widespread m-learning technologies" (p. 3). Further, a review of literature indicates that research in this field is still rather fragmented. The essence now would be to figure how best "to market, develop content to effectively promote, attract and sustain stakeholders' involvement in SMS-enabled distance education" and to explore if the use of SMS is viable as a stand-alone delivery tool or a component of "blended learning" (Ramos, 2005).

MOBILE LEARNING AT OUM

When the mobile learning project started in the January 2009 semester, one minute podcasts (audio and video) on selected content were developed for a compulsory first semester course, entitled "Learning Skills for Open Distance Learners." The decision was triggered by the findings from the mobile learning acceptance survey where 60 percent of the respondents indicated that they would like audio podcasts based on content from the modules. However, it was found that students were hesitant about downloading them into their mobile phones due to charges that would be imposed by their mobile service providers. The development of podcasts was then discontinued. However, this idea may be re-visited in the future.

In view of the above, OUM next explored the use of mobile learning through SMS texts. It was first implemented in the May 2009 semester with a compulsory first semester course on learning skills. However, given the general perception that using SMSes were less exciting than other more interactive mobile applications, it was a pleasant surprise to note that after a pilot run, the feedback from students and tutors were overwhelmingly positive. Hence, the project was continued in the September 2009 semester with three courses and six courses in the January 2010 semester. To achieve the learning outcomes of the course, the mobile learning team identified five categories of SMS content that could be sent to students twice or thrice weekly. The categories are: course management, content, forum, tips and motivation. Each category plays a unique role and together they were expected to enhance the current blended mode of learning at OUM. It was also expected that tutors will benefit from the SMSes sent as it would engage students in various parts of the learning process. In addition, the use of social media tools, in particular, Twitter and Facebook were included. Twitter was primarily used to function as an archive for SMSes sent while Facebook was used to support social interactions, academic discussions and learning. The design and implementation of mobile learning in general and how it enhances the blend of learning for one compulsory first semester course, OUMH1103 (Learning Skills for Open Distance Learners) in particular is described next.

Design of Mobile Learning

OUM's mobile learning initiative was designed to enhance its current blend of learning approaches

Figure 1. OUM's blended learning model as enhanced by mobile learning

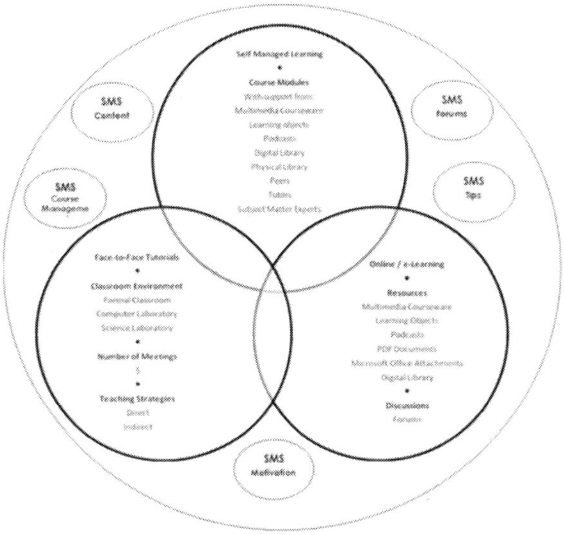

(see Figure 1). The current blended learning model comprises: self-managed learning, face-to-face learning and online learning.

As adult learners, students are expected to manage their learning independently. It means being able to balance and manage their time between work, family and study; monitoring their own performance as they go through the system; and ultimately achieving their academic goals. OUM provides students with print modules as a primary learning resource for every course offered. The module is a "tutorial in print," designed to engage the student in such a way that when he or she reads the module, it is as if the module is "talking" or "interacting" with the student in a personal way. In addition, several other forms of learning resources are made available for some of the courses. This includes multimedia courseware, learning objects and podcasts. In addition, students are encouraged to refer to books to supplement the print modules. These are available in the university library, in both print and electronic forms.

Face-to-face interactions are carried out during actual classroom sessions conducted at OUM

Learning Centres throughout the country. Known as *tutorials*, these classes are conducted five times per semester for a total of ten contact hours on weekends or evenings on weekdays. Tutorials are held fortnightly while students manage their own learning in between the tutorials using the learning resources provided and engaging in asynchronous discussions through online forums. During the tutorials, interactions include discussions of concepts and exercises from the module as well as discussions of course assignments. Discussions during the tutorials are continued virtually through interactions in the online forum discussions through the university's Learning Management System, myLMS. The objective of online discussions are to provide a collaborative learning environment where learners learn from each other and tutors monitor to ensure that learners are on track. In addition, the myLMS platform provides links within the course environment so that students can access the learning resources such as learning objects and podcasts.

The concept of mobile learning implemented by OUM neither duplicates the existing three modes, nor provides additional modes, but rather further enhances learning through the existing blended learning approaches. The objectives of mobile learning are to: (a) enhance the blend of learning modes at OUM, (b) increase the flexibility of learning offered to OUM learners, and (c) encourage and support ubiquitous learning (just in time, any time, anywhere) via mobile technologies. These are consistent with some of the recommendations reviewed in the literature review on the pedagogical strengths of mobile learning.

The first course offered through mobile learning was a compulsory course (OUMH 1103) on learning skills that is taken by new first semester students. OUMH1103 consists of 10 topics focusing on learning skills namely learning to learn, ICT and searching for information, time management, stress management, reading skills and so on. The course is aimed at the preparation of new students on how to cope as open and distance learners. Each topic in the course lists a set of learning outcomes which students need to achieve upon completing the topic. During tutorials, these learning outcomes are frequently referred to through quick feedback from short activities and questions. This is to continuously measure learners' progress and understanding of the topics discussed. This was the basis in constructing the content of the SMSes. There are three phases in the planning of SMS content for OUM's mobile learning course: (a) creating the SMS content, (b) scheduling the SMSes and (c) interacting on Facebook. These are described below.

PHASE 1: CREATING THE SMS CONTENT

It is important to define the purpose for each SMS sent to enhance the learning experience of the student. The five categories of SMS and purpose for each are described below.

Content

The purpose of this category is to help learners locate/remember important course facts easily and to review important content in the module. It supports the *self-managed learning* mode where learners are encouraged to refer back to the module whether to find the correct answers or for quick revision of what they have learned from the tutorial sessions.

Forum/Discussions

This category provides suitable stimuli for discussions. It supports the *online learning* mode or *virtual classroom* and contributes to an increase in learner participation in forum discussions.

Tips

Tips provide hints or strategies to learners on how to do well in their studies. These may or may not derive from the module. However, they help the learners manage their studies.

Motivation

Suitable motivational quotations are selected from a list of general quotations with regard to achieving success in life. The purpose is to motivate learners to persevere in the learning process and further contribute to retention of learners.

Course Management

For any course, there will be announcements pertaining to assignments, examination dates and others. This category provides timely announcements/reminders on these matters, thus helping the learners to remember important dates and actions to be taken related to the course.

The first two categories support self-managed learning. The SMS sent for the *content* category is usually in the form of statements or questions that will prompt students to perform a learning task such as reading or accessing an e-learning resource. SMS on *forum or discussions* are meant to prompt the students to engage in a series of postings in the myLMS online forums or Facebook. In the *tips* category, the SMS sent would inform the student on how best to remember a concept or formula. In the *motivation* category, motivational quotes are sent to inspire students to persevere, strive or stay on track. Typically, SMS sent for the *course management* category is one that is sent at the start, middle and end of the semester providing administrative details. The first SMS sent in the latter category is usually a Welcome message such as "Welcome to OUM. A series of free SMSes related to OUMH 1103 will be sent to you during the semester to help you in your studies." Another example of this category is, "Discussion on topics will be on Facebook. Register as a member and go to www.facebook.com/oumh1103 and register as a fan. Guide is available in myLMS." At the planning stage, that is, before the new semester starts, the entire SMS content for the course for the whole semester is created and charted in a template based on the semester's schedule. This will be emailed in advance to all tutors of the course.

Figure 2 is an example of how, based on the learning outcomes, the categories for SMS texts are identified and the respective contents are created. Figure 3 shows part of the entire SMS content by week and the dates the SMS will be sent to students.

PHASE 2: SCHEDULING THE SMSes

SMSes are scheduled to be sent during weekdays and after working hours. This avoids distractions while the students are at work or involved in family

Figure 2. Examples of SMS in Content and Forum categories for Topic 3: Reading for Information

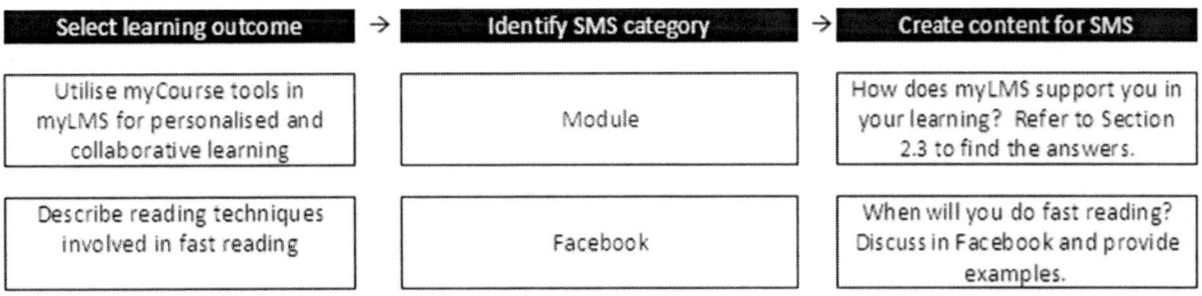

Figure 3. Example of SMS content and schedule for OUMH1103

DAYS	Week 3 (18 – 24 Jan 2010)	Week 4 (25 – 31 Jan 2010) (Topic 1 & 2)	Week 5 (1 – 7 Feb 2010) (Topic 1 & 2)
MONDAY	18	25 What are the strategies & advantages of the blended learning mode used in OUM? Discuss in Facebook.	FEDERAL TERRITORY DAY 1
TUESDAY	19 Welcome to OUM! A series of free SMSes related to OUMH1103 will be sent to you during the semester to help you in your studies.	26	2 How does myLMS support you in your learning? Refer to Section 2.3 to find the answers.
WEDNESDAY	20 Discussion on SMS topics will be in Facebook. Register as a member & go to www.facebook.com/oumh1103 to be a Fan. Guide is available in LMS.	27 Concentration is important to sustain learning. Identify causes of poor concentration and strategies to improve it. Refer to Topic 1.	3
THURSDAY	21 It is advisable to read the module before tutorials. You will understand better. Do read Topics 1 & 2 for Tutorial 1.	28	4 Have you participated in Facebook? If you haven't done so, refer to the guide in LMS or email for help to oumh1103@oum.edu.my.
FRIDAY	TUTORIAL 1 (22-24 Jan 2010) TOPIC 1 & 2	29 Setting goals and time management are important to your studies. What is your goal? How do you apply time management? Share in Facebook.	TUTORIAL 2 (5-7 Feb 2010) TOPIC 3 & 4
SATURDAY		THAIPUSAM 30	
SUNDAY		31	

activities on weekends. The SMSes in the *Content* and *Forum* categories are scheduled after the topics have been covered during the tutorials so as to enhance the students' learning and to reinforce what they covered in the face-to-face tutorials. On average, 25 – 30 SMSes are sent throughout the semester for OUMH1103. Every week, between two and three SMSes are sent around 7.30 pm.

To send the SMSes out, an SMS gateway is used. This is provided by an external vendor. Once the charting of SMSes is completed, as illustrated in Figure 3, a list of learners' and tutors' mobile phone numbers is obtained from the university's student database. The mobile numbers are keyed into Notepad (.txt file) and imported into the university's SMS delivery application, myWorkmate (See Figure 4). The next step in setting the

Figure 4. The SMS delivery application in the OUM ICT platform (myWorkmate)

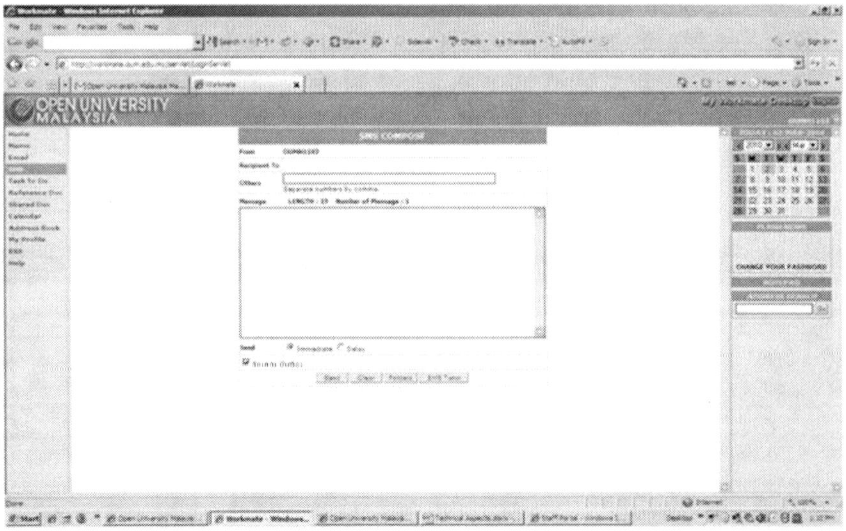

SMS for group messaging delivery is to copy the SMS text that was scheduled for delivery (of not more than 145 characters including spaces and not containing certain characters like inverted commas or quotation marks) into myWorkmate. The message may be sent immediately (as when a trial message is sent) or at a delayed time. Once the SMS has been uploaded, it would then be forwarded to an online SMS gateway provider responsible for registering the short code (which identifies OUM as the SMS content provider) with the various telecommunication service providers.

PHASE 3: INTERACTING ON FACEBOOK

SMSes that are designed to have students engage in the forum discussions are sent out once or twice a week. These discussions are to be managed or moderated by the mobile learning course coordinator. Students may be required to read their module before posting a message. The Facebook moderator is also a tutor teaching the course and will monitor the student's feedback while at the same time keep the discussions active, and as much as possible, give equal attention to all postings. The moderator's role is limited to the discussions pertaining to the content of the SMSes and not to replace the role of their designated face-to-face tutor who are available through the forums in myLMS. Figure 5 is an example of the discussions generated by two SMSes posted in the OUMH1103 Facebook.

Each semester, between 1200 and 1800 new students are registered for the OUMH1103 course. About 70 percent of the total number of new students typically registers as an OUMH1103 Facebook fan. It is not compulsory that students enrol as a fan. They will, however, need to interact in the myLMS forums with their own face-to-face tutor in the learning centre. Other than postings related to the SMS sent, the moderator also contributes some other relevant postings to keep the Facebook environment alive and meaningful. It was generally found that 20 – 30 comments are submitted by the students for each posting from the moderator. However, like any other social media platform, not all fans are actively interacting in the discussions. About 20 percent of the fans

Figure 5. Discussions in the OUMH1 103 Facebook based on SMS sent

make their presence felt each week. Some prefer to read the postings and some occasionally post opinion, comments or queries. The general page views are between 200 – 300 views per week. The number of views increases or decreases depending on holidays, assignment due dates, examinations, and so on.

Apart from using the Facebook as a discussion platform, the mobile learning team used Twitter to archive the SMSes for easy reference. Students who are not able to keep the SMSes in their mobile phones due to space limitation or did not receive it due to technical problems, may access all previously sent SMSes at http://twitter.com/oumh1103 (see Figure 6). The archive acts as a quick recap of SMSes based on selected topics or learning outcomes and may help the student when revising for the mid-semester and final examinations.

FEEDBACK AND EVALUATION OF THE MOBILE LEARNING INITIATIVE

To determine the success of the mobile learning initiative in meeting its objectives, a formative evaluation was conducted during the pilot in the May 2009 semester. A summative evaluation was next carried out at the end of the May and September 2009 semesters, respectively. These are reported below.

Formative Evaluation (May 2009 Semester)

In order to gain feedback on the pilot project, focus group discussions were held mid-way through the May semester. The discussions were held at six OUM learning centres representing the six geographical zones in Malaysia namely North, Central, South, East, Sabah and Sarawak. Four aspects were examined: (a) the students'

Figure 6. Use of Twitter to provide an archive of the SMSes sent to students

feelings about receiving SMS, (b) their actions upon receiving SMS, (c) their views regarding the frequency and timing of the messages, and (d) their perceptions of the usefulness of the SMS. A total of 51 students and 12 tutors participated in the discussions. Findings include:

1. Students felt good about getting the SMS, perceived that they were being cared for and that they were being given special treatment;
2. Majority was happy to receive the SMSes and looked forward to receiving SMS. Some reported that they promptly checked the SMS upon hearing the arrival tone at the particular time of the day when SMSes were sent;
3. They were not sure of what to do upon receiving the SMSes as they were not informed ahead about the initiative;
4. Upon receiving the messages, students said most of the time they acted as suggested (for example refer to a particular section in the module or go online to discuss certain issues) but their actions were delayed.
5. Some preferred the SMSes to be spaced out so that they did not get too anxious about their studies.
6. Perception of the usefulness of these SMSes varied. The majority of the learners were enthusiastic about SMS from the course management category, e.g. the scheduling of examinations, and reminders about deadlines for handing in assignments but were less keen on receiving SMSes on motivation and tips. Several said they could not understand the motivational quote sent because of the language and wanted simpler ones. An enthusiastic learner likened it to an alarm clock reminding them to prepare for tutorials.
7. A tutor reported that responses to the forum discussion seem to have increased tremendously and the postings showed that the SMSes extended the students' thinking as well as motivated them to share their thoughts online.

Based on the feedback obtained, implementation of the initiative for the second half of the semester was fine-tuned. This included (a) providing motivational quotes that were easier to understand, (b) sending tips related to the examinations, and (c) planning for more content-related SMSes to be sent near to the examination dates. The timing and frequency of sending SMSes were maintained. Other feedback such as the need for guidelines for learners and tutors were revisited in the second phase of the project during the following semester.

Summative Evaluation (May and September 2009 Semesters)

Summative evaluations were conducted at the end of the May and September semesters using a survey. Questionnaires were sent to all learning centres by post and they were administered by the tutors involved. The questionnaire was bilingual and comprised four sections with a total of 35 items including three open-ended questions. The four sections were: (1) Demographic data, (2) Perceptions of the mobile Learning project, (3) Impact on learning and (4) Overall impression. The Cronbach alpha coefficient obtained for the instrument during a pilot test was found to be .98 with the corrected item-total correlations ranging from + .48 to + .93. A five-point Likert scale was utilized for items on perception of usefulness and motivation, perceived effectiveness of the four categories of Content, Tips, Motivation and Course Management and perceived impact of mobile learning. The points on the scale were 1 for Strongly Disagree, 2 for Disagree, 3 for Neutral, 4 for Agree and 5 for Strongly Agree. Percentage agreement was calculated by adding the percentage obtained for Agree and Strongly Agree.

For the May semester, a total of 712 completed questionnaires were obtained out of a total of 1863 students who were involved in the project. For the September semester a total of 642 out of 1173 completed questionnaires were obtained.

Figure 7. Perceived effectiveness of the various categories of SMS (Percentage agreement)

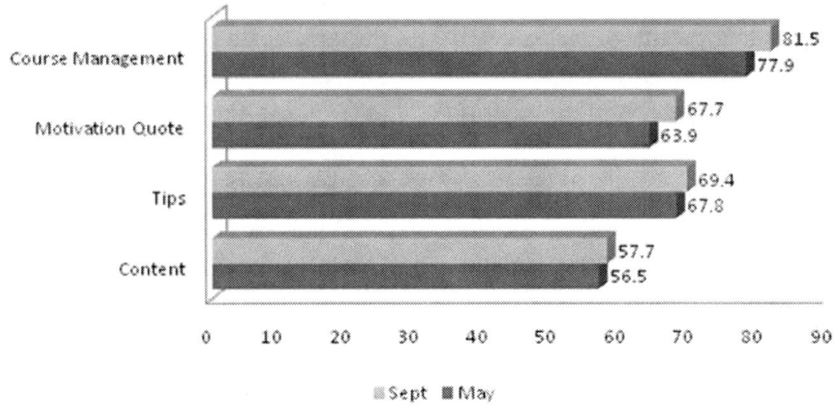

The number of respondents for both samples met a confidence level of 99 percent.

Findings from the surveys conducted at the end of the two semesters indicated that the majority of the respondents perceived mobile learning using SMS technology to have been useful in their learning with an average percentage agreement of 84.9 percent for May and 82.2 percent for September. It was also found that the SMSes had motivated them to learn (74.3 percent agreement for May and 69.4 percent for September).

The findings for perceived effectiveness of the various categories of SMS (Figure 7) showed that for the May semester, 56.5 percent of the respondents agreed that they remembered facts (Content) more easily because of the Mobile learning initiative as compared to 57.7 percent for the September semester. Meanwhile, 67.8 and 69.4 percent of the respondents for May and September respectively agreed that they got useful hints/strategies on how to proceed with their learning (Tips). As for the motivation quotes, 63.9 percent (May) and 67.7 percent (September) felt encouraged after reading the quotes (Motivation) and 77.9 percent (May) and 81.5 percent (September) reported that they were reminded of important details related to the course (Course Management). These findings are reassuring because learners seemed to be satisfied with the SMS, particularly with the Course Management text messages, of which 25.5 percent (May) and 27.5 percent (September) strongly agreed

As Figure 7 shows, the percentage of learners who found the various categories of SMS useful increased from the May to the September semesters. The range of percentage of agreement ranged from 56.5 percent to 77.9 percent for the May semester and 57.7 percent to 81.5 percent for the September semester. Of the five categories of SMS, learners reported that text messages that reminded them of important details related to their courses (Course Management) were found to be the most useful. At least a quarter of the respondents for both the May and September semesters had strongly agreed that the timely reminders helped them in their studies. Meanwhile, the Tips messages were perceived to be next most useful, followed by the Content messages.

As for the effectiveness of the SMS in getting learners to go online for discussions, it generally appears that after receiving the SMSes asking them to go into the LMS for discussion, the learners in the May semester did so more frequently than their counterparts who went into Facebook for discussions in the September semester.

As for the extent to which the initiative has impacted learners, the majority of learners (77.7 percent in May and 76.1 percent in September)

Figure 8. Learner Responses to SMSes on Online Discussions (Percent)

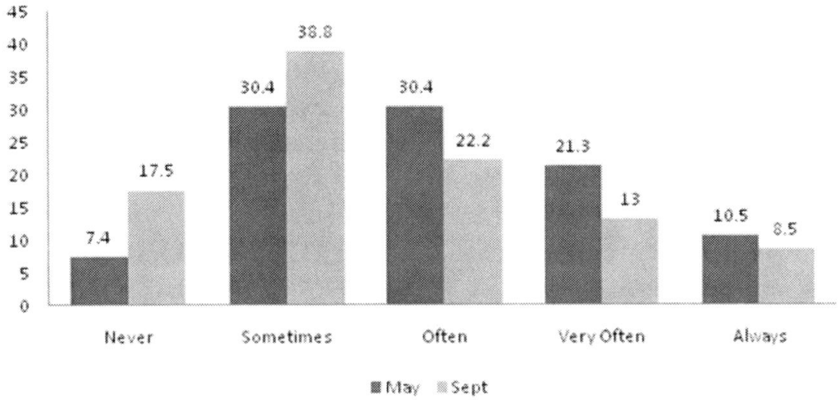

had felt it made their learning flexibile. In addition 77.5 percent (May) and 77.4 percent (September) of the respondents thought it had sustained their interest in the course. In addition, 76.3 percent (May) and 75.1 percent (September) of the learners felt encouraged to be focused in their learning. While 74.1 percent (May) and 75.5 percent (September) of the respondents agreed that the SMSes had added value to their learning experience, the majority 73.9 percent (May) and 72.9 percent (September) also believed that the SMSes had assisted them to become better self-managed learners.

Overall feedback from the questionnaire indicated that 81.4 percent of the learners for the May semester thought that the mobile learning experience had created a good impression of OUM as compared to 82.8 percent for the September semester. In addition for the May semester, 77.3 percent indicated that it had made a positive impact to their learning experience. The percentage agreement for this item increased to 78.1 for the September semester.

ISSUES AND CHALLENGES

During the first semester of mobile learning implementation for OUMH1103, discussions were conducted in the myLMS forum. Each forum was moderated by the face-to-face tutor in the respective learning centres according to the same tutorial grouping. At the same time, there was one other change for OUMH1103. Firstly, the award of marks for online participation that were previously given for learner participation (based on frequency and quality of postings) were discontinued. However, during the focus group interview in the middle of the semester, general feedbacks from students and tutors were encouraging. In response to whether the SMSes are helpful to them, most felt that the SMSes were 'good' and helpful. Although generally, tutors responded and supported the initiative, the level of interaction varied. Some forums were fairly active but others were less so. This was probably due to the fact that students no longer received marks for their discussing in the online forums. However, most agreed that the mobile learning should continue. They were pleased and wished to receive similar types of SMSes in future courses.

In the September semester, the mobile learning team tried something new. In addition to using the myLMS forum for discussions, the team tried out Facebook for discussions. Most students are Facebook account holders and are familiar with the interface. Unlike the individual forums in myLMS, Facebook allows interactions among students taking the same course throughout the country. As stated earlier, OUMH1103 is an entry level core course and these students are new to OUM and, most importantly, new to the learning environment. Open Distance Learning is a totally new world for them. It is expected that other academic related information from the courses, students need moral or social support from other students going through the same journey. The OUMH1103 Facebook provides a wider network of connections, not just course mated from the same learning centre. When usage of the Facebook was announced to the students before the first tutorial, the number of fans was low, about 5 percent of the student enrolment. This gradually went up to 50 percent by mid-semester. Discussions became more frequent when students who were in the Facebook started interacting about points made or messages received in the mobile. There was a marked increase after the mid-semester examination when several students commented that discussions in Facebook had helped them during the examination.

In addition, when first implemented, learners were questioning their need to go into the online discussions based on the SMS received because marks were no longer awarded for participation. However, the mobile learning team continued to push for discussions, explaining to learners the hidden benefit of participating actively online. When scheduling the SMSes for the semester, the "official" schedule was used comprising typical dates of the five tutorials. However, in reality, not all 61 of OUM learning centres kept the same tutorial dates. As such, the messages sent (particularly those related to content) may not match the learning outcomes of the course for those weeks. This was, however, not a major issue according to the tutors and students. There were three technical related issues however. This included:

1. Lack of response in getting learners to update their mobile phone numbers (mobile phone number is not a required field to be filled in during registration);
2. Receiving SMSes much later than the scheduled time;
3. Failed SMS delivery, particularly involving certain mobile operators.

SOLUTIONS AND RECOMMENDATIONS

Some students complained that they did not receive the SMSes. This could be due to the absence of their mobile phone numbers in the student database, failure to update numbers that were no longer in use or failure of the system in delivering the SMS to their mobile number, particularly if they are registered with certain mobile operators. Students are reminded through the myLMS and their tutors about the need to update the mobile numbers in the database and discussions with the SMS gateway provider has started to minimize the problem. It can be agreed that for any mobile learning projects to make an impact, implementation must include delivery of all information pertaining to the project and subsequently, content to all student mobile numbers.

In addition, all tutors who are supporting the face-to-face tutorials and online forums via myLMS need to understand how mobile learning works in tandem with the current blended learning modes. For this, the lead implementer of the project sent explanations via email regarding the objectives and how mobile learning will enhance learning, the benefits to students and how tutors can further support the initiative. However, some tutors still did not fully grasp the intent of the project. The best form of communication would have been via a face-to-face briefing and dialogue at each learning centre. However, it is not possible to cover all 61 learning centres to meet all tutors. This could be done on a regional basis but the expense involved to organize a regional meeting is not justifiable. One possibility is to create a set of guidelines on mobile learning and include FAQs (Frequently Asked Questions) in the guidelines. More detailed feedback should be collected from all tutors of mobile learning courses so as to produce a very useful set of guidelines so that, as tutors, they are able to optimize the students' learning opportunities through the blending of learning modes.

FUTURE RESEARCH DIRECTIONS

Mobile learning is a new experiment in some institutions. It has the potential to add value to existing modes of learning particularly in distance learning institutions where the number of students are larger per course and are scattered geographically. It provides standardized communication or interventions which can be designed to facilitate learner's understanding and help raise academic performance. Based on the fact that the attrition rate tends to be higher in distance learning institutions, mobile learning has the potential to reduce the attrition rate by reminding the receivers that they are enrolled in a tertiary institution and as such, have a few learning tasks to perform, examinations to study for and so on.

Hence, one valuable study in the future is to determine the relationship between mobile learning (based on the initiative by OUM) and attrition rate. It is believed that when mobile phones become more powerful but more affordable more students will have devices that will be able to do more. Coupled with affordable and the availability of greater bandwidth, mobile phones could be used to access resources on the internet, thus potentially adding more value to the learning.

CONCLUSION

The mobile learning initiative at OUM started by experimenting with one minute audio and video podcasts and later changed to use of SMS. The latter worked better than the podcasts and today, OUM has implemented mobile learning through the use of SMSes for a total of ten courses. The chapter highlighted the design and implementation of mobile learning for a compulsory first semester course on learning skills. The feedback from tutors and students has been positive. The initiative is an example of how SMSes can be designed to enhance learning and how it can be further supported by use of social media tools to enhance its blended learning model to produce the desired learning and help students achieve the learning outcomes. Specifically, the mobile learning initiative sought to: (a) enhance the blend of learning modes at OUM, (b) increase the flexibility of learning offered to OUM learners, and (c) encourage and support ubiquitous learning (just in time, any time, anywhere) via mobile technologies. Five categories of SMSes were formulated, each with a specific role to assist the student in understanding key concepts, prompt students to perform the necessary learning tasks, pace and manage their learning and provide useful tips and timely reminders. Since the decision to implement mobile learning came from top management, faculty members were supportive of the initiative. The challenges were thus mainly related to technical issues such as system delay in sending out the SMSes and failure of the SMSes to reach certain mobile numbers. However, the latter was overcome by using Twitter to act as an archive of all sent SMSes. It was also interesting to discover that Facebook can be used as a tool to bring students taking the same course (spread geographically across the nation) to a single virtual meeting place on Facebook and where a "friendly and safe" learning environment was created. The facilitator chosen to manage the interactions was a tutor whose friendly presence online made a difference to the students. The Facebook interactions are in addition to the usual student-tutor interactions via myLMS forums between the student and face-to-face tutor assigned to them. These are restricted to members of a tutorial group from the same learning centre. The mobile learning initiative is expected to grow and there are plans to continue offering mobile learning for other courses. Hence the total number of courses will increase. Meanwhile the mobile learning team will continue to fine tune the present implementation and explore other ways of increasing the effectiveness of mobile learning to further enhance OUM's current blended learning model to benefit its students.

REFERENCES

Abas, Z. W., Chng, L. P., & Mansor, N. (2009). A study on learner readiness for mobile learning at Open University Malaysia. In *Proceedings of IADIS International Conference on Mobile Learning* (pp. 151-57).

Caudill, J. G. (2007). The growth of m-learning and the growth of mobile computing: Parallel developments. *International Review of Research in Open and Distance Learning, 8*(2). Retrieved July 10, 2009, from http://www.irrodl.org/index.php/irrodl/ article/view/348/873

Devinder, S., & Zaitun, A. B. (2006). Mobile learning in wireless classrooms. *Malaysian Online Journal of Instructional Technology, 3*(2), 26–42.

Florida, R. (2005). *The world is spiky*. Retrieved July 13, 2009, from http://www.creativeclass.com/rfcgdb/articles /other-2005-The World is Spiky.pdf

Global, E. (n.d.). *Achievements of mobile learning today*. Retrieved July 10, 2009, from http://www.ericsson.com/ericsson/corpinfo/ programs/the_role_of_mobile_learning_in_european_education/products/ workpackage2.shtml

Hayes, P., Joyce, D., & Pathak, P. (2004). Ubiquitous learning: An application of mobile technology in education. In L. Cantoni & C. McLoughlin (Eds.), *Proceedings of World Conference on Educational Multimedia, Hypermedia and Telecommunications 2004* (pp. 1811-1816). Chesapeake, VA: AACE. Retrieved July 10, 2009, from http://www.editlib.org/p/12724

Herrington, A., & Herrington, J. (2007). *Authentic mobile learning in higher education*. Retrieved May 10, 2009, from http://www.aare.edu.au/07pap/her07131.pdf

International Telecommunication Union. (2010). *ITU Statshot, issue 1*. Retrieved February 17, 2010, from http://www.itu.int/net/pressoffice / stats/2010/02/index.aspx

Keegan, D., Kismihok, G., Mileva, N., & Rekkedal, T. (2009). *The role of mobile learning in European education*. Retrieved May 7, 2009, from http://www.ericsson.com/ericsson/corpinfo/ programs/the_role_of_mobile_learning_in_european_education/products/ workpackage4.shtml

Kirkwood, A. (1998). New media mania: Can information and communication technologies enhance the quality of open and distance learning? *Distance Education, 19*(2), 228–241. doi:10.1080/0158791980190204

Malaysian Communications and Multimedia Commission. (2010). *Facts and figures for cellular phone subscribers: Cellular phones in Malaysia*. Retrieved 23 February, 2010, from http://www.skmm.gov.my/facts_figures/ stats/ViewStatistic.asp?cc=51298166&srid =9247989

Markett, C., Sanchez, I. A., Weber, S., & Tangney, B. (2006). Using short message service to encourage interactivity in the classroom. *Computers & Education, 46*(3), 280–293. doi:10.1016/j.compedu.2005.11.014

Motiwalla, L. F. (2007). Mobile learning: A framework and evaluation. *Computers & Education*, *49*(3), 581–596. doi:10.1016/j.compedu.2005.10.011

Motlik, S. (2008). Technical evaluation report: Mobile learning in developing countries. *International Review of Research in Open and Distance Learning, 9*(2). Retrieved 10 July, 2009, from http://www.irrodl.org/index.php/irrodl/ article/view/564/1071

Nationmaster.com. (2010). *Media statistics: Mobile phones by country*. Retrieved February 17, 2010, from http://www.nationmaster.com/graph /med_mob_pho-media- mobile-phones#source

Nyiri, K. (2002). Towards a philosophy of m-learning. In *Proceedings of the 1st IEEE International Workshop on Wireless and Mobile Technologies in Education*, Växjö, Sweden, (pp. 121-124).

Parsons, D., & Ryu, H. (2006). *A framework for assessing the quality of mobile learning*. In Education (INSPIRE), Southampton Solent University, UK, 13 April 2006. Retrieved from http://citeseerx.ist.psu.edu/viewdoc/ download?doi=10.1.1.108.2612&rep=rep1 &type=pdf

Price, S. (2007). Ubiquitous computing. In Pachler, N. (Ed.), *Mobile learning: Towards a research agenda* (pp. 33–55). London, UK: WLE Centre, IoE.

Ramos, A. J. O. (2005). *The viability of mobile SMS technologies for non-formal distance learning in Asia*. Retrieved July 12, 2009, from http://[REMOVED HYPERLINK FIELD]www.pandora-asia.org/downloads/05-AAOU_Ramos.pdf

Ryu, H., & Parsons, D. (2009). *Innovative mobile learning: Techniques and technologies*. Hershey, PA: Information Science Reference. Retrieved April 7, 2009, from http://www.massey.ac.nz/~ryu/ acmelearn.pdf

Sharples, M. (2000). The design of personal mobile technologies for lifelong learning. *Computers & Education*, *34*, 177–193. doi:10.1016/S0360-1315(99)00044-5

Traxler, J. (2009). Current state of mobile learning. In Ally, M. (Ed.), *Mobile learning: Transforming the delivery of education and training* (pp. 9–24). AU Press, Athabasca University.

Uday Bhaskar, N., & Govindarajulu, P. (2008). Implications of mobile technology usage on learners in a learning process. *International Journal of Computer Science and Network Security*, *8*(5), 251–259.

ADDITIONAL READING

Alexander, B. (2004). Going nomadic: mobile learning in higher education. *EDUCAUSE Review, 39*(5), 28-35. Retrieved March 3, 2010 from http://www.educause.edu/EDUCAUSE+Review/EDUCAUSEReviewMagazine Volume39/GoingNomadicMobileLearningin Hi/157921

Cavus, N., & Ibrahim, D. (2009). M-learning: an experiment in using SMS to support learning new English language words. *British Journal of Educational Technology, 40*(1), 78-91. Retrieved February 22, 2010 from Ebscohost (Academic Source Premier).

Donnelly, K. (2009). Learning on the move: how m-learning could transform training and development. *Development and Learning in Organizations, 23*(4), 8-11. Retrieved February 22, 2010 from Emerald Management Xtra.

Duncan-Howell, J., & Lee, K. T. (2007). M-learning: Finding a place for mobile technologies within tertiary educational settings. In *ICT: Providing choices for learners and learning. Proceedings ascilite Singapore 2007*. Retrieved March 3, 2010 from http://www.ascilite.org.au/conferences/singapore07/procs/duncan-howell.pdf

Flowers, J. (2001). Online learning needs in technology education. *Journal of Technology Education,* 12(1), pp. Retrieved March 3, 2010from http://scholar.lib.vt.edu/ejournals/JTE /v13n1/flowers.html

Goh, T., & Kinshuk. (2006). Getting ready for mobile learning – adaptation perspective. *Journal of Educational Multimedia and Hypermedia,* 15(2), 175-198. Retrieved March 3, 2010 from http://www.editlib.org/index.cfm/files/ paper_6179.pdf?fuseaction=Reader.DownloadFullText&paper_id=6179

Huang, J. H., Lin, Y. R., & Chuang, S. T. (2007). Elucidating user behaviour of mobile learning: a perspective of the extended technology acceptance model. *The Electronic Library, (25)*5, 585-598. Retrieved March 3, 2010 from Emerald Management Xtra.

Laouris, C. Y., & Eteokleous, N. (2005, October 25-28). *We need an educationally relevant definition of mobile learning.* Retrieved March 3, 2010 from http://www.mlearn.org.za/CD/papers/ Laouris%20&%20Eteokleous.pdf

Lee, M. J. W., & Chan, A. (2007). Pervasive, lifestyle-integrated mobile learning for distance learners: an analysis and unexpected results from a podcasting study. *Open Learning, 22*(3), 201-218. Retrieved February 27, 2010 from Ebscohost (Academic Source Premier).

Low, L., & O'Connell, M. (2006). Learner-centric design of digital mobile learning. Paper presented at mLearn 2005 4th World Conference on mLearning: Mobile Technology: The Future of Learning in Your Hands, Cape Town, South Africa. Retrieved March 3, 2010 from http://www.eden-online.org/contents/publications/SMS/Ericsson. Mobile.A5.pdf

Nix, J., Russell, J., & Keegan, D. (n.d.). *Mobile learning/SMS (Short Messaging System) academic administration kit.* Retrieved March 3, 2010 from http://net.educause.edu/ir/library /pdf/ERM0532.pdf.

Pachler, N. (Ed.). (2007). *Mobile learning: towards a research agenda: The WLE Centre: Occasional papers in work-based learning 1.* London: The WLE Centre, Institute of Education. Retrieved March 3, 2010 from http://www.wlecentre.ac.uk/cms/files/ occasionalpapers/mobilelearning_pachler 2007.pdf

Porter, D. (2006, September 7). Innovations, trends and creativity in distance learning: using the social fabric of the web as a strategic lens to monitor trends and innovations. Paper presented at the 4th International Congress on Education and Technology – Inter-American University of Puerto Rico (UIPR). Retrieved March 3, 2010 from http://selma.bccampus.ca/Assets/BCcampus+Whitepapers/Innovations$!2c+Trends $!2c+and+Creativity+in+Distance+ Learning+report.pdf

Rozhan, M. Idrus, & Issham Ismail. (2008, April 16-18). Using SMS mobile technology for m-learning for physics distance learners at the Universiti Sains Malaysia. Paper presented at 3rd International Conference on Interactive Mobile and Computer Aided Learning, Amman, Jordan.

Scornavacca, E., Huff, S., & Marshall, S. (2009). Mobile phones in the classroom: if you can't beat them, join them. *Communications of the ACM, 52*(4), 142-146. Retrieved March 1, 2010 from Ebscohost (Business Source Premier).

Sharples, M. (Ed.). (2007). *Big issues in mobile learning: report of a workshop by the Kaleidoscope Network Excellence Mobile Learning Initiative.* Nottingham, UK.: Learning Sciences Research Institute. Retrieved March 3, 2010 from http://mlearning.noe-kaleidoscope.org /public/news/KALEIDOSCOPE%20 REPORT_07_Big_Issues_In_ Mobile_Learning.pdf

Sharples, M., Taylor, J., & Vavoula, G. (2007). A theory of learning for the mobile age. In Andrews, R., & Haythornthwaite, C. (Eds.), *The SAGE handbook of e-learning research* (pp. 221–247). Los Angeles, CA: SAGE Publication.

Somayajulu, B. K., & Ramakrishna, T. [2008]. *Distance learners and support services: current trends and prospects*. Retrieved March 3, 2010 from http://wikieducator.org/images/0/08/PID_25.pdf

Norbayah Mohd Suki, & Norazah Mohd Suki. (2007). Mobile phone usage for m-learning: comparing heavy and light mobile phone users. *Campus-Wide Information System, 24*(5), 355-365. Retrieved March 3, 2010 from Emerald.

Tarasewich, P., Gong, J., Nah, F. F. H., & DeWester, D. (2008). Mobile interaction design: integrating individual and organizational perspectives. *Information Knowledge System Management, 7,* 121-144. Retrieved February 27, 2010 from Ebscohost (Academic Source Premier).

Trondsen, E. (2005, February 24). Global perspective on mLearning. Paper presented at the Macromedia Higher Education Leadership Forum, San Francisco, California. Retrieved March 3, 2010 from http://www.sric-bi.com/news/ETrondsen-mLearning2005-02.pdf.

Wagner, E. D. (2005). Enabling mobile learning. *EDUCAUSE Review, 40*(3), 40-53. Retrieved March 3, 2010 from http://net.educause.edu/ir/library /pdf/ERM0532.pdf

Wang, M., Shen, R., Novak, D. & Pan, X. (2009). The impact of mobile learning on students' learning behaviours and performance: report from a large blended classroom. *British Journal of Educational Technology, (40)*4, 673-695. Retrieved March 1, 2010 from Ebscohost (Academic Source Premier).

Watkins, R., Leigh, D., & Triner, D. (2004). Assessing readiness for e-learning. *Performance Improvement Quality, 17*(4), 66-79. Retrieved March 3, 2010 from http://home.gwu.edu/~rwatkins/articles/PIQ2004.pdf

Wentzel, P., Van Lammeren, R., Molendijk, M., de Bruin, S., & Wagtendok, A. (2005). *Using mobile technology to enhance students' educational experiences: ECAR Case Study 2*. Retrieved March 3, 2010 from http://net.educause.edu/ir/library/pdf/ers0502/cs/ecs0502.pdf

KEY TERMS AND DEFINITIONS

Blended Learning Approach: A learning approach that comprises self-managed learning, face-to-face interactions with tutors and online learning (which includes learning using online resources as well as from online discussions with course mates and tutors).

Face-to-Face Learning: Learning that takes place in the physical classroom. Learner interact with tutors at designated times throughout a semester for tutorials and discussions.

Flexible Learning: The provision of flexibility to learners in the forms of approaches, methods, timing and media used. The learners are provided the opportunity to experience different or preferred modes of learning any time anywhere.

Mobile Learning via SMS: Short text messaging service whereby messages relating to content, course administration, and tips on how to manage learning, as well as suggestions on forum discussion and motivational quotes are sent to students via their mobile phones

Podcast: Short compressed video or audio files that are streamed via the Internet when downloaded. Viewable on personal computers or mobile devices.

Self-Managed Learning: Learning independently according to individual's time availability using both printed and digital learning materials and resources.

Social Media: Web-based platforms such as Facebook and Twitter that are used to support social interactions among members of an online community.

Ubiquitous Learning: Learning that occurs anywhere and at any time using mobile devices. Learners are not dependent on any physical set-ups, except for infrastructure that allows access to online or previously stored resources in their mobile devices.

Chapter 10
Promoting Learner Generated Content and Podcasting in Postgraduate Hospitality Education

Crispin Dale
University of Wolverhampton, UK

Ghislaine Povey
University of Wolverhampton, UK

ABSTRACT

The chapter aims to investigate the use of podcasting as a means of facilitating learner-generated content in hospitality management at post-graduate level. The research aimed to instigate critical skills development at this postgraduate level, which has been used in delivery at University of Wolverhampton, UK for two years, and addresses a gap in knowledge. Learner generated content is an innovation in this field, where students create 'learning objects' to share with their peer group and tutors. Theoretical perspectives of this technique are explored in the chapter and used to analyse the student experiences of generating a "mockcast" for a new gastronomic concept in a post-graduate hospitality management course. The chapter initially reviews podcasting technologies influence in education, before analysing and discussing the use of learner generated content in hospitality and the wider tourism and leisure subjects.

INTRODUCTION

The nature of the contemporary student has evolved significantly in recent years. Today's students have been referred to as digital natives who are technologically savvy, and process information differently from previous generations (Prensky, 2001). A society has emerged which reflects an "always-on" culture (Baird and Fisher, 2006) enabling a belief that interaction with

DOI: 10.4018/978-1-60960-511-7.ch010

information can occur, anytime, anyplace, and anywhere. This culture has largely been precipitated by the growth of new forms of social media and related technologies that enable engagement with networks of communities. As a phenomenon, social networking has arguably become a powerful means for interacting, communicating and learning (Barnes and Tynan, 2007). Indeed, its impact is proposed to have a pervasive effect across society, enabling people to engage as a shared community, whatever their age, gender or culture (Goodyear et al., 2003).

A generation of learners are entering higher education having been immersed in a range of electronic devices and gadgets, including personal computers, games consoles, personal music players, mobile phones and so on. These devices structure the daily lives of a generation who have been widely acknowledged as the "net generation" (Oblinger and Oblinger, 2005). The rapid emergence of learning technologies presents both opportunities and threats to a higher education system that has been predicated upon traditional lecture style approaches to learning and teaching (Bach et al., 2007). A traditional lecture is defined as a one-way process whereby the lecturer imparts information via speech and overheads, and students take notes. In this model, the lecturer is active and the student largely passive. Students can interact with the content by asking questions. Recent wide scale research into the adoption and use of learning technologies has noted how students have higher expectations and increasingly desire interactive experiences (Noss, 2008; Ito et al., 2008). Though Burgess and Mayes (2003, p.301) have acknowledged that "pedagogy will evolve to fit with the capabilities of the new technologies", challenges remain in the embedding of these technologies as part of the learning experience.

Higher education (HE) has been going through a period of significant change and has required teachers, educators and policymakers to be receptive to factors that are driving this change (Laurillard, 2008). Such factors have included educational policies concerning widening participation and the move towards blended modes of delivery (Motteram, 2006; Hughes 2007; Johnson 2007). Furthermore, the rising expectation of students, which has been driven by the payment of tuition fees, has placed an economic value upon education (Nulden, 2001). As a result the emphasis in HE has been towards enhancing the student experience and this is evidenced through the proliferation of evaluation and satisfaction surveys at module, course and institutional level (Douglas et al., 2006). In order to deliver an enhanced student experience, Higher Education Institutions (HEIs) have thus been empowered to become more dynamic and flexible in their approach (Laurillard, 2008). This has engendered a culture in some aspects of HE that explores new models of learning and teaching that "meet the needs of a generation of learners who seek greater autonomy, connectivity, and socio-experiential learning" (Lee and McLoughlin., 2007).

Central to these models has been the role of learning technologies to support learning delivery (Motteram, 2006). Electronic modes of communication including virtual learning environments and mobile and wireless devices have influenced learning engagement (Laurillard, 2006). These technologies have facilitated a process of "networked learning" (Steeples and Jones, 2002; Wise and Quealy, 2006) that involves the promotion of "connections" between learners, between learners and tutors and between the learning community and learning resources (Goodyear et al., 2006). However, according to Barnes and Tynan (2007) university teaching has fallen behind changes in the range of new technologies that have emerged.

PODCASTING IN EDUCATION

Podcasting was introduced by "The Podfather", an ex-MTV VJ, Adam Curry who sought a medium that would facilitate his audiences downloading broadcasts onto hand held, mobile technological devices such as iPods (Campbell, 2005). Since then, and despite mixed statistics concerning its growth, podcasting's use has grown radically, and for many downloading is now an everyday part of life. Neilsen/NetRatings (2006) research noted that there is a considerable percentage of the population who are not cognisant of podcasting technology. Despite this numbers of people downloading podcasts are projected to reach 56 million by 2010 (TDG, 2005).

There are many reasons for the growth in the podcasting phenomenon, according to Campbell these include:

1. The widespread adoption of internet usage worldwide
2. The facility to simply download large media files due to improvements in broadband infrastructure
3. The increasing growth and normalisation of personal computing in daily lives
4. A blurring of the distinction between downloading and streaming
5. The explosion in the uptake of iPod and mp3 devices (Campbell, 2005, p.38.)

When considering the user-creator viewpoint, a key benefit of podcasting is that it engenders a high level of independence, enabling the individual to take control and to be self-empowered. All internet users have the potential to create and distribute their podcasts on the Net simply and rapidly, and better still, without having to be conversant with sophisticated HTML code. Huann and Thong (2006) recognised this phenomenon, calling it the 'bottom-up' approach. In the academic context this gives the teacher or lecturer the opportunity to metamorphose into an 'educational broadcaster' with the opportunity to publish and circulate knowledge instantly, simultaneous to its creation. Jones (2006) identifies this podcast genre as 'coursecasting'. Huann and Thong (2006) illustrate this generation and publication of podcasts as a three-stage process. The content creation phase which includes the authoring of the podcast; the publication phase which includes the hosting of the podcast to a website; and finally the subscription phase which includes the use of aggregator software to capture the podcast.

With a rudimentary understanding of an appropriate technology it is reasonably simple to produce a podcast. These podcast episodes can be created using audio recording software such as Audacity or Garageband to record the dialogue being communicated by the lecturer/facilitator, which can be saved as an MP3 or similar audio file. This can then be either uploaded into a virtual learning environment (VLE) to enable student access, or transmitted through the medium of an RSS feed. The advantage of the RSS feed for podcasting is its immediacy. As soon as a subscription feed is established the RSS can instantaneously retrieve the podcast. This subscription feed is then linked to an aggregator such as iTunes, MediaMonkey, Doppler or Juice. The aggregators automatically search and retrieve any new updates from the Net when they are published. Our contemporary 'always on' culture ensures that the aggregator automatically searches for new feeds to which the user has subscribed (Hargis & Wilson, 2005). In terms of knowledge distribution once a subscription is established with an RSS feed, as soon as the host site has been updated, the subscriber will automatically get the information they desire (Ractham & Zhang, 2006).

EARLY STEPS IN EDUCATION

In 2004 Duke University in the United States used iPods with all their freshman students. This successful pilot highlighted how podcasting could be used as an innovative support for learning (Duke University, 2005). Since that debut iPods and podcasting have gone on to become a popular educational medium, taken up by numerous educational institutions worldwide (Blaisdell, 2006). It is seen by some to make a significant contribution to improving pedagogical approaches to aspects of both conceptual learning and information processing (Hargis & Wilson, 2005).

There are numerous educational uses for podcasting which have included student support for blended, online and distance learning, fieldwork, dissemination of lecture materials, assessment and feedback (Salmon & Edirisingha, 2008). Research into the use of podcasting in education has revealed some fascinating findings regarding the potential and rewards of podcasting, particularly looking at its benefits for use with a range of student learning styles and its value in the development of transferrable skills such as reflection and critical thinking, have been highlighted previously (Dale, 2007). Podcasting is also seen to be a valuable tool in engaging students from a diverse range of social backgrounds and cultures (Dale & Hassanien, 2008). The potential for enabling mobile learning has also been highlighted (Cooper *et al.;* Copley 2007; Evans 2008; Lee & Chan, 2007). This includes the potential for portability and flexibility of learning, the immediacy of communication, the ability to promote learner empowerment and active learning experiences and the anytime, anyplace connectivity (JISC, 2005). Further benefits of mobile learning can also include the enhancement of communication and collaborative learning, the engagement of reluctant learners and the raising of self esteem and confidence levels (Attewell, 2004; Liu et al., 2003; Naismithetal., 2004). Though it should be noted that students can reject the potential of listening to podcasts on the go in preference for listening on their home computers (Sutton-Brady et al., 2009).

Three broad types of educational podcast have been identified by Deal (2007). The first of these is the compiling of archives of lecture and seminar podcasts, either audio or video based as a resource for the students. Deal (2007)'s second category is the use of podcasting to deliver accompanying course materials to the students. Thirdly he identifies the role of podcasting as a student assessment method. During this review Deal (2007) concludes that the use of podcasting technology, and thus its value, is limited by the primary objective of its role as an aid to tutors and students in attaining their educational goals. Harris and Park (2008) divide educational podcasting into four categories; teaching-driven, where podcasting is used to create coursework and in the delivery of lectures; technology-driven, assisting technology adoption to aid teaching delivery methods; service-driven, delivering student support service information such as library resources and course details and lastly marketing-driven; where podcasting is used as a recruitment tool. In terms of podcast creation, the use of audio and multi-media broadcasts have found to develop academic creativity in teaching and learning (Dale, 2008; Lazzari, 2008; Middleton, 2009; Sutton-Brady et al., 2009).

One key attribute of the podcast is that it is easily shared amongst academic groups, and they have helped to facilitate the development of collaborative learning (Alexander, 2005; Ractham & Zhang, 2006). According to Lim (2005) it is particularly good for use with students who either prefer or have to learn 'on the go'. The podcasts put the students in control, enabling them to choose the time and place of learning (Baird & Fisher, 2006), thus reaching individuals with less traditional learning styles (Alexander, 2005). In fact, Cebeci & Tekdal (2006) assert that podcasting can enhance the appeal of learning, and engender inclusivity. It has shown that it can potentially satisfy the needs of individuals who require additional learning support, because for example they may have specific learning difficulties or do not have English as a first language (Schmidt, 2009; Sutton-Brady et al., 2009). Podcasting can also promote inter-cultural exchange (Lee, 2009).

It has been purported that podcasting offers benefits over written forms of communication (Abt & Barry, 2007). Research has also shown that podcasts can enhance student engagement and reflection (Baird & Fisher, 2006). It has even been proposed that the use of podcasting can generate a shared learning environment and thus lead to spontaneous creation of conceptual thoughts and ideas (Hargis & Wilson, 2005). Podcasting has also been shown to enable development of key skills including critical and analytical thinking, problem solving, communication and time management (Huann & Thong, 2006). An analysis of the influence of podcasting on language learning concluded that it augments accustomed behaviours, as students are familiar with the technology and its use (Thorne & Payne, 2006). Miller and Piller (2005) argue that courses who supply course materials that include audio elements have students who experience higher satisfaction. The quality of podcasts is noted as a concern as misinformation can result (Anderson, 2009). This therefore requires the use of information searching skills so as to ascertain the validity, relevancy and legitimacy of podcasts (Anderson, 2009).

Podcasting has also been found to alleviate student's anxiety levels regarding the subject matter (Ragusa et al., 2009), however the counter argument could be used, as the use of the new technologies has the potential to increase stress (Chan & Lee, 2005). Lane (2006) conducted research at University of Washington, where he found that podcasting enabled students to more fully understand teaching materials. Whilst processing sophisticated conceptual information via audio media is challenging for learners it enhances understanding of general arguments and opinions (Chan & Lee, 2005). It is also appealing for those learners who prefer to take a 'bite size' instrumentalist approach to study (Dale & McCarthy, 2006). Abt and Barry (2007) reported positive outcomes in exam performance when podcasts were included in materials for an undergraduate exercise physiology class.

Some studies into the use of podcasting in education have had mixed results. It has been reported however that some students were not clear about podcasts and their use, or how to access them (Copley, 2007). This study also found that using podcasts did not influence a student's propensity to attend lectures, as this provided interaction and a 'structured learning environment' (Copley 2007, p.398). Though, Fernandez et al (2009) note that podcasting has the potential to increase the feeling of permanent contact between tutors and students thus increasing their motivation.

LEARNING THEORY

Social constructivism acknowledges the importance of culture and context in forming understanding (McMahon, 1997; Wise and Quealy, 2006). Learning is driven through the social and cultural contexts through which the knowledge was constructed (Hung, 2001; Palmer, 2005). The word social is used as the construction of meaning is developed through interaction with others (Azzarito & Ennis, 2003; Hung, 2001; Hung et al., 2003) including adults and peers, in order to negotiate meaning (Palmer, 2005). The "truth or reality" of a situation emerges from the consensus of the constructions that emerge in a social grouping (Adams, 2006). Tutors are viewed as being pivotal in supporting learners and their social construction of meaning. The social constructivist school of learning builds upon the work of Vygotsky (1978). Vygotsky argued that learning is not a solitary activity but is "grounded in a system of social relations" (Macellan, 2005, p.139). He argues that tools, such as language, are used for social interaction and knowledge construction (Jordan et al., 2008). McGregor (2007) notes how social constructivists such as Vygotsky, "value and support the development of dialogic exchange because it is seen as pivotal in transforming cognitive activity into a more tangible form" (p. 56). The internalisation of this process enables learning to occur (Vygotsky, 1978) and critical skills are developed through the internalisation of dialogical argumentation (Ravenscroft, 2003).

Bruner (1987) used the term "scaffolding" to describe how a person's learning can enable them to enter the Zone of Proximal Development (ZPD) (Vygotsky, 1978). The ZPD "is the distance between the actual level as determined by independent problem solving and the higher level of potential development as determined through problem solving under adult guidance or in collaboration with peers" (Vygotsky cited from McGregor 2007, p.56). Learners can therefore co-construct extensions to their existing knowledge (Moran 2008) and the ZPD acts as a space where learners and teachers interact to develop knowledge (Adams, 2006, Jordan et al., 2008). It is argued that via the interaction with more advanced learners, students can develop their learning far more than if they were on their own (Palmer, 2005). Collaborative problem solving learning therefore becomes a social activity which develops a students learning (Neo, 2003). Furthermore, from a Vygotskian perspective, learning leads development and the teacher plays a more proactive role in the development of learning (McGregor, 2007).

Previous research into eLearning has proposed that scaffolding techniques are required for students to develop deeper learning skills (Allen, 2005; Cotton & Gresty, 2007; Dalgarno, 2001; Hughes and Daykin, 2002; Willet, 2007). Scaffolding can be achieved through the use of differentiated learning techniques (McGregor, 2007) and podcasting has the potential to develop a student's ZPD. Zembylas (2005) also acknowledges the role of "emotional scaffolding" and assisting students in their ability to cope in these learning situations. Constructivist learning theories have acknowledged how play can promote learning. Indeed, play can generate ZPD and develop actual to higher levels of learning (Bodrova, 2008). Indeed, Prentice (2000) notes how technologies have blurred the boundaries between work and play. Though it is acknowledged that students have clear boundaries when using their mobile devices (such as iPods) for leisure or work/learning (Lee and Chan, 2007). This is in contrast to recent research that suggests the blurring between work and play is a key factor in promoting creativity in learning and teaching (Dale, 2008).

It is acknowledged that traditional learning theories have emerged within the context of conventional methods of learning and teaching. They therefore do not fully take account of the influence of new technological innovations on learning engagement (Barnes and Tynan, 2007). These innovations are argued to have developed a generation of "digital natives" (Prensky, 2001) who are characterised as digitally literate, connected, prefer immediacy and experiential learning, are social, prefer team-based work that is structured and are visual and kinesthetic in learning style (Oblinger & Oblinger, 2005).

In forwarding his theory of connectivism, Siemens (2005) argues that traditional learning theories are concerned with the actual process of learning, not with the value of what is being learned (Siemens, 2005). Connectivism is based upon the premise that learning starts with the connections that students make and not with a fixed body of content (Barnes & Tynan, 2007). Learning is, therefore, actionable knowledge that can reside outside of ourselves and is facilitated through non-human appliances (Siemens, 2005).

Connectivism acknowledges the influence of chaos and networking theory in the process of learning. Siemens (2005) contends that learning is a process that "occurs within multiple overlapping environments of dynamic core elements that support the amplification of learning, knowledge and understanding through the extension of a personal network". The focus is on social interaction, connection and collaboration, as opposed to just the learning processes involved with the individual (McLoughlin & Lee, 2008). In the same vein as socio constructivist theory, the tutor's role is to act as a mediator in the learning process. However, in this context learning is not purely content driven, but begins and is maintained through the connections that students make (Barnes & Tynan, 2007).

Learner Generated Content

Constructivism tends to emphasise students' engagement in creating personally meaningful knowledge (Adams, 2004) and learner autonomy (Ewing, 2000). Technologies have been argued to promote opportunities for autonomy in learning (Downing, 2001). The development of learner-generated content in many respects takes a radical constructivist perspective (Perkins cited from Delgarno, 2001). Learners are given the autonomy and freedom to generate knowledge with the tutor acting as a facilitator of this knowledge (Dalgarno, 2001; Hunter, 2008). Thus, in contrast to traditional learning paradigms, Siemens (2005) argues that "learning is no longer an internal, individualistic activity".

According to Lee and McLoughlin (2007) the educational consumer has become a "prosumer" producing the knowledge that they consume. This articulation of understanding is argued to provide students with the development of academic learning (Laurillard, 2002). Through the development of Web 2.0 technologies, which have facilitated the creation of content, learners have become familiar with this approach thanks to commercial social sites such as Facebook, Wikipedia and YouTube. Students' potential to learn is greater when using learner-generated content, (Lee *et al.,* 2005) increasing the intellectual capital of the group as well as that of the individual.

A key tenet of connectivism is the potential for learners to generate knowledge (McLoughlin and Lee, 2008). Paavola and Hakkarainen (2005) contend that cognitive and social processes alone cannot account for the development of expertise and propose an alternative "knowledge creation view". Basing their research upon Paavola and Hakkarainen's (2005) three metaphors of learning; the acquisition metaphor; the participation metaphor and the knowledge creation metaphor, Lee et al. (2008) found that students developed greater potential for knowledge building when using learner-generated content. Lee et al comment that students "are there not to simply participate in activity and acquire skills, but also to produce shared outcomes and advance the intellectual capital of the group" (p. 510). From a constructivist perspective, podcasts offer the potential for students to challenge their existing beliefs and assumptions (Bullard, 2003; Lonn & Teasley, 2009). Students are able to construct meaning from the development of podcasts that are made with peers and this acts as a basis for scaffolding a students learning.

Power relationships between students and tutors have also been challenged, with learners generating and sharing knowledge with their peer group. This is in contrast to traditional written 'end products' of formative and summative assessment work which are not normally shared with peers. Yet there are many benefits to this process (Sener, 2007). Students, are required to differentiate between good content and that which is less valid (Dale and Lane, 2004). In this context, and consistent with social constructivist and connectivist perspectives, the role of the tutor becomes one of facilitator guiding and developing the students capability in determining the validity of knowledge. Furthermore, it is important for the tutor to facilitate the creation of knowledge, otherwise feelings of disempowerment amongst learners could potentially result (Lee et al., 2008).

THE CASE STUDY

The research focused on two annual iterations of a postgraduate module called 'Gastronomy'. This is a core module for students on the MA Hospitality Management or an option for those taking MA Tourism Management. The module teaches theoretical aspects of gastronomy and consumer behaviour in the international restaurant context. The modules ran over a twelve week period with taught material being delivered in the first half of the module and students using the second half to develop and generate their assessment materials. The students produced a 'mockcast' of a new gastronomic concept they had developed for the intended audience of potential investors worldwide.

Student views on the process and the assessment itself were gained by use of eportfolio blogs and focus groups. In addition to the analysis of the data screen grabs of the student's podcast creation are illustrated. The module ran over a twelve week period with the subject material and critical skills being delivered in the first part. In the second part of the module and to meet the assessment needs as outlined below, students were asked either in a small group or working individually to create a podcast to promote a new gastronomic concept to a group of potential investors. The idea was that this could be played by a potential investor, to help as a 'hook', to interest them in the concept proposal. The module incorporated a field trip to a trade Food Show, where students could meet potential investors in an educational setting to enable an understanding of their needs and wants when assessing new gastronomic proposals. Following the fieldtrip the students produced an embryonic 'mockcast' (that is a dummy podcast) of a concept they had encountered to enable them to prepare for the assessment, and use the software.

In creating their podcast students followed the same procedure as that delineated by Dale (2007). Using the Apple programme GarageBand (see www.apple.com/ilife/garageband/) the students created an enhanced podcast. GarageBand enables the user to embed audio, visual and weblink materials to the podcast. The students were given hands on guidance on how to use the programme during workshop sessions as a cohort. They also had the opportunity to see the tutor on a one-to-one basis for individual support. Prior to production students created a storyboard of their podcast. They were also expected to produce a mood board (See Figure 1) of the concept to help set the scene at the podcast presentation, where they were able to show the podcast. The students had to understand and use the underpinning concepts of gastronomy, as well as gather market intelligence data pertinent to their proposed concept. The innovative ideas proposed included Polish/Indian fusion, Steamed Food, Fusions of Types of Indian food, Just Eggs and many others.

An individual reflective blog was also completed by students, exploring their thoughts and

Table 1. Learning outcomes and assessment for the module

Learning Outcomes
It is expected that master's students will gain knowledge and develop understanding in the following areas:
• Mastery of the concepts of gastronomy in the context of the contemporary hospitality industry, communicating these complex concepts in an appropriate way. • Creatively apply gastronomic knowledge demonstrating mastery of research and enquiry into hospitality management at strategic and operational levels. • Demonstrate mastery of the gastronomic creative process, and critically analyse information from a range of sources, and synthesise an innovative gastronomic concept.
The assessment
Working either on your own or in small groups you should create a new gastronomic concept for a food or hospitality based business of your choice. You need to prepare a 30 minute presentation to a group of potential investors, who are also gastronomes. At this presentation you need to explain how you developed your concept and justify that choice with reference to theory.
You should also present a podcast to represent the concept to the audience. The podcast should include pictures and narrative, it may include background music and internet links if you wish. At the presentation you should provide written materials which back up your justifications, and clearly relate to underpinning theories and concepts.

Figure 1. Mood Board

feelings as they progressed through the process of creating their podcast. This research tool was adopted as according to Cobanoglu (2006) blogs can effectively generate student reflection on apposite topics. The students were directed to consider various aspects of the process when completing their blogs. It should be noted that the podcasts themselves were not marked, but students were asked to premiere their work at a formal summative presentation, which required them to present a logical rationale for their innovative concept, based in gastronomic theory, and with appropriate use of market intelligence data. These presentations were followed by focus group interviews, with each of the participant groups, which were recorded and consequently transcribed. A thematic content analysis was used on the ten focus group interviews and the twenty eight individual student blogs generated.

Evaluation

An enhanced podcast that could be played on hand held technological devices was created by each of the student groups (See Figure 2). This process enabled them to understand the value of

Figure 2. Example of an enhanced podcast

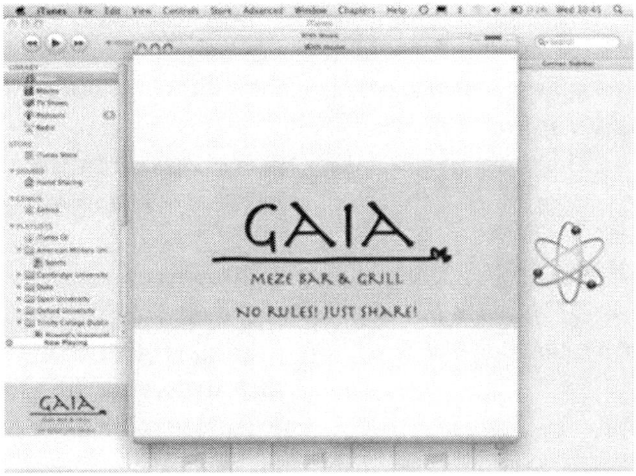

podcasting in the hospitality context, and also enabled them to learn from the innovative strategies of their peer group. The analysis of empirical data showed that the students were very satisfied with this type of learning activity as a whole. A number of themes emerged from the analysis including student's initial thoughts on the podcast assessment, the extent to which deeper learning was developed, podcasting's value as an assessment method, and the level of support that was needed. The postgraduate group comments were noticeably more critical of the process than those of undergraduates that have been studied previously (Dale & Povey, 2008).

Initial Impressions

The group had a full spectrum of feelings and thoughts when this assessment was first introduced. Despite a general familiarity amongst the group with the use of podcasts in their normal lives, they voiced a range of emotions and thoughts about creating a podcast as part of their assessment on the module. Students had not encountered podcasting at all in their previous study experiences. There was some concern about the use of Apple operating systems however many of the students had used iPod technology before. Some students expressed fear about the topic and subject matter as much as the assessment process itself. Blog comments included:

I was terrified and I was worried that I would not cope as I'm not a kitchen person and this was about the restaurant function, I'm front office and housekeeping.

My thoughts and feelings about developing a podcast for the assessment in this module are good. I'm looking forward to producing a podcast with my group.

When I first found out that we were going to be doing a podcast, I was excited about the process, as it was something different, compared to the reports, essays and presentations that we usually have to do for our assignments.

My thoughts on the outcomes of the podcast are generally good. I'm looking forward to preparing a podcast as all the aspects you have to create. For example, images, videos, voices and background music etc. It's a new experience in general for an assignment.

Scaffolding Learning

The students generally reported that having to produce the podcast made them think more deeply about the subject matter. From a social constructivist perspective, the selection of the groups was based upon mixed abilities and this gave the opportunity for students to interact and problem solve collaboratively with more advanced learners as the following comments suggest:

We worked within the gastronomy group together, and all kept a sense of team cohesion, spirit, goals and to take positive criticism constructively, with the more technically competent helping those of us who were less brave.

Though collaborative learning using technology can be a frustrating experience for some, as the following comment suggests;

Major difficulties of working with a partner, some wanted to be less responsible.

The experience developed the student's ability to enter their ZPD via a sense of play enabling them to be more creative in their learning and elevating their understanding of the subject matter. It was perceived to be more fun method of learning on the whole. There was a general consensus that this had motivated them to do more research and learning than other assessment methods as the following comments highlight;

It was good to be creative.

Podcast assessment is a very creative type of assessment, consequently, I had to use my imagination before creating the podcast

By doing podcast get practical experience of what were doing with music and emotions, much better than just doing PowerPoint slides.

It also promoted the students ability to conceptualise the problem under investigation;

It made me realise that it's important to have a gastronomic concept and how many hotels have poor restaurants that don't actually have any underpinning concept and direction. (Focus Group)

A number also said that the experience had motivated them to engage in visits to trade shows and hospitality venues to widen their gastronomic experience and further extend their knowledge;

We went to explore similar restaurants to experience the styles, and collect our pictures and videos that we could add to our podcast.

Furthermore this assignment has encouraged me to visit restaurants and food events that I would not have visited of my own accord, and it has reminded me how enjoyable it is to learn about new cultures. Moreover we were also able to take reading materials away with us from the food fairs, that not only gave us some useful information for our podcast but also allowed us to read further about how the restaurant is run (Student Blog)

The process enabled students to generate and build upon knowledge by making connections with further sources of information, on and offline, as the following comments suggest;

I went back to the books and the recipes involved and it greatly added to my knowledge, I neglected to understand gastronomy before.

I thought about different restaurants and even food fairs, and how to use podcast to promote them. I looked in the internet in order to find some podcasts.

Assessing Learning

When contrasted with other modes of assessment that were experienced, students found the novelty of producing a podcast to be a motivating factor:

At this point in the process the expectations of the podcasting process is still positive, finding the collection of information and images for the podcast to be motivating to learn and do the work. When putting the podcast together, the podcasting tool was still seen as a positive assessment to do.

Because of the fact that this assessment was different to other assessments, I was more curious and wanted to learn more about podcasting as well as about use of podcasting in hospitality and tourism.

However, it is acknowledged that the experience may have been influenced by the novelty effects of the adopted method, as the following focus group comments suggests;

The use of the podcast assessment did enhance my motivation to learn more because if it was something new. If I was just given a report I would have just left it till last minute, like I usually do, whereas the podcast motivated me to do it earlier because it was a new experience.

Further longtitudinal research should explore the extent to which this type of assessment impacts upon student learning and motivation. Indeed, some students offer contrasting opinions on their learning experience of the assessment;

Doing the podcast as part of the assessment it encouraged me to learn about gastronomy and it's development. However I'm not sure if I learnt any more or less than I would have done, if I was doing a report. (Focus Group)

Facilitating Learning

Students in some cases were critical of aspects of the process. This was reflected in their accessibility and use of the computer equipment. However, this in itself became part of the learning experience as the following comments insinuate;

'A' had a first go at the Apple Mac to figure out how we could record and add pictures on it. As she gradually got the grip of it, I followed her foot steps to learn. I found the technology on the Apple Mac confusing and difficult to comprehend. Getting acquainted with this new kind of technical skill was very challenging and quite a bit of work. We were ready to share knowledge, information and individual skills to kick off the big game! I think it worked as our major strength in the long run.

We started to encounter problematic situations like small errors while speaking our script eloquently which consumed time as work had to be repeated individually. Delays as the Apple Mac technology was confusing and new to us but eventually we anticipated overcoming these barriers by taking appropriate measures as we went through more practices on the Apple Mac. Our pro -active attitude towards learning the whole concept was optimistic and re paying at last as we managed to compile the work together to make it a success. Everyone worked hard towards this process and contributed to maximum ability and created productivity to achieve the objectives. I know for sure that I have learnt and developed a new skill but certainly I do need more practices to get completely familiarise with the equipment. This is a good way to assess a student's capability in a form of an assessment as it is challenging, thrives exictment, innovative, modern concept and dwells a skill in us. These are the modern techniques which we will be requiring in enhancement of careers.

Problems recording voices was difficult at times as we don't have a studio, we were in the learning centre and other students were larking about and we couldn't get the podcast done.

Facilitating the learning experience is an important part of students being able to develop their knowledge. Indeed, students requested further guidance in the area as the following comments suggest;

There were classes on how to use the podcasting, however obviously from the errors that kept being made. We could have done with more assessment guidance on it.

We needed a lecture on all the artistic things we could do on a podcast

CONCLUSION

The research conducted for this chapter investigated the use of student-generated "mockcasts" and the podcasting process as a form of assessment. Analysis of the findings implies that the experience of developing learner-generated content enhances the learning experience. The student's prospective employability is improved as the process enables them to develop a wide range of practical skills and theoretical understanding. Creating a podcast is a skill that the students can offer to a potential employer, which will add to their value as a candidate. The participants generally find that they undertake a process of wider theoretical research to deepen conceptual understanding of the subject matter and were also motivated to undertake more visits to real life situations and thus add to their practical experience. Some students initially perceived the software to be a barrier to engagement, but in all but a small number of cases this was overcome once they had hands on experience.

Whilst this study is small scale it generally agrees with the findings of other research by the author's in this area (Dale & Povey, 2009). The novelty of this assessment did skew the results as the uniqueness from the student perspective was seen as a key factor to enhancing their experience and as mentioned this requires further investigation longitudinally. Theoretically, the research contributes to the body of knowledge on constructivist and connectivist learning. In addition, it furthers our understanding of learner-generated content through the use of podcasting and how this can scaffold a student's learning.

An essential factor for all tutors considering using student-generated content in the learning and teaching strategies is the key need to make adequate support available to their groups, and that students need access to the relevant software and hardware. The postgraduate students were more strategically critical of the process than was found in research with undergraduate groups, and more appreciative of the opportunity to be creative. For some students this was the first time they had been able to incorporate any artistic or creative elements to their work, which for most added greatly to the experience. It was particularly pertinent as a medium to meeting the learning outcome for the module.

REFERENCES

Abt, G., & Barry, T. (2007). The quantitative effect of students using podcasts in a first year undergraduate exercise physiology module. *BioScience Education Journal*, 10. Retrieved June 1, 2008, from http://www.bioscience.heacademy.ac.uk/journal/vol10/beej-10-8.aspx

Adams, P. (2006). Exploring social constructivism: Theories and practicalities. *Education*, *34*(3), 243–257.

Alexander, B. (2005). Podcasting and the liberal arts. *The Newsletter of the National Institute for Technology and Liberal Education*, *4*(3). Retrieved on November 1, 2006 from http://newsletter.nitle.org/v4_n3_summer 2005/podcasting.php

Allen, K. (2005). Online learning: Constructivism and conversation as an approach to learning. *Innovations in Education and Teaching International*, *42*(3), 247–256. doi:10.1080/01587910500167985

Anderson, M. C. (2009). The medium is the messenger: Using podcasting to deliver consumer health information. *Journal of Consumer Health on the Internet*, *13*(2), 119–128. doi:10.1080/15398280902896428

Attewell, J. (2004). *Mobile technologies and learning*. Learning and Skills Development Agency. Retrieved from http://www.mlearning.org/docs/The%20m- learning%20project%20 %20technology% 20update%20and%20project%20 summary.pdf

Azzarito, L., & Ennis, C. D. (2003). A sense of connection: Toward social constructivist physical education. *Sport Education and Society*, *8*(2), 179–198. doi:10.1080/13573320309255

Bach, S., Haynes, P., & Lewis Smith, J. (2007). *Online learning and teaching in higher education*. Maidenhead, UK: McGraw Hill, Open University Press.

Baird, D. E., & Fisher, M. (2006). Neomillennial user experience design strategies: Utilizing social networking media to support always on learning styles. *Journal of Educational Technology Systems, 34*(1), 5–32. doi:10.2190/6WMW-47L0-M81Q-12G1

Barnes, C., & Tynan, B. (2007). The adventures of Miranda in the brave new world: Learning in a Web 2.0 millennium. *ALT-J Research in Learning Technology, 15*(3), 189–200.

Blaisdell, M. (2006). Academic MP3s: Is it time yet? *Campus Technology*. Retrieved on November 1, 2006, from http://campustechnology.com/article.asp ?id=18001

Bodrova, E. (2008). Make-believe play versus academic skills: A Vygotskian approach to today's dilemma of early childhood education. *European Early Childhood Education Research Journal, 16*(3), 357–369. doi:10.1080/13502930802291777

Bruner, J. (1987). *Actual minds, possible worlds*. Cambridge, MA: Harvard University Press.

Bullard, J. (2003). Constructivism: Does your practice match your conceptual framework? *Journal of Early Childhood Teacher Education, 24*, 157–162. doi:10.1080/1090102030240304

Burgess, H., & Mayes, A. S. (2003). Tutoring primary trainee teachers through e-conferencing. *Mentoring & Tutoring, 11*(3), 285–305. doi:10.1080/1361126032000138337

Campbell, G. (2005). Podcasting in education. *EDUCAUSE*, Nov/Dec 05.

Cebeci, Z., & Tekdal, M. (2006). Using podcasts as audio learning objects. *Interdisciplinary Journal of Knowledge and Learning Objects, 2*, 7–57.

Chan, A., & Lee, M. (2005). *An MP3 a day keeps the worries away-exploring the use of podcasting to address preconceptions and alleviate pre-class anxiety amongst undergraduate information technology students*. Student Experience Conference, Charles Stuart University. Retrieved on November 1, 2006, from http://www.csu.edu.au/division/studserv/ sec/papers/chan.pdf

Cobanoglu, C. (2006). An analysis of blogs as a teaching tool as perceived by hospitality management students. *Journal of Hospitality, Leisure, Sport and Tourism Education, 5*(2), 83–88. doi:10.3794/johlste.52.138

Cooper, S., Dale, C., & Spencer, S. (2009). A tutor in your back pocket: Reflections on the use of iPods and podcasting in an undergraduate popular music programme. *British Journal of Music Education, 26*, 85–97. doi:10.1017/S0265051708008280

Copely, J. (2007). Audio and video podcasts of lectures for campus-based students: Production and evaluation of student use. *Innovations in Education and Teaching International, 44*(4), 387–399. doi:10.1080/14703290701602805

Cotton, D. R. E., & Gresty, K. A. (2007). The rhetoric and reality of e-learning: Using the think-aloud method to evaluate an online resource. *Assessment & Evaluation in Higher Education, 32*(5), 583–600. doi:10.1080/02602930601116920

Dale, C. (2007). Strategies for using podcasting to support the learning of tourism students. *Journal of Hospitality, Leisure, Sport and Tourism Education, 6*(1), 49–57. doi:10.3794/johlste.61.155

Dale, C. (2008). iPods and creativity in learning and teaching: An instructional perspective. *International Journal of Learning and Teaching in Higher Education, 20*(1), 1–9.

Dale, C., & Hassanien, A. (2008). Podcasting in tourism education. *Journal of Teaching in Travel & Tourism, 7*(3), 43–60. doi:10.1080/15313220801909361

Dale, C., & Lane, A. M. (2007). A wolf in sheep's clothing? An analysis of student engagement with virtual learning environments. *Journal of Hospitality, Leisure, Sport and Tourism Education*, *6*(2), 100–108. doi:10.3794/johlste.62.156

Dale, C., & McCarthy, P. (2006). I like your style: The learning approaches of leisure, tourism and hospitality students studying generic modules. *Journal of Hospitality, Leisure, Sport and Tourism Education*, *5*(2), 48–58. doi:10.3794/johlste.52.116

Dale, C., & Povey, G. (2008). Learner generated content and podcasting. *Journal of Hospitality, Leisure, Sport and Tourism Education*, *8*(1), 117–123. doi:10.3794/johlste.81.214

Dalgarno, B. (2001). Interpretations of constructivism and consequences for computer assisted learning. *British Journal of Educational Technology*, *32*(2), 183–194. doi:10.1111/1467-8535.00189

Deal, A. (2007). *A teaching with technology white paper: Podcasting.* (Document ID: CSD5003). Retrieved October 4, 2007, from http://connect.educause.edu/files/CMU_Podcasting_Jun07.pdf

Douglas, J., Douglas, A., & Barnes, B. (2006). Measuring student satisfaction at a UK university. *Quality Assurance in Education*, *14*(3), 251–267. doi:10.1108/09684880610678568

Downing, K. (2001). Information Technology, education and healthcare: Constructivism in the 21st century. *Educational Studies*, *27*(3), 229–235. doi:10.1080/03055690120076510

Duke University. (2005). *Duke digital initiative.* Retrieved on November 1, 2006, from http://www.duke.edu/ddi/

Evans, C. (2008). The effectiveness of m-learning in the form of podcast revision lectures in higher education. *Computers & Education*, *50*, 491–498. doi:10.1016/j.compedu.2007.09.016

Ewing, J. (2000). Enhancement of online and offline student learning. *Educational Media International*, *37*(4), 205–217. doi:10.1080/09523980050210394

Fernandez, V., Simo, P., & Sallan, J. M. (2009). Podcasting: A new technological tool to facilitate good practice in higher education. *Computers & Education*, *53*(2), 385–392. doi:10.1016/j.compedu.2009.02.014

Goodyear, P., Asensio, M., Jones, C., Hodgson, V., & Steeples, C. (2003). Relationships between conceptions of learning approaches to study and students' judgments about the value of their experiences of networked learning. *ALT-J. Research in Learning Technology*, *11*(1), 17–27.

Hargis, J., & Wilson, D. (2005). *Fishing for learning with a podcast net.* University of North Florida. Retrieved on November 1, 2006, from http://www.unf.edu/dept/cirt/tech/podcast /HargisPodcastArticle.pdf

Harris, H., & Park, S. (2008). Educational usages of podcasting. *British Journal of Educational Technology*, *39*(3), 548–551. doi:10.1111/j.1467-8535.2007.00788.x

Huann, T. Y., & Thong, M. K. (2006). *Audioblogging and podcasting in education.* Edublog.net. Retrieved on November 1, 2006, from http://edublog.net/astinus/mt/files/docs/ Literature%20 Review%20on%20audio blogging%20and%20 podcasting.pdf

Hughes, G. (2007). Using blended learning to increase learner support and improve retention. *Teaching in Higher Education*, *12*(3), 349–363. doi:10.1080/13562510701278690

Hughes, M., & Daykin, N. (2002). Towards constructivism: Investigating students' perceptions and learning as a result of using an online environment. *Innovations in Education and Teaching International*, *39*(3), 217–224. doi:10.1080/13558000210150036

Hung, D. (2001). Theories of learning and computer-mediated instructional technologies. *Educational Media International, 38*(4), 281–287. doi:10.1080/09523980110105114

Hung, M. C. M., Mui, W. S. W., Wah, W. Y. L., & Ching, L. M. Y. (2003). Moving from a personal to a social constructivist view of learning: Implications for support in the student teaching experience. *Asia Pacific Journal of Education, 23*(1), 85–98. doi:10.1080/0218879030230106

Hunter, J. L. (2008). Applying constructivism to nursing education in cultural competence: A course that bears repeating. *Journal of Transcultural Nursing, 19*(4), 354–362. doi:10.1177/1043659608322421

Ito, M., Horst, H., Bittanti, M., Boyd, D., Herr-Stephenson, B., & Lange, P. … Robinson, L. (2008). *Living and learning with new media: Summary of findings from the digital youth project.* The John D. and Catherine T. MacArthur Foundation Reports on Digital Media and Learning. Retrieved from http://digitalyouth.ischool.berkeley.edu/report

JISC. (2005). Innovative practice with e-learning, a good practice guide to embedding mobile and wireless technologies into everyday practice. *JISC.* Retrieved from http://www.jisc.ac.uk/whatwedo/programmes /elearning_pedagogy/elp_practice.aspx

Johnson, G. M. (2007). Learning style under two Web-based study conditions. *Educational Psychology, 27*(5), 617–634. doi:10.1080/01443410701309159

Jones, D. (2006). *Enhancing the learning journey for distance education students in an introductory programming course.* Central Queensland University. Retrieved on November 1, 2006, from: http://cq-pan.cqu.edu.au/david-jones/ Publications/Papers_and_Books/react1.pdf

Jordan, A., Carlile, O., & Stack, A. (2008). *Approaches to learning: A guide for teachers.* Maidenhead, UK: McGraw Hill/Open University Press.

Lane, C. (2006). Podcasting at the UW: An evaluation of current use. *Catalyst.* Retrieved on November 1, 2006, from http://catalyst.washington.edu/projects/ podcasting_report.pdf

Laurillard, D. (2002). *Rethinking university teaching: A conversational framework for the effective use of learning technologies* (2nd ed.). London, UK: Routledge. doi:10.4324/9780203304846

Laurillard, D. (2006). E-learning in higher education. In Ashwin, P. (Ed.), *Changing higher education: The development of learning and teaching.* London, UK: Routledge.

Laurillard, D. (2008). Technology enhanced learning as a tool for pedagogical innovation. *Journal of Philosophy of Education, 42*(3-4), 521–533. doi:10.1111/j.1467-9752.2008.00658.x

Lazzari, M. (2008). Creative use of podcasting in higher education and its effect on competitive agency. *Computers & Education, 52*(1), 27–34. doi:10.1016/j.compedu.2008.06.002

Lee, L. (2009). Promoting intercultural exchanges with blogs and podcasting: A study of Spanish-American telecollaboration. *Computer Assisted Language Learning, 22*(5), 425–443. doi:10.1080/09588220903345184

Lee, M., & Chan, A. (2007). Pervasive, lifestyle integrated mobile learning for distance learners: An analysis and unexpected results from a podcasting study. *Open Learning, 22*(3), 201–218.

Lee, M. J. W., & McLoughlin, C. (2007). Teaching and learning in the Web 2.0 era: Empowering students through learner-generated content. *International Journal of Instructional Technology and Distance Learning, 4*(10). Retrieved on 24 January, 2008, from http://www.itdl.org/Journal/Oct_07/article02.htm

Lee, M. J. W., McLoughlin, C., & Chan, A. (2008). Talk the talk: Learner-generated podcasts as catalysts for knowledge creation. *British Journal of Educational Technology*, *39*(3), 501–521. doi:10.1111/j.1467-8535.2007.00746.x

Lim, K. (2005). *Now hear this–exploring podcasting as a tool in geography education*. Nanyang Technological University. Retrieved on November 1, 2006, from http://homepage.mac.com/voyager/brisbane_kenlim.pdf

Liu, T. C., Wang, H. Y., Llang, J. K., Chan, T. W., Ko, H. W., & Yang, J. C. (2003). Wireless and mobile technologies to enhance teaching and learning. *Journal of Computer Assisted Learning*, *19*, 371–382. doi:10.1046/j.0266-4909.2003.00038.x

Lonn, S., & Teasley, S. D. (2009). Podcasting in higher education: What are the implications for teaching and learning? *The Internet and Higher Education*, *12*(2), 88–92. doi:10.1016/j.iheduc.2009.06.002

Maclellan, E. (2005). Conceptual learning: The priority for higher education. *British Journal of Educational Studies*, *53*(2), 129–147. doi:10.1111/j.1467-8527.2005.00287.x

McGregor, D. (2007). *Developing thinking: Developing learning. A thinking skills guide for education*. Milton Keynes, UK: Open University Press.

McLoughlin, C., & Lee, M. J. W. (2008). The three p's of pedagogy for the networked society: Personalization, participation, and productivity. *International Journal of Teaching and Learning in Higher Education*, *20*(1), 10–27.

McMahon, M. (1997). Social constructivism and the World Wide Web-a paradigm for learning. Ascilite Conference, December 7-10, Retrieved on 1 October, 2009, from http://www.ascilite.org.au/conferences/ perth97/papers/Mcmahon/Mcmahon.html

Middleton, A. (2009). Beyond podcasting: Creative approaches to designing educational audio. *ALT-J. Research in Learning Technology*, *17*(2), 143–155.

Miller, M., & Piller, M. (2005). Principal factors of an audio reading delivery mechanism – evaluating educational use of the iPod. In P. Kommers & G. Richards (Eds.) *Proceedings of World Conference on Educational Multimedia, Hypermedia and Telecommunications 2005*, Chesapeake, VA: AACE. (pp. 260-267).

Moran, S. (2008). After behaviourism, navigationism? *Irish Educational Studies*, *27*(3), 209–221. doi:10.1080/03323310802242138

Motteram, G. (2006). Blended education and the transformation of teachers: A long-term case study in postgraduate UK higher education. *British Journal of Educational Technology*, *37*(1), 17–30. doi:10.1111/j.1467-8535.2005.00511.x

Naismith, L., Naismith, P., Lonsdale, G., Vavoula, G., & Sharples, M. (2004). *Literature review in mobile technologies and learning*. Future Lab. Retrieved from http://www.futurelab.org.uk/resources/ publications_reports_articles/literature_reviews/ Literature_Review203/

Neo, M. (2003). Developing a collaborative learning environment using a Web-based design. *Journal of Computer Assisted Learning*, *19*, 462–473. doi:10.1046/j.0266-4909.2003.00050.x

Nielsen/NetRatings. (2006). The 21st century UK digital consumer. *Nielsen/NetRatings MegaPanel UK Digital Consumer Survey*.

Noss, R. (2008). *Education 2.0? Designing the Web for teaching and learning*. Teaching and Learning Research Programme, www.tlrp.org/pub/documents/TELcomm.pdf

Nulden, U. (2001). Education: Research and practice. *Journal of Computer Assisted Learning*, *17*, 363–375. doi:10.1046/j.0266-4909.2001.00193.x

Oblinger, D., & Oblinger, J. (2005). Is it age or IT: First steps toward understanding the Net Generation. In D. Oblinger & J. Oblinger (Eds.), *Educating the Net Generation*. Educause 2005. Retrieved on 24 January, 2005, from http://www.educause.edu/ educatingthenetgen/

Paavola, S., & Hakkarainen, K. (2005). The knowledge creation metaphor—an emergent epistemological approach to learning. *Science and Education, 14*, 535–557. doi:10.1007/s11191-004-5157-0

Palmer, D. (2005). A motivational view of constructivist- informed teaching. *International Journal of Science Education, 27*(15), 1853–1881. doi:10.1080/09500690500339654

Prensky, M. (2001). Digital natives, digital immigrants. *Horizon, 9*(5), 1–6. doi:10.1108/10748120110424816

Prentice, R. (2000). Creativity: A reaffirmation of its place in early childhood education. *Curriculum Journal, 11*(2), 145–158. doi:10.1080/09585170050045173

Ractham, P., & Zhang, X. (2006). Podcasting in academia-a new knowledge management paradigm within academic settings. *SIGMIS-CPR, 06*(April), 13–15.

Ragusa, A. T., Chan, A., & Crampton, A. (2009). iPods aren't just for tunes: Exploring podcasting as a socio-cultural technology facilitating student experiences of higher education in rural Australia. *Information Communication and Society, 12*(5), 678–690. doi:10.1080/13691180802471471

Ravenscroft, A. (2003). From conditioning to learning communities: Implications of fifty years of research in e-learning interaction design. *ALT-J. Research in Learning Technology, 11*(3), 4–18.

Research, T. D. G. (2005). *Podcasting users to approach 60 million US consumers by 2010*. Retrieved November 1, 2006, from http://www.tdgresearch.com/press044.htm

Salmon, G., & Edirisingha, P. (2008). *Podcasting for learning*. Maidenhead, UK: Open University Press.

Schmidt, J. (2009). Podcasting as a learning tool: German language and culture every day. *Unterrichtspraxis, 41*(2), 186–194. doi:10.1111/j.1756-1221.2008.00023.x

Sener, J. (2007). In search of student-generated content in online education. *e-mentor, 21*, 1-8.

Siemens, G. (2005). Connectivism: A learning theory for the digital age. *International Journal of Instructional Technology and Distance Learning, 2*(1). Retrieved on 24 January, 2008, from http://www.itdl.org/Journal/Jan_05/ article01.htm

Steeples, C., & Jones, C. (Eds.). (2002). *Networked learning: Perspectives and issues*. London, UK: Springer.

Sutton-Brady, C., Scott, K. M., Taylor, L., Carabetta, G., & Clark, S. (2009). The value of using short-format podcasts to enhance learning and teaching. *ALT-J. Research in Learning Technology, 17*(3), 219–232.

Thorne, S., & Payne, S. (2006). Evolutionary trajectories, Internet-mediated expression, and language education. *CALICO Journal, 22*(3), 371–397.

Vygotsky, L. (1978). *Mind in society*. London, UK: Harvard University Press.

Willett, R. (2007). Technology, pedagogy and digital production: A case study of children learning new media skills. *Learning, Media and Technology, 32*(2), 167–181. doi:10.1080/17439880701343352

Wise, L., & Quealy, J. (2006). At the limits of social constructivism: Moving beyond LMS to re- integrate scholarship. Paper presented at the 23rd Annual ASCILITE Conference: Who's learning? Whose technology? Retrieved on 1st February, 2008, from http://www.ascilite.org.au/conferences/ sydney06/proceeding/pdf_papers/p158.pdf

Zembylas, M. (2005). Three perspectives on linking the cognitive and the emotional in science learning: Conceptual change, socio-constructivism and poststructuralism. *Studies in Science Education*, *41*, 91–116. doi:10.1080/03057260508560215

ADDITIONAL READING

Alonso, F., López, G., Manrique, D., & Viñes, J. M. (2005). An instructional model for web-based e-learning education with a blended learning process approach. *British Journal of Educational Technology*, *36*(2), 217–235. doi:10.1111/j.1467-8535.2005.00454.x

Ausburn, L. J. (2004). Course design elements most valued by adult learners in blended online education environments: an American perspective. *Educational Media International*, *41*(4), 327–337. doi:10.1080/0952398042000314820

Bull, M. (2005). No dead air! The iPod and the culture of mobile listening. *Leisure Studies*, *24*(4), 343–355. doi:10.1080/0261436052000330447

Condie, R., & Livingston, K. (2007). Blending online learning with traditional approaches: changing practices. *British Journal of Educational Technology*, *38*(2), 337–348. doi:10.1111/j.1467-8535.2006.00630.x

Deepwell, F., & Malik, S. (2008). On campus, but out of class: an investigation into students' experiences of learning technologies in their self-directed study, *ALT-J. Research in Learning Technology*, *16*(1), 5–14.

Denton, P., Madden, J., Roberts, M., & Rowe, P. (2008). Students' response to traditional and computer-assisted formative feedback: A comparative case study. *British Journal of Educational Technology*, *39*(3), 486–500. doi:10.1111/j.1467-8535.2007.00745.x

Downing, K. J., Lam, T., Kwong, T., Downing, W., & Chan, S. (2007). Creating interaction in online learning: a case study, *ALT-J. Research in Learning Technology*, *15*(3), 201–215.

Jelfs, A., Nathan, R., & Barrett, C. (2004). Scaffolding students: suggestions on how to equip students with the necessary study skills for studying in a blended learning environment. *Learning, Media and Technology*, *29*(2), 85–96.

Oravec, J. A. (2003). Blending by Blogging: weblogs in blended learning initiatives. *Learning, Media and Technology*, *28*(2), 225–233. doi:10.1080/1358165032000165671

Rosell-Aguilara, F. (2007). Top of the Pods - In Search of a Podcasting "Podagogy" for Language Learning. *Computer Assisted Language Learning*, *20*(5), 471–492. doi:10.1080/09588220701746047

Shim, J. P., Shropshire, J., Park, S., Harris, H., & Campbell, N. (2007). Podcasting for e-learning, communication, and delivery. *Industrial Management & Data Systems*, *107*(4), 587–600. doi:10.1108/02635570710740715

KEY TERMS AND DEFINITIONS

Connectivism: Theory that learning starts with the connections that learners make rather than the body of content, in the context of action knowledge being outside the learner, and can reside in non-human technologies.

Learner Generated Content: The development of media files that can be shared amongst learners through social media.

Podcasting: The process of broadcasting audio and/or multi-media files that can be downloaded to a personal computer or mobile device.

Social Constructivism: Theory that our perception of meaning is shaped by our culture and social environments, and our interactions with other people.

Social Networking: The building of a community between two or more users online.

Chapter 11
Closing in on Vocabulary Acquisition:
The Use of Mobile Technologies in a Foreign Language Classroom

Carly J. Born
Carleton College, USA

Andrea Lisa Nixon
Carleton College, USA

Christopher Tassava
Carleton College, USA

ABSTRACT

Vocabulary acquisition is one of the critical building blocks in acquiring foreign language fluency. While a number of studies have focused on effective vocabulary learning techniques for second language learners, several confounding factors complicate the practical application of this research in a classroom. For instance, faculty, pressed for time and results, frequently find it too cumbersome to explore new variations in their teaching and opt for standard methods of providing students with vocabulary lists which the student are expected to study on their own using their own methods. This tactic falters when the students are unaccustomed to second language learning and have not yet identified effective learning strategies suited to their own learning styles. This chapter will discuss one attempt to resolve this problem through the use of mobile devices as digital flashcards. This technological intervention may address the need to help students study vocabulary more effectively and do so in practical, sustainable ways that do not increase work loads for faculty, students, or academic technical support staff. Based on the results from a small-scale study, the authors make recommendations about this pedagogical approach and the technology used, aiming toward the goal of creating a pedagogically sound and scalable application of mobile devices in foreign language learning.

DOI: 10.4018/978-1-60960-511-7.ch011

INTRODUCTION

Carleton College is a small residential liberal arts institution for undergraduates in Northfield, MN. Our institution requires foreign language study, which can be met through study in any of five language departments which offer a total of ten different languages (Department of Asian Languages & Literatures: Chinese, Japanese; Department of Middle Eastern Languages: Arabic, Hebrew; Department of Classical Languages: Greek, Latin; Department of Romance Languages & Literatures: French, Spanish; Department of German & Russian). For instance, a student can complete their foreign-language requirement in the French & Francophone Studies Department by completing four classes – French 101, 102, 103 and 204. Students should achieve intermediate fluency in the target language at the end of the sequence. Classes are conducted five days a week over three 10-week terms for the first academic year of the language sequence. Students are expected to reach an intermediate level of language control by the third term, progressing to a high level proficiency by the end of the required sequence in their second year. To achieve this level of achievement, students are expected to learn a large amount of material very quickly, finishing the entire elementary book in two 10-week terms. Given this time pressure and the pedagogical goals of the sequence, our instructors are constantly exploring new ways to meet the goals of their curriculum, such as new technologies that can help students acquire language basics more quickly. This allows students and instructors to spend more class time on communicative activities designed to increase students' ability to become high-functioning speakers of French.

It goes without saying that a student must acquire a large amount of vocabulary in order to achieve a level of proficiency in the target language. At the college level, vocabulary is not frequently tested discretely, but embedded into grammar, writing, and speaking exercises in which students must demonstrate the ability to use both the pertinent vocabulary and the grammatical and culturally appropriate knowledge. These activities tend to be part of "an explicit learning/teaching paradigm" (Ma, 2009, p. 114) in which students also study vocabulary separate from other aspects of language, such as grammar forms or syntax. Instructors expect students to learn vocabulary through drilling as well as other, implicit contextual techniques, such as dialogs or short written pieces that include new vocabulary. However, in many cases, faculty shy away from suggesting specific strategies for vocabulary acquisition, instead trusting students to discover methods that work best for themselves.

For students, then, vocabulary study can become a monotonous and time-consuming area of study. Many students attempt to learn new words through a wide variety of methods, including rote memorization or creative mnemonics (Ma, 2009). Even as students struggle to find more effective ways to acquire more vocabulary, instructors strive to provide students with more opportunities to use their new knowledge, hoping that greater opportunities for comprhensible input (Krashen, 1981, 1985) and production of the foreign language will lead to a higher rate of achievement (Krashen et al).

It is at this intersection of needs that both faculty and students may turn towards technology. Students hope that technology will make it easier for them to acquire new language skills, while foreign language faculty hope that it can help streamline the content delivery so that class time can be focused on interactive and communicative tasks. One of the most common myths believed by both students and faculty is that the use of technology – any technology – will automatically meet the need to learn more information more rapidly. However, since "technology is theoretically and methodologically neutral" (Blake, 2008, p.11), the mere adoption of the newest language learning package will rarely meet these goals. It becomes critical to examine not only the theoretical methodology behind a particular implementation of technology, but also the effects that this implementation may have on a variety of learners.

This chapter explores one particular use of mobile technology as a supplementary tool to classroom learning of French as a second language. We examine the theoretical framework of the project, the results of a descriptive analysis of student achievement after the implementation, and a number of practical issues associated with implementing large-scale mobile technology support for students at a residential college.

BACKGROUND

Technology has become an integral part of our daily lives, and as such has begun to infuse the modern college curriculum almost out of necessity. Aside from helping to achieve intended pedagogical goals, one of the most interesting side effects of computer-mediated assignments is the incidental technology skills training that students receive. These technology skills can arguably be cited as just as important as the stated content goals of an assignment. In today's world, our students will face even greater expectations to know how to effectively manipulate a wide variety of technologies for personal and professional uses. This, coupled with the volumes of research suggesting greater academic gains with the appropriate use of technology, suggests that instructors cannot afford to shy away from incorporating technologies into our teaching.

The models for incorporating educational technology vary nearly as widely as the disciplines that are taught. Indeed, it is the particular discipline and the intended audience that help to drive the exploration of new technological applications. Initial efforts to take advantage of new technologies primarily focused on task-level interactions, giving language students computerized grammar drills to practice or asking elementary students to hand in typewritten compositions. Distance education in particular has enjoyed a renaissance as it has moved almost completely away from the traditional correspondence course model to full online delivery, complete with asynchronous and synchronous interactions and nearly immediate feedback. Recent advances in "voice over IP" (VoIP) technologies such as Skype have allowed professors to make themselves available via 'virtual office hours' to their students, even at residential colleges.

With such a vast sea of technologies available to educators the pedagogical possibilities are literally endless. Yet educational technology as a field of study is still young and many educators have not had the opportunity to pursue training or study in this area to inform their own teaching. Faculty at colleges and universities are often left to explore the use of educational technology on their own, taking little time to fully understand the wide variety of applications that could further their pedagogical goals. As a result, many applications of technology in higher education classrooms are tentative explorations of new lecture delivery tools (such as PowerPoint or podcasts), or as digital versions of the same assignment types typical to their discipline (e.g., electronic submission of a paper that the professor prints to grade).

The nature of foreign language education raises a number of additional issues that make it both difficult and critical for educators to explore technology. Students of a foreign language are generally expected to become familiar with all aspects of language use, including reading, writing, speaking, and aural and cultural comprehension. As such, foreign language educators have long used a variety of media to instruct students in these different modes of language, and have rejoiced in the new multimedia capabilities of computers in recent years. Audio components have always been helpful, but the recent proliferation of web cameras has opened up a new realm for foreign language teaching by allowing students to not only hear the spoken word, but see a speaker's facial and body language cues.

The embrace by language educators of technology has led to the birth of a subfield of study in second language learning research: Computer Assisted Language Learning. CALL is a growing field of academic inquiry into the effects of technology in teaching foreign languages. To date, CALL research has explored a wide variety of topics and spawned a number of peer-reviewed journals devoted to this research. Much of the ongoing research in CALL focuses on computer-mediated communication facilitated by chat programs, email, VoIP applications, virtual worlds or asynchronous text-based bulletin boards. These methods have obvious appeal for foreign language learners who are frequently far removed from native speakers of the language of study. Educators continue to explore these technologies as vehicles for introducing students to authentic language (Johnston, 1999) and to increase students' motivation to achieve greater success in language learning (Gardner, 1985).

But the history of CALL also includes the use of technology as supplementary tools in explicit language instruction, in the form of vocabulary or grammar drills, spelling and grammar checking, or electronic dictionaries. Computers are well-suited to these kinds of basic reference tools or closed-response drills (multiple-choice exercises which can be graded by a computer). And since the overall aim of foreign language instruction is to foster the development of communicative skills, there is still a place for explicit exercises in a well rounded instructional model (Ma, 2009). Examples of explicit instruction include the so-called drill-and-kill exercises commonly found in beginning language sequences, such as fill-in-the-blank grammar or conjugation activities.

Learner Differences and Mobile Assisted Language Learning (MALL)

Adding technology, including mobile learning technologies, to language learning environments creates a rich, not to say tangled, environment for research into pedagogy. Three particularly compelling areas of study are the utility of mobile learning technologies to provide multimodal delivery of content, to attend to different learning styles, and to increase learner motivation.

Soo (1999) nicely synthesizes research on learning styles in the context of CALL learning environments, noting that "multimedia lessons appear to be able to address the modalities of a large number of learning styles simultaneously" (p. 299). The ability to deliver aural, visual, and kinesthetic cues through a single piece of instructional material – for example, a video clip – makes multimedia learning objects appeal to a large number of students in a single classroom. Extending this idea to a mobile learning environment, students can not only use materials appropriate to their learning style, but take those materials, on the mobile device, with them wherever they happen to be – potentially increasing the frequency with which students access the material.

However, as Shih (2007) points out in his research comparing mobile language learning to comparable online learning modules, students must take time to adapt to the capabilities of a mobile learning environment, a process which can sometimes be frustrating enough for them to give up on the medium. He also suggests that instructional materials need to be delivered in smaller chunks to suit the capabilities of mobile devices. Additionally, Peters, Weinberg and Sarma (2009) found, in their research on the perceptions held by students in Canadian French classes, that students will only continue to prefer using technologies if they perceive the usefulness of the activities (p. 888). This confirms that technologies must be chosen both on the functionality appropriate to the pedagogical goals of the curriculum and the ease of use and design.

Motivation and Mobile Assisted Language Learning (MALL)

Ellis (1994) in his review of the field of second language acquisition points out that a learner's motivation to learn the second language "is influenced by external factors" (p. 36) in additional to factors intrinsic to each individual. However, motivation is a broad category that has been the subject of study in psychology and second language acquisition for decades. Gardner (1985), Oxford and Shearin (1994), and others have all posed varying explanations about different types of motivation that affect the success of foreign language learners. All agree that learner motivation is a key factor to academic success, even as – or because - a learner's motivation can change over time.

Specifically, a students' *instrumental motivation* (incentive-driven motivation, e.g. receiving a reward for achievement) can change over time to a more *resultative motivation* (i.e. a learner's achievement motivates them to learn more). (For more on motivation in second language acquisition, see Ellis 1994.) When mobile devices are involved, students may feel an instrumental motivation to study because using the mobile device gives them joy (because it's new, it makes them look cool, etc). However, educators know that instrumental motivation centered on technology rarely lasts long and must be replaced by resultative motivation or other intrinsic motivations for the student to achieve success. Thus, students' perceptions of usability and usefulness play a key role in their willingness to adopt and continue to use the mobile platform.

A growing body of research into the use of mobile devices in language education examines both the modality of input (e.g., Sydorenko, 2010) and the effects of the delivery platform of learning materials. In a study just published, Stockwell (2010) studied the effect of platform on the use of vocabulary exercises for students of English as a second language in Japan over the course of three years. Students were given the freedom to choose between completing the exercises on their mobile phone or on their desktop computer and allowed to change freely throughout the term. Stockwell found that adoption of the mobile platform over the desktop computer platform increased significantly in the third year of his study, even though full use of the mobile platform required students to take significantly more time to complete the activities. He offers one possible explanation for the jump in usage of the mobile platform by suggesting that students had started to take advantage of the mobile medium in the time and places that made it most useful (when commuting on a train to school, for example).

iPOD TOUCH PILOT PROGRAM DESIGN

Based on these theories and research, and on our own experiences working with undergraduate students at our institution, in the spring of 2009, we began a small pilot program to investigate the usefulness of mobile devices as a supplement to the existing French curriculum at the college.

In the spring of 2009, we acquired a stock of iPod Touch devices, 20 of which were allocated to the foreign language academic technologist for use in an experimental study of the use of the devices for digital flashcards. We chose to use the devices to deliver digital flashcards for vocabulary study for a number of reasons. First and foremost, the flashcards would require almost no adjustment to the existing, successful French curriculum and would therefore not detract from current practices. The French faculty already encouraged students to make use of paper flashcards, so this was a natural extension of a pre-existing technique to the digital realm.

Second, it seemed likely that vocabulary acquisition could be a natural fit for the hand-held, inherently mobile iPods. We expected that students would carry the iPods with them most of the time, using them for entertainment as well as for French study. Given this expectation, we hoped that students would take advantage of even brief opportunities to study their flashcards while standing in line or waiting for class to start. While most of the research into input frequency centers on grammar acquisition (Ellis, 1994), we hope this increased frequency of contact with French vocabulary could potentially impact their ability to retain vocabulary.

Third, vocabulary acquisition is an area that is easily transferrable to any language learning class. If our project was successful, we would easily be able to expand the program to encompass any of the other 10 languages taught on our campus.

Our initial work included the establishment of a relationship with the developer of a digital flashcard application called Mental Class. Based in part on the feedback from our pilot study, the software developer designed an educational version of the software that supported features facilitating easy creation and distribution of large flashcard sets. In fall 2009, we took advantage of these new features to manage a larger number of participating students. With as many as 50 iPods available to us, we were able to offer the devices to all students enrolled in Elementary French 101.

The Mental Class application was chosen in particular for its ease of use and pedagogically sound design. In addition to supporting the ability to maintain separate lists (known as *cases*) per chapter, the software provided students with a "Lesson" option that included one or more cases. When the "Lesson" option is enabled, Mental Class prompts students to study the marked cases at a spaced-repetition algorithm, similar to the well-known Leitner system of flashcard study. In addition to a randomizing feature and the ability to reverse the direction of the flashcards (e.g., show English prompts rather than French), Mental Class allowed students to mark individual cards in a study session to be randomly repeated later in the session until they were marked as learned.

Working in conjunction with professors teaching Elementary French 101, we developed a program to explore the impact of digital flashcards on test scores for the students who participated in the study. Faculty identified the required vocabulary, and the academic technologist prepared digital versions of the lists in a format compatible with Mental Class, so that students could download the lists directly to their iPods. In this iteration of the program, flashcards were created with French on one side and the English translation on the other.

Participating students were asked to complete both a short background survey on their language learning history and an online version of the Learning Style Survey developed by Cohen, Oxford, and Chi (2006). In a pre-test/post-test format, students completed the first two chapters of the textbook in the first half of the term without the iPods, after which iPods were distributed for the second half of the term. At the end of the term, the iPods were collected from the students and usage data for the Mental Class flashcard application was collected directly from the iPod backup files. The usage data includes both application launch counts and number of minutes the application was used. Test scores from four chapter tests, two before the iPods were distributed and two after, were also collected for all students.

Student participation in the study was optional and was made available to a total of 45 students enrolled in French 101. The total number of students included in the study who returned both surveys was 25 (56% of students in French 101), while only 10 (22% of students in French 101) chose to make use of the iPods in the second half of the term. Prior to discovering how to obtain application usage statistics from the devices themselves, students were asked to complete study logs for one week in the first half and one week in the second half of the term. The return rate for the study logs was rather low, and was a likely deterrent for participating in the study. In future versions of this program, study logs will not be required.

Initially, instructors and students expressed concern about the amount of technology training and support that would be required to ensure students would be able to use the iPods without incident. Students were told to bring all technical issues to the academic technologist who authored the flashcards, and that malfunctioning devices would be reset to factory default and reconfigured with Mental Class. Prior to the launch of the program, the technologist also prepared a study guide for students including links to iPod Touch how-to articles on Apple.com and a home-made screencast on how to use the software. As a result, only one student experienced technical issues with their iPod and all ten participating students were able to learn Mental Class software without individual training.

STUDY ANALYSIS

We first conducted a descriptive analysis of the data gathered from the devices and through the four chapter exams. Of the 25 students who completed the Language Learning Style Survey, 15 students showed a preference towards visual learning, at least in part. Six students showed a preference towards auditory learning, and seven students showed a preference towards tactile/kinesthetic learning. It is important to note that the survey allows students to express a preference for more than one learning style.

The authors then grouped students into two groups, 10 students who used the iPods ('iPod Group') and 29 students who did not ('No iPod Group'). Figure 1 shows mean test scores of the group of students who opted to use the iPod Touch devices and those that did not, showing the distribution of individual test. The group that used the mobile devices in fact have lower mean scores for the first and second tests. In contrast,

Figure 1. Student test performance on Test 1, 2, and 3 with and without iPod Touches

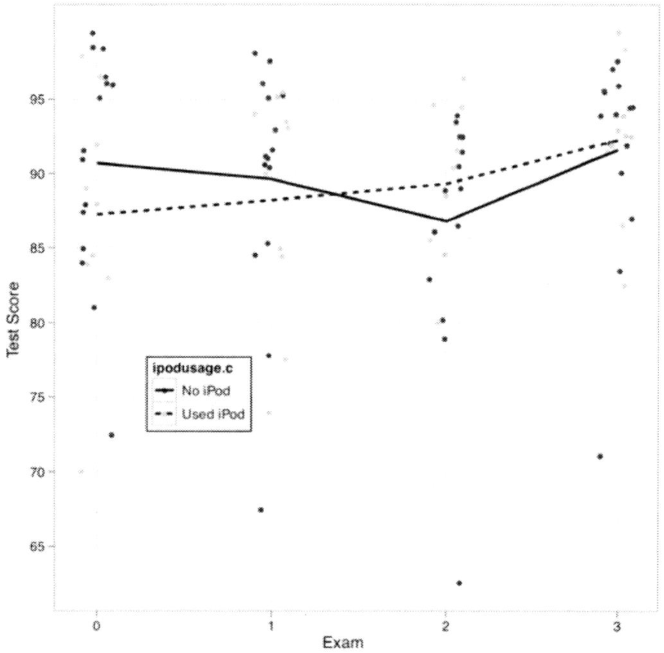

the mean scores for non-iPod users decreased for the third test and iPod users increased. iPod usage began after the second test so this difference is particularly interesting. Finally the mean scores of iPod users were slightly higher than those of other students on the final exam.

These general statements can be illustrated by looking at the individual cases, of Students 9 and 23. Figure 2 depicts individual change curves for each student. Students 9 and Student 23 can be identified as students who initially struggled, both scoring below 80% on the first test. Student 9, who began using the iPod after Test 2, demonstrates a significant increase in test scores for Tests 3 and 4. Student 23, who did not use the iPod, scored a few percentage points higher than Student 9 in Test 1, but did not make consistent progress in test scores throughout the term, ending with a score in Test 4 nearly identical to that of Test 1. The scores for students 9 and 23 belie the variation underlying mean group scores.

Following the preliminary descriptive analysis, the authors then used linear mixed effect regression techniques (LMER) to analyze the data. LMER techniques are particularly appropriate for repeated measures over time and are suited to examine mean changes over time that account for both group-level and subject-specific effects. In other words, LMER is suited to studies in which measurements are taken from the same subjects over time and there is no assumption of independent measures (Fitzmaurice et al, 2004; Long, in press). The authors analyzed polynomial models to determine the best fitting model was intercept only, linear, and quadratic. An intercept model would imply that the group differences in test scores appear in the first exam and are consistent throughout. A linear model implies that there is not only a difference in mean scores for the first exam but that there are differences in group mean scores over time and that those differences are linear. Finally the quadratic model accounts for

Figure 2. Individual test scores for Tests 3 and 4 with and without iPods

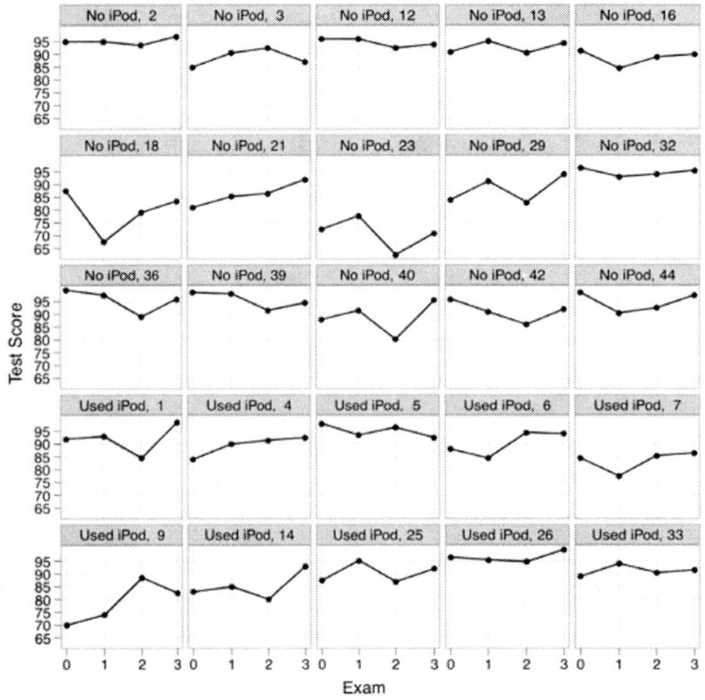

differences in group mean scores initially and over time but those differences are curvilinear in nature. In other words, this analysis determined whether the mean changes in test scores of iPod and noniPod users were best explained by statistical models that were intercept, linear, or quadratic. The following notation depicts these three models.

$$Y_{ij} = (\beta_0 + b_{i0}) + e_i$$
$$Y_{ij} = (\beta_0 + b_{i0}) + (\beta_1 + b_{i1})Test_{ij} + e_i$$
$$Y_{ij} = (\beta_0 + b_{i0}) + (\beta_1 + b_{i1})Test_{ij} + (\beta_2 + b_{i2})Test_{ij}^{\,2} + e_i$$

Akaike Information Criterion (AIC) is a metric used to determine among statistical models which is the best fit and a variant of this metric, AICc, is particularly well suited to studies with small sample sizes. The AICc scores of each of the above models were compared and the quadratic model was the best fit and most plausible with an AICc score, 638.9, approximately 4 points smaller than competing models. Having established that the quadratic model had the greatest explanitory power, the remainder of the inferential analysis was conducted using quadratic models reflecting the curvilinear change curves.

In a continuation of the analysis, we also use a multimodal analysis approach to simultaneously test the ten working hypotheses listed in Table 1. The first model hypothesizes that differences associated with iPod usage or learning style preferences were negligible in terms of mean changes over time in test scores. The second model suggests that students who opted to use iPods wanted to improve their initial test scores and that the iPods helped do so in the third and fourth tests. The remaining eight models relate to ways in which learning style preferences may mediate the effect of iPod usage. For each learning style preference, there are two models: the first examines whether iPod adoption is affected by a student's learning style preference as it relates to the physical senses, while the second examines the degree to which any associated mean increases in test scores are mediated by a learning style preference. Again, all of these models were quadratic.

The multimodel analysis employed information criteria to identify the best fitting of the ten models identified. This involved calculations of the weights of evidence and evidence ratios. Figure 3 clearly depicts that Model 2 is the best fitting model with a probability of 37% that it is the best fitting relative to other models in the study. Figure 4 contains a graph of evidence ratios.

Table 1. Ten working hypotheses that guided the analyses of the data

Model
LMER.1 iPod usage differences and differences associated with preferences for learning styles as they relate to physical senses are negligible.
LMER.2 iPod users on average have lower initial scores than other students but the magnitude of the difference changes over time.
LMER.3 iPod usage is mediated by a preference for a tactile learning style at least partially.
LMER.4 iPod users with preferences for tactile learning have higher gains over time.
LMER.5 iPod usage is mediated by a preference for a kinesthetic learning style at least partially.
LMER.6 iPod users with preferences for kinesthetic learning have higher gains over time.
LMER.7 iPod usage is mediated by a preference for a visual learning style at least partially.
LMER.8 iPod users with preferences for visual learning have higher gains over time.
LMER.9 iPod usage is mediated by a preference for a auditory learning style at least partially.
LMER.10 iPod users with preferences for auditory learning have higher gains over time.

Figure 3. Weight of evidence for each model

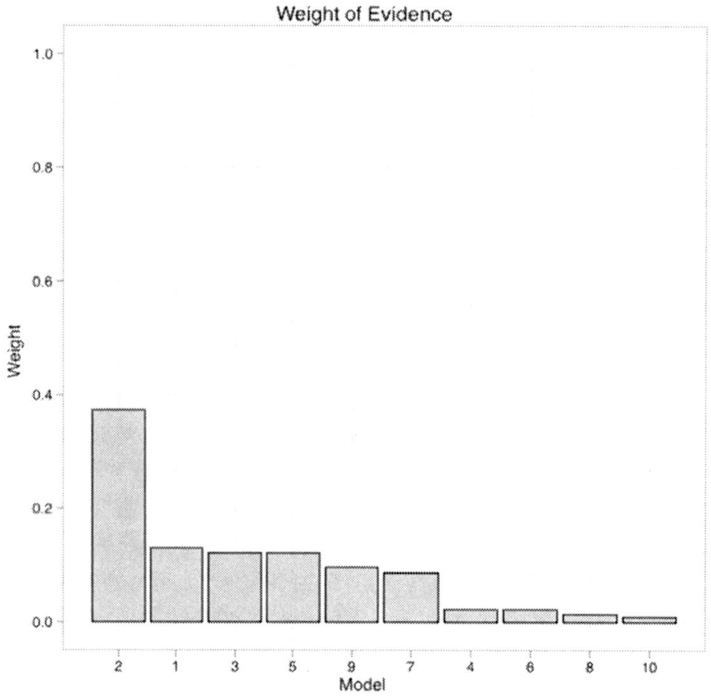

Figure 4. Evidence ratio for each model

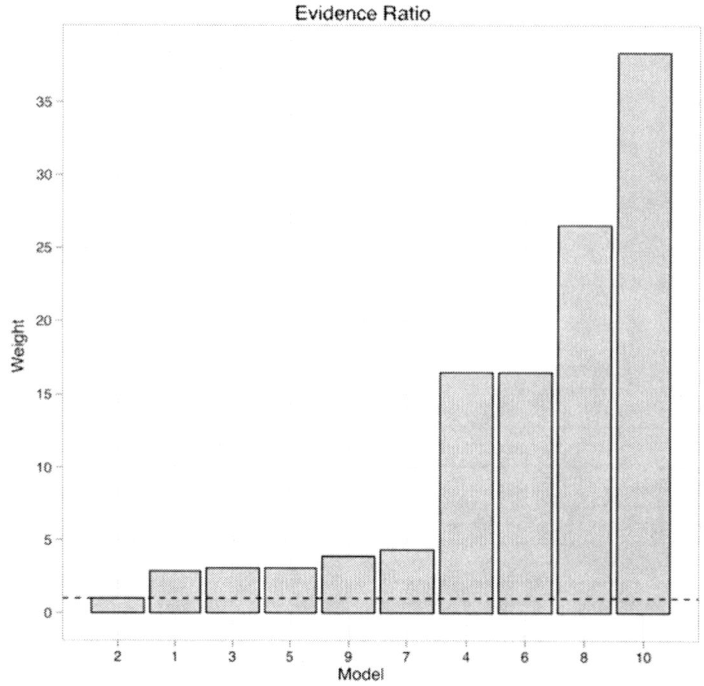

Models 1, 3, and 5 had approximately a 3 to 1 odds of being the best fit upon replication of the study. Models 9 and 7 both had approximately 4 to 1 odds of being the best fit upon replication. Models 4, 6, 8, and 10, which examined slope effects of learning styles, had odds as low as 16 and as high as 38 to 1 of being the best fit upon replication and are elimnated from consideration.

Given that Model 2 has greatest predictive value it is important to examine it in greater detail. This model examined the hypothesis that iPod users on average have lower initial scores than other students but that the magnitude of the difference changes over time. This is a zero mediation model insofar as the effect of iPod usage is not mediated by learning style preferences.

Table 2 shows the fixed effects estimates for the model. The negative 3.8 estimate indicates that students who opted to use iPods had initial mean test scores almost four points below the mean score of noniPod users. The estimates for the fixed effect for Exams and $Exams^2$ indicate that for each subsequent test score there is a mean decrease of 4.3 points but that the aggregate change curve is convex. This means that across all subjects, following a decline in mean test scores there is an increase in mean test scores. As a group, the iPod users had a mean increase of 4.5 points greater in each successive test relative to the rest of the class. The change curve for iPod users is modestly concave as that group's relative increase in scores begins to taper.

The standard errors associated with iPod usage, 3.1, and iPod usage over time, 2.6, indicate that there was variation within this subgroup's initial scores and in scores over time. The t ratio for the higher-order terms $Exams^2$ is 2.76 strong, e.g. greater than 1.96, but the modest t absolute value of the interaction term of iPod Usage and $Exams^2$, 1.16, might be negligible. In other words, Model 2 shows that the mean test scores of iPod users began lower and then increased at a greater rate than the non-iPod users. The initial magnitude of the difference in scores shifted as the mean iPod user scores surpassed the non-iPod users but the final tapering of this curve resulted in mean scores that were similar for the two groups.

While the sample size is very small in this study, the results indicate an interesting trend that merits future study. In general, the students who used the iPods enjoyed greater gains in test scores than the students who did not use the iPods. Though the beginning mean score of the iPod user group was lower than the control group, by the end of the term the mean test scores for both groups was nearly the same.

*Table 2. Linear mixed model fit by maximum likelihood (Formula: test ~ exams * iPodusage.f + I[exams²] * iPodusage.f + [exams | id])*

Fixed effects:			
	Estimate	Std. Error	t-value
(Intercept)	91.1675	1.9618	46.47
exams	-4.3992	1.6666	-2.64
iPodusage.fYes	-3.8325	3.1019	-1.24
I($exams^2$)	1.4625	0.5308	2.76
exams:iPodusage.fYes	4.5467	2.6351	1.73
iPodusage.fYes:I($exams^2$)	-0.9750	0.8393	-1.16

RECOMMENDATIONS

Based on this initial program we have a number of recommendations, both pedagogical and logistical, for similar applications of mobile devices to language programs.

Pedagogical Considerations

Design pedagogically useful activities for mobile delivery: Mobile devices, such as the iPod Touch, are excellent ways to get students' attention, but as mentioned earlier this 'wow factor' fades quickly. All students were excited to try out new technology, especially since they were able to keep the device for the whole term and there were no restrictions as to what they could load onto the iPods. But students will discard the technology quickly if they find that it does not fit in with their study habits or advance their academic goals. Students who expressed interest in audio capabilities, for example, tended to make less use of the iPods for studying because the software lacked a desired feature. Additionally, the technology itself could easily distract the student from the pedagogical goals, rather than advance them. It is only with truly pedagogically sound applications of the technology that students will continue to engage in the prescribed activity and recognize academic gains.

Ensure that the mobile activities blend well with the existing curriculum: Many language programs, including our own, attempt to create a partial immersion program that gives students the maximum amount of contact with the French language. Therefore, using French-English flashcards is somewhat counter to the methods of the curriculum and could detract from efforts to teach students to think in French rather than in translation. The technology could be used to deliver flashcards that included picture representations of vocabulary and/or examples of vocabulary use opposite to the French side of the flashcard, rather than simple bilingual translations. In fact, we considered this in first designing this program, but ultimately abandoned the idea after judging it be to too time consuming to identify appropriate graphics for each and every vocabulary item. Ideally, the flashcards would be carefully designed with this consideration in mind far enough in advance of the course to allow adequate development time.

Take advantage of multimedia to accommodate different learning styles: To take better advantage of the multi-modal nature of technology, audio recordings of the French pronunciation could also be added to assist students developing their 'ear' for the French language and guide their own French pronunciation. While this feature is not yet available in the production version of Mental Class (as of March 2010), it is already a feature of the next major revision and could be implemented as soon as the 2010-2011 academic year. As with the graphical material mentioned above, this would significantly increase the development time needed to prepare the flashcards, as audio recordings would need to be developed for each and every vocabulary item. Again, proper advance planning for materials and staff time would be critical to developing curriculum-appropriate multi-modal flashcards.

Create interactive activities to encourage engagement: Finally, if students had camera- and microphone-enabled mobile devices the flashcards could be generated, at least in part, by the student themselves. Professors could provide flashcards with the French side only, and instruct students to complete the opposite side of the flashcard by taking a picture of the object, or by typing or recording an aural explanation of the item (in English or French depending on the level of study). This method, though more time consuming for students, would ensure they engaged with the vocabulary items fully for complete comprehension. To ensure students would complete the tasks, they could then be instructed to share their versions of the flashcards with their classmates via a course management system or other class web space.

Do not assume the students already know the technology: Though our so-called millennial students have grown up with computers, their experiences are largely centered on the use of technology for entertainment. Faculty often mistake students' casual treatment of technologies as mastery, assuming that students will have the skills necessary to complete technology-enhanced tasks with little to no instruction. As a result, faculty who attempt to use technology in a blended learning environment often experience mixed results, finding that students were sometimes unable to complete the tasks or did not understand the pedagogical goal of the exercise and therefore circumvented the intended process to more quickly achieve the stated outcome. Experiences like these can lead faculty to question the validity of the incorporation of technology into assignments, along with creating exasperation at how to manage the many ways technology can go wrong.

However, students' extensive use of technology for personal entertainment does not often allow them to transfer skills to the academic setting. As we are reminded by Peters, Weinberg and Sarma (2009), many students are resistant to using new technologies in their learning because they have a "need for competence in technical skills rather than just language skills." In short, instructors and support staff must be sure to include technology training materials for their students. We recommend making tutorial handouts, or even better screencasts, available to students online and linked directly from the content they are expected to use.

Logistical Considerations

Consider how students will have access to and get support for mobile devices: Distributing electronic devices to students always comes with risks. There is always the chance that the device will fail, or that the student will lose or break it. The institution is then responsible for financing and providing technical support for these high-risk devices.

On the other hand, requiring students to provide their own mobile device creates alternative concerns. Students will not always purchase the recommended equipment, and therefore the institution must provide content materials that are flexible enough to work on a number of platforms. At Carleton College, ownership of the device could be irrelevant, putting the College in the position of needing to provide technical support for all the mobile devices that students use. In this scenario, the institutional commitment and financial liability are even greater in supporting devices that it does not actually own, requiring a greater number of staff trained in a wider variety of technologies.

From a technological support perspective, it is far easier to provide students with one, supported device chosen by the institution. Though this requires significant funding, the benefits in terms of tech staff support time and the predictability of the environment for students and faculty will go a long way toward smoothly integrating the technology into the curriculum and the lives of students, staff, and faculty. In this scenario, broken or lost devices can easily be replaced by identical devices (assuming that support staff have adequate swap units available), thereby minimizing the loss of study time for students. Training in using the devices is also streamlined, and faculty can be better prepared to integrate a known technology into their pedagogy.

Bring Technical Support Staff into the Project Early: Everyone benefits when technical support staff are well informed of new applications of technology in the curriculum. In many cases, technical support staff can be helpful in identifying appropriate and supportable technologies that meet the pedagogical goals identified by faculty. The earlier that support staff can be brought into the development of a new project, the more easily all parties to the project can identify and address potential issues before they block student learning. Many institutions are now recognizing the importance of this collaboration between educators and support staff, and are specifically hiring technologists with background and training in educational technologies.

FUTURE RESEARCH DIRECTIONS

In general, more research is needed in the effects of mobile learning, and specifically on foreign language learning. As new technologies continue to evolve and offer new ways of incorporating multimodal, dynamic, and interactive content, educators and researchers will need to continue investigating how these materials impact student attitudes and success rates in learning. Devices like the Apple iPod Touch and the underlying operating system offer a great deal more flexibility than traditional PCs in that development of pedagogically appropriate learning applications can be done quickly and, in some cases, with minimal programming skills. The devices can then be quickly deployed with these custom pieces of software to meet a particular learning goal. In addition, the devices are both far more mobile than even the smallest laptop computers, and very familiar to undergraduate users – though not necessarily as learning tools. All of these characteristics should be tested, qualitatively and quantitatively, in future research projects.

In particular, software and interface designers of educational content should pay closer attention to current research on learning styles and how to match them with new learning objects. Educators also need to attend to these issues by becoming both better designers of pedagogically appropriate materials and educated consumers of materials sold by software vendors. Certainly, these two parties to the learning experience can and should collaborate closely to develop new learning tools, especially in the ecosystem of, say, Apple devices, which allows for relatively "bottom-up" development of new applications – which can include features designed to allow or accelerate assessment of student learning and of the tool's efficacy.

Finally, consideration must be given for the technical support infrastructure available to faculty and students before embarking on large scale implementations of mobile and/or blended learning technologies. The most pedagogically sound projects can have disastrous results if the technical support staff of the institution are not well informed enough to provide appropriate levels of training and technical support for both faculty and students. This suggests a need for technologists to have a closer tie to the academic curriculum, and the ability to collaborate with faculty on developing new applications of technology. These collaborations can and should occur at every level of the curriculum, from the most general consideration of where mobile learning tools can best be used (introductory courses? advanced courses? in the classroom via exercises? outside the classroom as students study?) all the way to the most specific deployment of particular assignments ("flashcards" going from English to the target language at one point in a course, but in the other direction later). Only further research will determine what, if any, principles should shape the nature and direction of these collaborations.

Having already carried out the project described here, and possessing a good stock of iPod Touch devices, Carleton intends to explore some of these issues in future research projects, including some that have been conducted during the writing of this article.

CONCLUSION

Though focused on the problem of vocabulary acquisition in a foreign language course, this project yielded encouraging results that suggest new and effective ways to deploy mobile technologies in contemporary classrooms. The flexibility and extensibility of mobile devices such as iPod Touch computers can, under the right circumstances, be used to refine older pedagogical models and to develop new ones.

The use of such devices is not a revolutionary change to language learning, but rather a clear development or enrichment of earlier forms of computer-assisted language learning. In particular, the ability of students to use the devices wherever and whenever they choose seems to be an improvement over earlier forms of CALL, which were usually tied to IT infrastructure such as a language lab.

Similarly, the relatively open and straightforward process of developing new applications for devices like the iPod Touch means that users (including language teachers as well as IT professionals) can readily create and modify software to meet new learning goals or to capitalize on the devices' inherent characteristics, which include the ability to display still or video images, to play audio files, to record sound, and to accept relatively natural input via the touchscreen.

For all these reasons, we feel confident in asserting that the use of mobile devices in language learning is pedagogically appropriate and likely to grow in importance over the coming years.

REFERENCES

Blake, R. J. (2008). *Brave new digital classroom*. Washington, DC: Georgetown University Press.

Cohen, A., Oxford, R., & Chi, J. (2006). Learning style survey: Assessing your learning styles. In *Styles- and Strategies-Based Instruction: A Teacher's Guide*. Center for Advanced Research on Language Acquisition.

Ellis, R. (1994). *The study of second language acquisition*. Oxford, UK: Oxford University Press.

Fitzmaurice, G. M., Laird, N. M., & Ware, J. H. (2004). *Applied longitudinal analysis*. Hoboken, NJ: Wiley-Interscience.

Gardner, R. C. (1985). *Social psychology and second language learning: The role of attitudes and motivation. The Social Psychology of Language, 4*. London, UK: E. Arnold.

Johnston, B. (1999). Theory and research: Audience, language use and language learning. In Egbert, J., & Hanson-Smith, E. (Eds.), *CALL environments: Research, practice, and critical issues* (pp. 55–64). Alexandria, VA: Teachers of English to Speakers of Other Languages, Inc.

Krashen, S. D. (1981). *Second language acquisition and second language learning*. Oxford, UK: Pergamon Press.

Krashen, S. D. (1985). *The input hypothesis: Issues and implications*. London, UK: Longman.

Long, J. D. (in press). *Longitudinal analysis for the behavioral sciences using R*. Thousand Oaks, CA. *Sage (Atlanta, Ga.)*.

Ma, Q. (2009). *Second language vocabulary acquisition*. Bern, Germany: Peter Lang AG.

Oxford, R., & Shearin, J. (1994). Language learning motivation: Expanding the theoretical framework. *Modern Language Journal, 78*(1), 12–28. doi:10.2307/329249

Peters, M., Weinberg, A., & Sarma, N. (2009). To like or not to like! Student perceptions of technological activities for learning French as a second language at five Canadian universities. *Canadian Modern Language Review, 65*(5), 869–896. doi:10.3138/cmlr.65.5.869

Shih, Y. (2007). *Dynamic language learning: Comparing mobile language learning with online language learning*. Unpublished doctoral dissertation, Capella University.

Soo, K. (1999). Theory and research: Learning styles, motivation and the CALL classroom. In Egbert, J., & Hanson-Smith, E. (Eds.), *CALL environments: Research, practice, and critical issues* (pp. 55–64). Alexandria, VA: Teachers of English to Speakers of Other Languages, Inc.

Stockwell, G. (2010). Using mobile phones for vocabulary activities: Examining the effect of the platform. *Language Learning & Technology*, *14*(2), 95–110.

Sydorenko, T. (2010). Modality of input and vocabulary acquisition. *Language Learning & Technology*, *14*(2), 50–73.

ADDITIONAL READING

Blake, R. J. (2008). *Brave New Digital Classroom*. Washington, DC: Georgetown University Press.

Egbert, J., & Hanson-Smith, E. (Eds.). (1999). *CALL environments: Research, practice, and critical Issues*. Alexandria, Va: Teachers of English to Speakers of Other Languages.

Ellis, R. (1994). *The Study of Second Language Acquisition*. Oxford, England: Oxford University Press.

Journal, C. A. L. I. C. O. available online at https://calico.org/page.php?id=5

Language Learning & Technology journal, available online at http://llt.msu.edu/intro.html

Lightbown, P. M., & Spada, N. (1999). *How languages are learned.* Oxford, Engiand: Oxford University Press.

Mitchell, R., & Myles, F. (1998). *Second Language Acquisition Theories*. New York: Oxford University Press.

KEY TERMS AND DEFINITIONS

Asynchronous Learning: Earning that occurs outside of direct interactions between two or more students, or between an instructor and a student.

Computer Assisted Language Learning: Foreign language learning enabled by some form of computing technology, usually in addition to conventional forms of learning via classroom instruction and textbooks.

Second Language Learning: Learning a foreign language, especially via formal instruction, of a language in addition to one's own native language(s).

Mobile Assisted Language Learning: Language learning enabled by mobile computing technology such as cell phones, iPods, tablet computers, or other such portable devices.

Vocabulary Acquisition: A core element of second-language learning, the mastery of new words and terms.

Section 3
Extending Mobile Learning

Chapter 12
Augmented Reality and Mobile Technologies

Grant Potter
University of Northern British Columbia, Canada

ABSTRACT

Unlike Virtual Reality (VR) that attempts to replace the perception of an immediate environment with an artificial one, Augmented Reality (AR) applications aim to enhance a person's perception of their immediate environment. A blend of both the virtual and the real, AR application interfaces on mobile devices display information that is dependent on users' time and location. AR applications are not necessarily an entirely new technology and have been emerging in various sectors over the past 5 years. For example, in aviation, AR in the form of 'heads-up-displays' has been used to display important data to pilots for decades. As mobile devices diversify in their speed, power consumption needs, network connectivity, and locative functions, developers are able to port AR applications to next generation mobile handsets, opening a wide range of utility and potential across public and private sectors.

INTRODUCTION

Mobile AR technologies offer a unique, context aware means of content delivery with great potential for enhancing the effectiveness and attractiveness of teaching and learning for students in real-time situations. The technology provides a unique means of progressing through teaching and learning in situated environments. Many of the criticisms leveled at VR environments revolve around their struggle to effectively establish a meaningful sense of presence and connection between multiple users in an environment. A key purpose of any educational environment is to promote social interaction among users located in a shared space. With AR applications, multiple users access a shared space populated by virtual objects, while remaining grounded in the real world. This technique is powerful for both marketing and educational applications when users are co-located and use proximal means of communication (e.g. gestures, speech), but can also interact with location-aware data.

AR applications offer advantages over manual-based and VR models as users can see and touch the actual objects presented while at the same time receiving contextual, interactive, guided informational elements. AR applications are able

DOI: 10.4018/978-1-60960-511-7.ch012

to sequence and highlight specific objects in a users' field of view, depending on the task and context. Additional functionality and contextual information can be integrated into AR applications through possible live interaction with a remote expert to providing further assistance by further updating informational elements displayed by the system application. This chapter will examine the potential and limitations of AR through profiles of projects utilizing mobile AR applications across the domains of business, tourism, and education through an examination of how emerging AR technologies converge mobile content, integrated global positioning system (GPS) functionality, and the affordances of hardware and software in mobile devices.

BACKGROUND

Augmented reality (AR) is an emergent field of computer research developing hardware and software capable of blending situated, real world experience with computer generated data. To date, most AR research involves integrating processed live video imagery which is "augmented" through the addition of graphics and textual information. Advanced research includes the use of pattern recognition via digital optics, motion tracking, and the generation of controlled environments via sensors.

AR is generally aligned opposite of VR such that rather than attempting to immerse a user into a entirely computer generated environment, the goal of AR is to augment a user's immediate environment with information access and management capabilities. Augmented reality is considered a specific element of the more general concept of computer mediated reality (Med-R). MR pursues the development of technologies that effectively filter a users' vision of their immediate environment through digital overlays on a display placed within a users' field of view. AR applications growing in popularity spurred by the rapid devel-

Figure 1. Technologies employed by AR

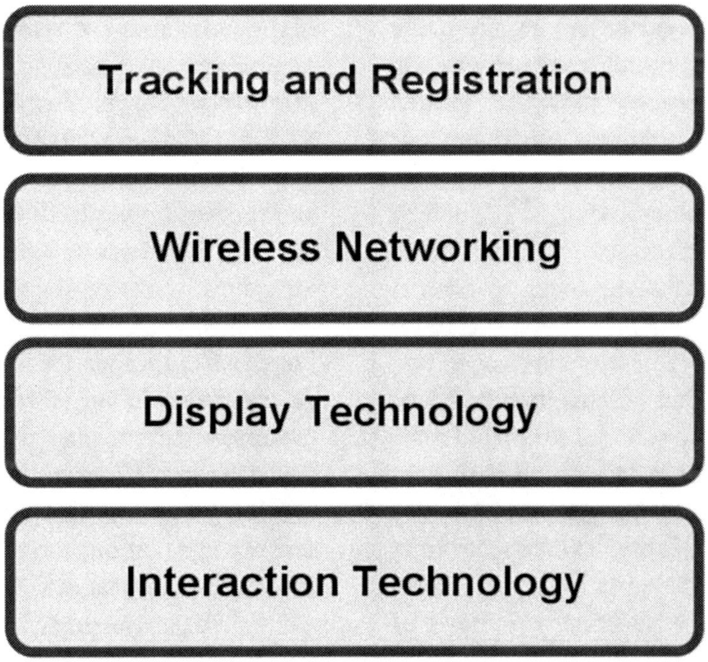

opments in personal mobile technologies offer placement of virtual elements in digital overlays and are more augmentations than direct mediations of users' environmental experience. One of the key characteristics of AR is the manner in which a users' locus of interaction is no longer limited to a specific zone of influence, but also encompasses a user's immediate and extended environment. Interactivity is not defined as a screen to face connection, but rather a more fluid dynamic between a user, environmental feedbacks, surrounding objects and spaces, thereby facilitating access to information flows that are no longer exclusively a directive and intentional act.

AR research and development has parallels with research in ubiquitous computing models (UC) and elements of wearable computing. Weiser's idea of "embodied virtuality", which he defined before coining the term "ubiquitous computing", aims to widen the scope of virtual reality developments (Weisner, 1991). While UC does not deal with the absence of conscious and intentional interactions with information systems like AR, UC devices maintain the idea of direct, intentional interaction. Humanistic Intelligence theory (HI), however, also challenges this model. HI is defined as intelligence arising from the activity of human interaction in a feedback loop of a computational process in which human activity is required, yet not necessarily with conscious, intentional effort or thought. In this way, HI shares much in common with AR. Much of emergent research operates on this definition of AR and has been articulated as such by Azuma who defines augmented reality systems as those that combine real and virtual interactives in real time (Azuma, 1997). This combination both real and virtual is often also referred to 'mixed reality environments' (MREs). Drascic and Milgram (1996) describe them as: "between the extremes of real life and Virtual Reality lies the spectrum of Mixed Reality, in which views of the real world are combined in some proportion with views of a virtual environment" (p. 123).

Applications typically identified as virtual reality encompass a broad range of developments companies and researchers use to define their work. The coining of term "virtual reality" is attributed to Jaron Lanier, founder of VPL Research and one of the first companies to commercialize virtual reality equipment and systems. The term was is defined as "a computer generated, interactive, three-dimensional environment in which a person is immersed." (Aukstakalnis & Blatner, 1992)

This definition consists of three important elements. As virtual environments are rendered by computers in three-dimensional scenes, they require robust computing power and high performance graphics to provide rich visual detail and realism. Virtual reality environments are meant to be responsive to user presence and activity requiring effective real-time feedback responses from the system. The third element of virtual reality environments is that they are immersive such that all activity and feedback takes place within the simulated environment. Research labs have developed complex installations to simulate immersive experiences for many years and although they have yielded commercial applications such as flight simulators and technical training environments, popularization and growing sophistication of both personal computers and gaming platforms have led to an exponential growth in virtual reality developments. Advancements in graphics and processing capabilities of gaming platforms have led to mass commercialization of a range of virtual reality environments spanning game-based systems to collaborative, online meeting spaces. For these consumer accessible systems, the idea of immersion in virtual reality is rooted more in the notion of roleplay and online identity and is a departure from the large scale installations involving systems that enclose users in multi-sensory installations to the exclusion of external stimuli. By contrast, AR systems require users to maintain a sense of presence in their surroundings my merging generated data visual and textual data with real-time video feeds of the physical environment. This generated data must be accurately registered with the immediate surroundings in all dimensions as deviations between data displayed and the users's surroundings will interfere with a user's ability to form direct relationships between the generated informational elements and their immediate environment. The challenge in object registration occurs as the user moves within their physical environment creating possible delays or discrepancies in data registration. Maintaining accurate registration is crucial in ensuring changes in the rendered AR scenes match the perceptions of the user's immediate environment. Without timely, responsive representation of data in AR, utility of the information is reduced leading to possible confusion and disorientation, making the system completely unusable. Responsive visual capture and representation affords such systems a greater influence in user perception (Welch, 1978).

Milgram (Milgram & Kishino, 1994; Milgram, Takemura et al. 1994) established a taxonomy identifying the relationship between augmented reality and virtual reality applications. This relationship places a user's perception of their immediate environment and a simulated virtual environment at opposite ends of a continuum.

The region between these two points is identified as Mixed Reality (MR). Defining the realm of MR further, Milgram identifies three dimensions for categorizing these systems: Extent of World Knowledge, Extent of Presence Metaphor, and Reproduction Fidelity. The Extent of World Knowledge dimension deals with the degree to which accurate registration of simulated objects are rendered with representations of the immediate environment. This requires detailed understanding of the relationship between the frames of reference between the user, the physical environment, and the optical systems receiving the imagery. With AR installations in lab conditions, the relationships between these elements are limited and well defined, making referencing far less complex than instances encountered by AR systems functioning in changing environments and situations. Reproduction Fidelity refers to the quality of the computer generated images that can range basic textual data to rendering of photorealistic items. Limitations in processing power combined with the constraints on responsiveness of real-time computer generated imagery forces most AR applications developed for mobile devices to the lower end of the Reproduction Fidelity domain. AR systems are also situated low within the Extent of Presence Metaphor domain. This domain measures a user's degree of immersion within a displayed AR scene and is tied directly to display technology formats being used. When considering mobile AR applications, part of a users' field of vision is the direct view of the immediate environment. Specifically in the case of mobile handsets, AR displays act more as augmented 'AR windows' Overall, AR applications reside closer to the 'real world' end of the

Figure 2. Milgram's Virtuality Continuum

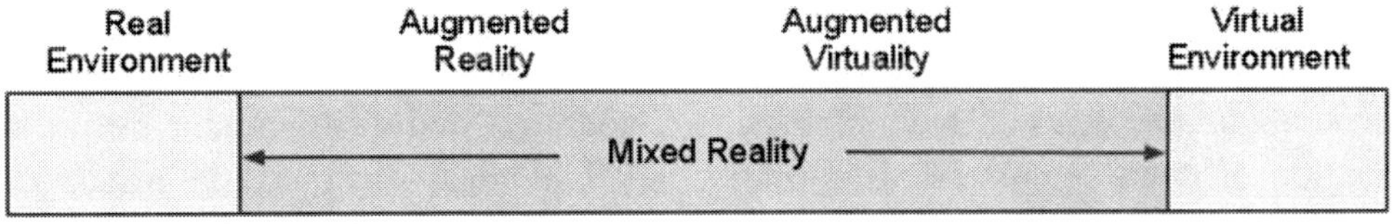

span as a user's predominate perception is based on experience of their immediate physical environment which is augmented by computer generated data. As advancements in graphics and processing power further refine optics and computer generated imagery the distinction between synthetic elements and the scenes within the actual environment will become less distinguishable.

Research on the benefits of AR applications include improving collaborative working and planning (Fjeld et al., 2002), enhancing learning (Underkoffler & Ishii, 1998), and enhancing user experiences (Camarata et al., 2002; Schnadelbach et al., 2002). Research exploring how and why augmenting real life with digital elements produces desirable effects has yielded a range of hypothesis. It has been proposed that manipulating familiar, tactile artifacts or acting in physical spaces when interacting with digital information provides greater embodiment for a user - embodiment implying more concrete presence, as opposed to more abstract representations of data or information. (Dourish, 2001). In essence, interactions experienced in augmented reality fit more naturally with the way we interact with the world by taking advantage of our familiarity and experience with the immediate physical world, in particular, our ingrained skill sets involving physical actions (e.g. pushing, holding).

AUGMENTED REALITY IN BUSINESS

Recent advances in AR have reduced development costs and opening up new advertising applications previously not economically viable. For years, the ability to integrate 3D graphics in mobile delivery systems was been limited to projects with large budgets, such as film productions from major studios. Advances in technologies and proliferation of more sophisticated mobile devices has improved the ability of marketers to effectively scale quality, complexity, and and interactive elements with within a reasonable cost/benefit ratio for advertisers, publishers, and content providers. Marketers are increasingly leveraging the possibility of AR in promotional and advertising campaigns by combining the novelty of AR with sharing of branded user-generated AR content to generate viral popularity phenomenon indicative of Web 2.0 web services. Campaigns offering interactive experiences via AR applications that allow for sharing of branded user-generated content online have demonstrated the potential to yield considerable growth in brand visibility. Among the added parameters offered by AR marketing initiatives are the ability to explore detailed product information, visualization elements, and in some cases the ability to interact with extended features of the product.

Much of the development around promotions and advertising utilizing mobile applications have been through the creative use specialized barcodes called Quick Response codes (QR codes). A QR code is essentially a barcode - a mechanism to capture information without the need to enter letters or numbers into a system or application using a keyboard. Denso Wave, a Toyota subsidiary, developed QR codes in 1994 as a means to track vehicle parts during manufacturing processes. The QR code design enables characters to be stored in a format not possible with the traditional single vertical black strip barcodes and allows characters to be placed anywhere within a rectangle storing data in patterns in both horizontal and vertical directions. The ability to store more characters in a machine readable code or label has enabled the storage of a much broader range of information including: web site addresses, telephone numbers, text messages, contact cards (e.g VCards), geolocative information, and images. Although Denso Wave owns the patent rights on this technology, it has chosen not to exercise them, leaving QR code use and development free of any license. While Japan is by far the largest user of QR codes for advertising and marketing and have broadly adopted their use in the Japanese mobile market, Australia and North America are showing signs of similar directions and developments. Mobile handsets feature a wide range of QR code readers which utilize integrated digital cameras to scan the codes on packages, signs, or public displays and billboards. For example, McDonalds restaurants in Japan use QR codes on the sides of food containers to direct customers to web resources outlining the nutritional information of food. In North America (McDonalds, 2007), and Google launched a QR code promotional campaign in 2009 to promote their 'Favorite Places on Google' feature by sending decals to over 100,000 businesses in the US indexed by their search appliance as being the most sought after businesses in their search and map services. (Google Favorite Places, 2010) Scannable QR codes on these decals take visitors directly to the business's website on the mobile device. The codes also direct visitors to reviews, contextual site-specific information, and special offers. The project also plans to incorporate the ability for mobile users to submit reviews and ratings of the business directly from their mobile devices. NHL hockey team, the Detroit Red Wings use QR codes to extend their promotion and marketing print campaigns by redirecting visitors to multimedia features and have tracked mobile access to their content as the foremost medium fans use to access promotional videos and materials. (Detroit Red Wings Make Game Programs Interactive, 2010)

In all of these examples, QR codes are used as highly visible labels used to trigger mobile internet browser access to video, text, and images from websites. In an effort to generate market interest in their new product lines, snack company Frito-Lay explored the potential to move beyond traditional image and text media by developing a campaign that integrates QR codes patterns with product packaging to trigger 3D multimedia AR. (King, 2009) Launched initially in Brazil, this project marked an move towards using mobile devices and codes to diversify the range of consumer accessible AR interactive media. A code much like a QR code on product packaging was designed to trigger a 3D toy-like character that is superimposed on a video stream from a users' device. A range of interactive features extended the generated animation including sharing the photos, videos, and games involving the generated characters via social networks. The promotion generated significant viral marketing spinoffs, adding to online product visibility.

Children's toy manufacturer Mattel tool this approach to product promotion further by positioning AR as an essential component of an merchansie advertising campaign for movie studio Twentieth Century Fox's action adventure film "Avatar". The toy line, developed in partnership with Twentieth Century Fox featured characters, creatures, and vehicles from the film. Each toy in the product line has an symbol marketers have labelled a 'i-TAG' that works much like a QR code. When this symbol is scanned with an application via a mobile device animated, interactive 3-D models are generated for the user. According to promotional literature from Mattel, placing two AR generated characters next to each other results in the generation of animated battles. Increasing numbers of corporations are recognizing the rising popularity of mobile devices as media access points for potential customers and the potential for AR applications as delivery systems. Like the Frito-Lay project, some campaigns are AR specific multimedia promotions but AR is also being leveraged an effective complement to online, print, television, and mixed media campaigns.

The integration of AR features in the December 2009 edition of Esquire magazine is one example of AR being used to extend print-based publishing and content deployment. In addition to featuring an AR enabled cover that generates a video of actor Robert Downey Jr. introducing the issue, there are several special features and supplements accessible on Esquire's website via an application that interprets AR markers within the magazine layout by holding the magazine facing a webcam. The extended features range throughout the magazine, featuring a fashion section that when tilted in different directions controls weather surroundings and clothing of models. Features also include elements that change according to user access time - access the AR component after midnight and the content delivered differs from that offered during the day. AR features were leveraged to generate curiosity, promote magazine market visibility and were limited users possessing a copy within a two week period. Esquire's editor-in-chief has expressed further interest in AR for adding further features such as audio versions of the magazine and its enhanced advertising delivery potential. Print publishers are realigning their business models as generations shift their media habits to mobile, digital access methods, AR applications offer a possible means of binding users to print editions, especially when AR supplements offer unique and exclusive access to content.

AR in marketing and promotion is being adopted as a novel and effective means to complement print and online advertising campaigns, attracting customers to product campaigns in compelling ways previously not possible due to cost and technology barriers. Combined with the potential for AR applications to generate viral 'word of mouth' product visibility in user-generated content communities and social networks, expect to see more AR functionality released in mobile devices as core features in the coming years.

Although marketers have embraced AR's potential for value-added services and features, a great number of these developments revolve around the entertainment industry. As mobile devices have diversified, more applications have started to focus on providing contextual information to assist users with inquiries situated in their immediate environment such as restaurant reviews, locations, and visualizations of products and services. Building on marketing and promotional potential of AR applications, tourism initiatives are increasingly leveraging AR to offer tourists access to location-aware, historically accurate experiences and perspectives.

AUGMENTED REALITY IN TOURISM

The Augmented Reality-based Cultural Heritage On-site Guide (Archeoguide) was the first mobile augmented reality guide developed for outdoor archaeological sites. (Vlahakis V., Ioannidis N., Karigiannis J.) This European Union Information Technology Societies funded project pioneered on-site, real-time access to archaeological multimedia data, facilitated the reconstruction of ancient monuments, and aided education regarding the history and context of ancient sites. The project was the first of its kind to envision the use of personalized mobile devices with navigation and interactivity features in place of kiosks and conventional guidebooks. Like many augmented reality applications emerging on fully features mobile phones, Archeoguide consisted of two systems working in close collaboration: a mobile client running on a handheld device linked via a wireless local area network to a central server. Being an early prototype, this system functioned on a local area network rather than a wide area network tied to internet accessible resources due to limits on bandwidth and connectivity during its development. The system did allow for scalability through an implementation of a client-server model allowing for the use of multiple devices and expandability. The application server consisted of a multimedia database where all information regarding a particular site was archived. Multimedia elements consisted of photographs and architectural drawings, 3D reconstruction models of monuments, text, audio, and video informational elements. All of these pieces were stored along with associated metadata capable of relating each item to a historically accurate former geographic location. The metadata embedded in these items made it possible to achieve efficient search and retrieval of the multimedia elements. The server featured a suite of graphical authoring tools that can be used to create new content and database applications used to organize thematic and geographic information. Early prototypes were successfully tested on sites in Greece and the project began commercialization of the technologies in 2004.

This model of augmented reality application development has been adopted by a range of companies aiming to provide value-added tourism experiences driven by a range of revenue models. France, China, Switzerland, and Germany are beginning to invest research and development resources to develop technologies that allow vistors to historic sites to access reconstructed images of past landmarks and sites overlapping the current environment giving the sense of peering through windows in time.

Historic sites in China suffered a great deal of damage in the 20th century due to armed conflict and cultural upheaval. One such site, the imperial Garden of Perfect Brightness in Beijing, built in the late 18th and early 19th centuries, was destroyed in 1860 during the Opium Wars. With no photographic records of the landscape, vistors to the site relied on reproductions of paintings and sketches to gather insights in the how gardens once appeared. A project team at the Beijing Institute of Technology created a virtual reconstruction of the gardens using geo-locative data, 3D models, and viewing devices that allow tourists to scan the existing garden site with the layer of reconstructed historical imagery superimposed on the view.

France's Cluny Abbey has developed and implemented augmented reality applications as a key component to their historical site reconstruction. The resultant device stands as possibly the single most used augmented reality device in the world to date (Joscelyne, 1994). Eschewing the small form factor of the smartphone, the device resembles a large window pane that visitors are able to rotate and pivot. The direction and orientation of the device is related to high resolution imagery depicting the abbey as it stood hundreds of years ago. What began as a single installation soon grew to a number of installations throughout the site based on volumes of positive feedback from visitors on the device and the degree to which it added to their appreciation of the architecture and site's historical importance. A similar project is in place at by the DNP-Louvre museum lab. Utilizing a custom mobile tablet PC with location aware technologies, the device does not situate visitors in the past, but does detect vistors' location in the museum and offers access to contextual information and histories for to provide insights into the collections when visitors draw near or point the device in the direction of collection pieces. The lab is currently working on the next phase of this device which will be more compact and offer multi-lingual support. In Berlin, IGD Fraunhofer and Instant Reality have devised a digital imagery system allowing visitors to use their own smartphones to capture images of historic landmarks and then see how the sites have changed over periods of time. This is done by submission of the users' image to a central server housing a database of photographs covering decades of the city's growth. Images from smartphones include GPS metadata allowing central servers to retrieve archival images from the visitor's exact location.

This example from Berlin is indicative of current trends towards embedding augmented reality applications in personal mobile devices rather than installations or specialized, single purpose devices. A new generation of augmented reality applications are being developed and released on smartphones and portable gaming platforms that aim to put a broader range of AR in consumer electronics. Increasingly sophisticated mobile AR experiences are being made possible by the emergence of mobile devices equipped with GPS functions, tilt sensors, cameras, high speed cellular connectivity and, perhaps the key component for situating a user in their environment, a digital compass. This last item is vital, and until recently it was the one missing element with mobile hardware platforms. Apple, Google, Nokia, Palm, Sony, and a number of other mobile developers all released devices in 2009 with this key feature spurring a flurry of AR application development. The combination of accelerometers, tilt sensors,

GPS, and a compass enables a mobile device to determine where it is, its orientation relative to the ground, and the direction it is being pointed in. An integrated digital camera provides optical functions for both the user and processing algorithms, and the high speed cellular access allows applications to retrieve relevant information relating to its surroundings, displayed as an amalgam of a live view from the camera and displayed on the screen.

Wikitude (Wikitude, 2010) is a smartphone application that has taken this approach to mobile augmented reality and combined it with user-generated, social networking elements popularized by the phenomenon labelled Web 2.0. Wikitude, as the name implies, draws information from Wikipedia, the online encyclopedia, by pouring through the vast arrays of data to locate entries listing a longitude and latitude. Using the application with a mobile device, a tourist can move through the streets of a city and view the names of the landmarks in the vicinity by simply holding their device in the direction of landmarks and locations. The full Wikipedia entry on any landmark can be retrieved from Wikipedia by selecting markers that appear on the live display. It is estimated that Wikipedia contains more than 600,000 entries that include longitude and latitude co-ordinates. As this collection of locations and data points are constantly being authored and updated by millions of users, the richness and depth of data available to the application is growing exponentially and at a rate that far surpasses a scale possible by any single directory. Reliability of user-generated data sources like Wikipedia is often raised as an impediment to developing trustworthy applications, yet the an investigation reported in the journal Nature suggested that for scientific articles Wikipedia came close to the level of accuracy in Encyclopedia Britannica with a very similar rate of "serious errors". Such studies indicate that information systems enabling collaboration on a massive scale provide for scalable checks and balance the yield data sets of high reliability (Giles, 1995).

Mobile platform Layar functions much like Wikitude such that it superimposes data 'layers' over realtime video streams from users' mobiles, but differs in both its business and development model. Mobile AR platform Layar allows publishers, brands, developers, and producers to engage users with a wide range of AR experiences based on curated datasets as well as vast arrays of data filtered and accessed via social networks by offering an application development interface (API) to third-party developers. Layar is the first company to reach out to third-party developers to create interactive data layers on top of their platform. Although these developments and methods are fairly recent, in 2009 alone more than 400 distinct layers in the Layar AP application were published in categories ranging from real estate, health care, and transportation to entertainment, tourism, and social networks with more than 1200 layers in development. In addition to a developer API, Layar producers and publishers aim to offer methods for both free an paid access to AR models supporting global payment methods with established online payment systems. These synergies are an attempt to provide an AR platform that allows producers to focus on content development and leave distribution, cross-platform compatibility, and financial administration to integrated platform supports. A further advantage of the Layar platform model is that it allows promoters, marketers, and advertisers to effectively curate and filter which real-time data layer users will be presented with on their mobile applications.

Mobile AR applications like RobotVision exposes latent online information about about a users' surroundings by drawing on massive databases like the Microsoft Bing Local Search. Using services like Bing Local search to acquire location and business review data, RobotVision capitalizes on access to popular aggregate search engine features. This key feature combined with the ability to draw on microblogging service

Figure 3. An augmented reality application

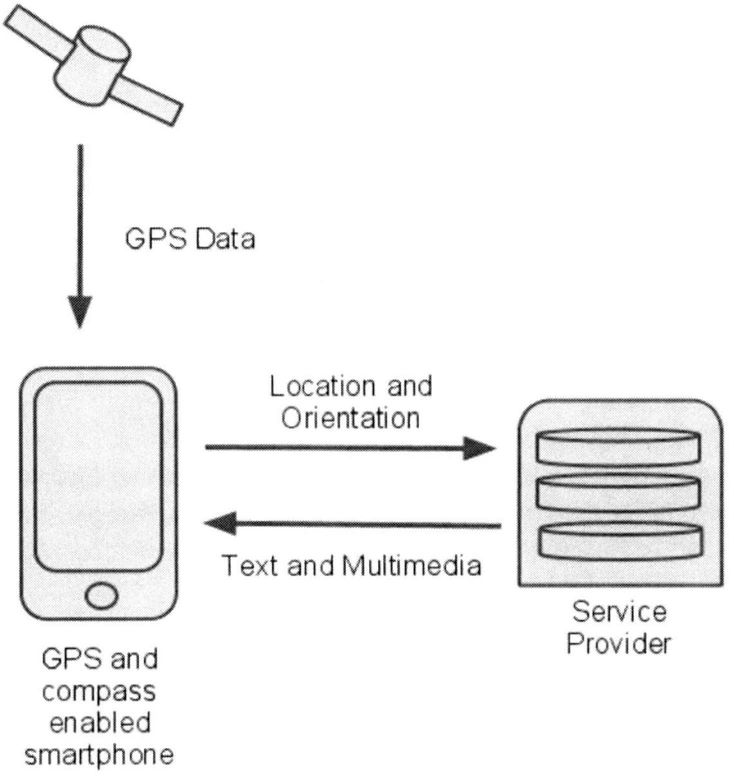

Twitter and photo sharing service Flickr provides users with a highly localized AR experience that is potentially highly responsive to localized, time dependent updates for sites and special events. (Microsoft Adds "Augmented Reality" to Bing Maps, 2010) The most common feature of early mobile augmented reality applications is the overlay of GPS metadata on top of a mobile phone's camera viewer, a feature offered by Wikitude, Layar, and Robotvision. the distinguishing feature yet to emerge in these applications will how these platforms go about collecting, displaying and interacting with layers of web-based data. The accuracy and usefulness of these data layer overlays is also highly dependent on the integrated GPS functions of mobile devices. All integrated GPS on mobiles take differing periods of time calibrate each time they are engaged. In addition, users' locative data refreshes at different rates depending on mobile carrier and bandwidth, resulting in data overlays mapped to relative locations such as nearest street intersection instead of exact location. Furthermore, devices without clear access to GPS satellites can lead to further inaccuracies. Such inaccuracies in mobile AP applications lead to data overlay results that indicate objects or informational items that are not actually present in a user's environment or failure to provide information for items that are present. As elements of these two technologies improve in performance and reliability, expect AP applications to become more robust, inexpensive, and common on mobile devices.

Not only are major web search engines being used as AR data sources, but they are beginning to recognize the end-user potential of these applications and in turn investing considerable time and effort in the research and development of their own feature sets. Microsoft has developed AR prototypes and revealed a promise of what is to come in beta releases like Bing Streetside Photos. This feature of the Bing search engines gathers digital imagery and video segments and inserts them into the Bing Map's street-level view perspectives using a combination of geolocation information drawn from manually entered geo-locative metadata, data from GPS-enabled devices, and imaging-matching algorithms. The cumulative effect of this creative combination of data sources and image processing technology results in an AR map application that provides a user with a perspective of a location with either a given image of a particular location or a user-submitted photo image sharing and social networking sites like Flickr and Twitpic rather than a static photo. The applications for tourism and marketing for such an application are considerable as up to date images and perspectives can be provided for locations and sites of interests. A large crowd gathered for a special event equipped with such an application would be capable of providing unique, data-rich perspectives of events thereby serving as a powerful user-generated marketing force. The project developer of this AR system, Agüera y Arcas, was lead developer of another image-stitching application called Microsoft Photosynth. Early prototypes of Microsoft Bing AR applications have incorporated Photosynth technologies allowing AR applications to recombine thousands of photos with detailed attention to overlap into 3D perspectives. This image recognition and stitching algorithm compensates for deviations in GPS locative data allowing image matches to combine within within a few inches of connection points. (Blaise Auguera y Arcas, 2009) These developments take AR beyond the 2D mobile data overlay experience and allow for multiple perspectives in a 3D environment. The commercial potential of such rich user experiences for tourism and hospitality industries are only beginning to be tapped as projects like these emerge from prototypes to versions capable of functioning on mobile devices offered by consumer telecommunications providers.

AUGMENTED REALITY IN EDUCATION

Driven by commercial interests in mobile content and the ability to reach and interact with new customers, mobile devices continue to emerge at more affordable, accessible price points without sacrificing complex functions and feature sets. Mobile devices are already widely adopted by students as part of their personal lives and they will bring resultant preferences and related skills sets to their educational contexts derived from their adoption and integration of these technologies. Curriculum and instructional designs capable of incorporating the capabilities and advantages of AR offer unique potential for building on these preferences and emerging skills sets. Drawing on both information systems and students' immediate, social environments, AR applications can offer multiple sensory experiences, amplify and highlight elements of a physical space, facilitate both simultaneous online and face-to-face collaboration, and incorporate elements of sensory, spatial, and kinesthetic domains of learning.

Developers and educators are beginning to identify best practices for developing effective AR curricula in a number of innovative projects. Supported by a grant from the U.S. Department of Education targeted at enhancing literacy and mathematics skills in urban schools, Harvard University, the Massachusetts Institute of Technology (MIT), and the University of Wisconsin at Madison have developed The Handheld Augmented Reality Project (HARP) curriculum. HARP utilizes wireless enabled mobile devices to enhance teaching and learning through activities that integrate AR functions that enable interpretation of students' surroundings. HARP uses GPS technology to track student movement, offering students multimedia elements relating to scenarios and problem-solving exercises once they have reached designated points in a physical space.

HARP's first prototype exercise, "Alien Contact," offers a alien Earth invasion scenario, and challenges students to work through literacy and mathematics problems related to the narrative. Students are presented with the following scenario: Aliens have landed on Earth apparently preparing for an unknown course of action. Some proposed intentions include invasion, peaceful relations, or simply exploration. The exercises organize students in groups of four who work together to explore the world rendered by the AR application which include collection of digital artifacts, interviews with virtual characters, and literacy and mathematics related problems leading to clues that point to alien intentions. The GPS and geo-locative features of the handsets are central to the arc of the narrative and exercises as the AR application provides evidence to students based on information gathered at 'hot spots'. Once within a defined radius of these 'hot spots' situated in a physical space, students are presented with activities and information by the AR application that connect and extend the narrative. Team members are assigned roles: linguist, chemist, linguist, computer scientist, and government agent. The roles define the AR generated artifacts and elements seen by each student. In order to successfully navigate the AR enabled environment students must collaborate both within and between the teams by devising hypotheses based on their interpretation of the collected data. The unit culminates when students present their conclusions to the class using their collected data from the AR application to support their position.

HARP's 'Alien Contact' project was a deliberate attempt to create a place-independent AR based curriculum. MIT's Teacher Education Laboratory has developed a range of curricula enabling students to collaborate within scenarios that are place-dependent implementations of AR. For example, their Environmental Detectives outdoor game was designed to challenge participants to work in teams to respond to a simulated oil spill on the MIT campus. Team equipped with mobile devices with AR applications were required to seek out and explore campus locations using GPS waypoints and gather related information provided by the AR. Problem sets include virtual character interviews and datasets indicating site specific levels of toxicity. MIT's Reliving the Revolution (RtR) is another example of place-dependant AR curricula. Designed for middle and high school social studies programs, it aims to engage students in civic literacy and an awareness of local histories. Situated in Lexington, Massachusetts, site of the Battle of Lexington of the American Revolution, AR-based curricula challenges students explore the events of the battle in order to deduce who may have fired the first shot of the battle - a topic that remains controversial to this day among historians. Students are asked to explore areas situated in Lexington Common using GPS enable mobile devices that register designated locations and relay multimedia elements as students arrive at the locations. Students are offered multimedia interviews with key figures such as Paul Revere and receive contextual information of both existing and simulated structures as they navigate the physical environment. The contextual information provided by the mobile devices are both digital scans and recreations of actual documents from the period such as newspaper articles, photographs, maps, diaries, and letters. These media elements offer subjective accounts of events that preceded the firing of the first shot at the battle. Students are challenged to use this information to support or refute their narratives. Situating students among physical artifacts on the historical site while providing discoverable AR elements adds a rich, engaging dimension to the long practiced exercise of the class field trip.

While such site-specific AR curricula are good examples of AR for teaching and learning, their reliance and design based around a specific site exclude them from being easily repurposed and adapted for off-site use. AR projects like HARP's 'Alien Contact!' essentially allow educators to design an AR enabled field trip irrespective of physical locality. The Alien Contact! pilot used GPS enabled PDAs to correlate physical locations mapped to GPS waypoints to virtual locations within the game's narrative. As the students move through a designated area, such as a sports field or playground, a map on their mobile device displays multimedia elements. It is important to note that at the time of HARP's 'Alien Contact!' pilot, MIT provided the required mobile devices. The scale of maintenance and management of customized equipment of this nature is a highly prohibitive for barrier for most schools. As with any specialized technology, such designs are time consuming, prone to obsolescence and present a significant cost of ownership to a school budgets. As telecommunication providers upgrade and enhance networks and mobile devices to gain market share in the highly competitive wireless market, many of the feature sets required for such AR curricula are emerging on mobile cell phones. This migration from custom hybridized PDA/GPS technologies to commercially available mobile phone technologies with both wireless and cellular capabilities will allow educators to take advantage of the affordances of using ubiquitous, feature-rich devices students could bring to classrooms.

With these possible affordances come challenges and controversies that revolve around the role of personal technologies in schools. Adoption of AR enable mobile devices for teaching and learning will require a considerable shift in perceptions among educators and administrators regarding the presence of mobile phones in classrooms. Much of the debate and controversy around mobile phones in the classroom revolves around the perception that mobile phones will only serve as a barrier to instruction, citing their potential to distract and a means to facilitate cheating on tests and assessments as justification for their exclusion from classroom utility. School systems will need to address the evolution of mobile phones not simple as a telephony device, but a powerful, fully featured, commercially supported information infrastructure already paid for and maintained by students. The powerful opportunities presented by AR for teaching and learning is poised to move school systems closer to rethinking approaches to mobile devices in schools.

Projecting the outcomes of emergent research and development in AR applications leveraging faster mobile networks, improved registration algorithms, and advancements in mobile hardware, AR stands not only to be powerful force in education, but a disruptive one. During the 2010 Mobile World Congress, Google CEO Eric Schmidt demonstrated a prototype of visual search application named Google Goggles — an experimental AR application capable of interpreting and translating text captured in photos. This prototype uses technologies emerging from Google's research and development of machine translation and image recognition.(Brian, 2010) Although browser based machine translations and optical character recognition (OCR) functionality have been available for a number of years, integration of these features with a mobile device aims to provide immediate access to additional layers of context. The prototype enables a user take a photo using a mobile device which interprets and encodes recognizable characters contained in the image using a integrated OCR algorithm. The application maximizes performance and efficiency by passing this data via a wireless broadband connection to servers capable of interpreting this OCR generated data quickly and returning relevant translations and contextual information. There are considerable implications of such portable translation functionality for ESL language instruction. As this technology matures, the ability to readily translate and represent foreign texts in a native language on mobile devices with emerge. This feature stands to be a significant disruptive technology, mirroring in some respects the introduction of calculators in mathematics instruction. The speed with which such an application will be likely to render a translation will engender the same type of debates still ongoing among mathematics educators: what does it mean to assess, what forms of assessment are most important, what assessment practices are inadequate measures of learning, what new practices are to be proposed, and what are the curricular implications of changes in assessments. (Bright, et al, 1993)

CONCLUSION

A perfect storm of hardware, computing power, locative technology, and intuitive user interface design have converged in new generations of mobile devices enabling rapid growth in consumer AR applications. These applications are capable of retrieving, filtering, and organizing context specific information in increasingly novel and useful ways. These applications offer business new ways of delivering content and services to customers, provide rich informational systems for tourism and museum applications, and offer a means to situate students in authentic learning environments with interactive data delivered with multiple modalities. The expense of mobile devices featuring the unique hardware combinations required for augmented reality applications put these devices out of reach of the majority of consumers. As Moore's Law ("Moore's Law", 2010) evolves manufacturing processes of mobile devices and reduces price points, the unique capabilities that facilitate augmented reality will become standard features on all mobile devices. This process is already underway and is evidenced by the emergence and proliferation of commercial AR applications. The mass popularization of the Internet has led to an unprecedented growth in both access and dissemination of information. The ability to readily access rich, context specific information for collaborative and creative problem solving is critical for effective teaching and learning. This process has always been a part of teaching and learning from the Gutenberg Revolution to the emergence of the Internet, a notion that is addressed in Rheingold's concept of 'network awareness'. (Rowell, 2010) Rheingold asserts that societies are network oriented and suggests that the emergence of Internet enabled technologies is simply a natural evolution in global information ecosystems and should be considered an important part of digital literacy. Personal, mobile technologies will continue to be a disruptive technology in educational institutions, challenging policy and practice while presenting new issues and opportunities.(Christensen, 2008) As applications emerge, educators must consider how best to incorporate AR functions into pedagogy, and how to best to support AR through a reconsideration of the role of students' personal mobile technologies in schools. In 1996, Blair MacIntyre forecast that wearable see-through displays could emerge to be the Sony Walkman of the early 21st century (MacIntyre & Feiner, 1996). Augmented reality applications are poised to become a pervasive feature of ever expanding information ecosystems and fulfill this prophecy in the coming decades.

REFERENCES

Aguera, B. (2007). *Arcas demos Photosynth*. Retrieved April 10, 2010, from http://www.ted.com/talks/blaise_aguera_y_arcas_demos_photosynth.html

Aukstakalnis, S., & Blatner, D. (1992). *Silicon mirage-the art and science of virtual reality*. Berkeley, CA: Peachpit Press.

Azuma, R. T. (1997). A survey of augmented reality. *Presence (Cambridge, Mass.)*, *6*(4), 355–385.

Brian, M. (2010). Google Goggles releases an exciting new feature. Lets you translate…the world. Retrieved June 1, 2010, from http://thenextweb.com/mobile/2010/05/06/google-googles-releases-exciting-feature-lets-translatethe-world/

Bright, G. W., Waxman, H. C., & Williams, S. E. (1993). *Impact of calculators on mathematics instruction*. Lanham, MD: University Press of America.

Camarata, K., Yi-Luen Do, E., Gross, M. D., & Johnson, B. R. (2002). *Navigational blocks: Tangible navigation of digital information*. Minneapolis, MN: ACM Press.

Caudell, T. P., & Mizell, D. W. (1992). Augmented reality: An application of heads-up display technology to manual manufacturing processes. *Proceedings of 1992 IEEE Hawaii International Conference on Systems Sciences*, (pp 659-669).

Christensen, C. M., Johnson, C. W., & Horn, M. B. (2008). *Disrupting class: How disruptive innovation will change the way the world learns*. Toronto, ON: McGraw-Hill.

Dourish, P. (2001). *Where the action is: The foundations of embodied interaction*. Cambridge, MA: MIT Press.

Drascic, D., & Milgram, P. (1996). Perceptual issues in augmented reality. In *Stereoscopic Displays and Virtual Reality Systems III* (pp. 123–134). San Jose, CA: SPIE.

Esquire Magazine. (2010). *Augmented reality issue*. Retrieved January 6, 2010, from http://www.esquire.com/the-side/augmented-reality

Feiner, S. K. (2002). Augmented reality: A new way of seeing: Computer scientists are developing systems that can enhance and enrich a user's view of the world. *Scientific American*, 34–41.

Fjeld, M., Lauche, K., Bichsel, M., Voorhorst, F., Krueger, H., & Rauterberg, M. (2002). Physical and virtual tools: Activity theory applied to the design of groupware. *Computer Supported Cooperative Work*, *11*, 153–180. doi:10.1023/A:1015269228596

Giles, J. (2005). Internet encyclopaedias go head to head. *Nature*, *438*, 900–901. doi:10.1038/438900a

Google. (2010). *Favorite places*. Retrieved April 10, 2010, from http://www.google.com/help/maps/favoriteplaces/gallery/#los-angeles-ca

Joscelyne, A. (1994). VR means virtual reconstruction: Cluny Abbey has been rebuilt. *Wired Magazine*, *2*(1), 114–115.

King, R. (2009, November 3). Augmented reality helps sell the product. *BusinessWeek*. Retrieved April 10, 2010, from http://www.businessweek.com/technology/content/nov2009/tc2009112_198019.htm

MacIntyre, B., & Feiner, S. (1996). Future multimedia user interfaces. *Multimedia Systems*, 4.

Mashable.com. (2010). *Detroit Red Wings make game programs interactive with QR codes*. Retrieved February 15, 2010, from http://mashable.com/2010/02/10/red-wings-qr-codes/

Milgram, P., & Kishino, F. (1994). A taxonomy of mixed reality visual displays. *IEICE Transactions on Information Systems E*, *77-D*(12), 1321–1329.

Milgram, P., Takemura, H., et al. (1994). Augmented reality: A class of displays on the reality-virtuality continuum. *SPIE Proceedings: Telemanipulator and Telepresence Technologies*. (pp. 282-292).

MIT HARP. (2010). Retrieved February 10, 2010, from http://www.isites.harvard.edu/icb/icb.do?keyword=harp&pageid=icb.page69619

Moore's law. (2010). In *Wikipedia, the free encyclopedia*. Retrieved June 14, 2010, from http://en.wikipedia.org/wiki/Moore's_law

O'Shea, P., Mitchell, R., Johnston, C., & Dede, C. (2009). Lessons learned about designing augmented realities. *International Journal of Gaming and Computer-Mediated Simulations*, *1*(1), 342–349.

Perez, S. (2009). Matell's new Web enabled toys will offer augmented reality. Retrieved April 10, 2010, from http://www.readwriteweb.com/archives/mattels_new_web-enabled_avatar_toys_will_offer_augmented_reality.php

Rowell, L. (2010). An interview with Howard Rheingold. Retrieved March 28, 2010, from http://www.elearnmag.org/subpage.cfm?article=111-1&section=articles

Schnadelbach, H., Koleva, B., Flintham, M., Fraser, M., Izadi, S., & Chandler, P. ... Rodden, T. (2002). *The Augerscope: A mixed reality interface for outdoors.* CHI'02, ACM Press, Minneapolis, USA, (pp. 9–16).

Technology Review. (2010). Microsoft adds augmented reality to Bing maps. Retrieved February 15, 2010, from http://www.technologyreview.com/blog/editors/24800/?ref=rss%20target=&a=f

Underkoffler, J., & Ishii, H. (1998). Illuminating light: An optical design tool with a luminous-tangible interface. In *CHI'98 Proceedings* (pp. 542–549). New York, NY: ACM Press/Addison-Wesley Publishing Co.

Vlahakis, V., Ioannidis, N., & Karigiannis, J. (2002). Archeoguide: Challenges and solutions of a personalised augmented reality guide for archaeological sites. *Computer Graphics in Art, History and Archaeology. Special Issue of the IEEE Computer Graphics and Applications Magazine, 22*(5), 52–60. doi:10.1109/MCG.2002.1028726

Wall Street Journal. (2009). *Esquire flirts with digital reality.* Retrieved January 6, 2010, from http://online.wsj.com/article/SB10001424052748704222704574501122991439500.html

WARC. (2009). *Augmented reality attracts big customers.* Retrieved December 13, 2009, from http://www.warc.com/News/TopNews.asp?ID=25974&Origin=WARCNewsEmail

Weisner, M. (1991). The computer for the 21st century. *Scientific American, 265*(3), 94–104. doi:10.1038/scientificamerican0991-94

Welch, R. B. (1978). *Perceptual modification.* New York, NY: Academic Press.

Wellner, P., Mackay, W., & Gold, R. (Eds.). (1993). Special issue on computer augmented environments: back to the real world. *Communications of the ACM, 36*(7), 212–231.

Windowshop from your mobile. (2009). Retrieved January 6, 2010, from http://googleblog.blogspot.com/2009/12/explore-whole-new-way-to-window-shop.html

ADDITIONAL READING

Barfield, W., & Caudell, T. (Eds.). (2001). *Fundamentals of Wearable Computers and Augmented Reality.* Mahwah, NJ: Lawrence Erlbaum.

Billinghurst, M., Kato, H., & Poupyrev, I. (2001, November). The MagicBook: A Transitional AR Interface. *Computers & Graphics*, 745–753. doi:10.1016/S0097-8493(01)00117-0

Brooks, F. (1995). Augmented Reality, http://www.cs.unc.edu/Research/graphics.

Caudell, T. P. (1994). *Introduction to Augmented Reality. SPIE Proceedings (Vol. 2351).* Telemanipulator and Telepresence Technologies.

Cawood, S., & Fiala, M. (2007). *Augmented reality: A practical guide.* Raleigh, NC: Pragmatic Bookshelf.

Digital reconstruction of Yaunmingyuan garden using AR view (2010). [Video file] Retrieved April 10, 2010 from http://www.youtube.com/watch?v=N-9tF4dJISs

Dunleavy, M., C. & Mitchell, R. (2008). Affordances and Limitations of Immersive Participatory Augmented Reality Simulations for Teaching and Learning. *Journal of Science Education and Technology*, *18*, 7–22. doi:10.1007/s10956-008-9119-1

Durlach, N. I., & Mavor, A. S. (Eds.). (1995). *Virtual Reality: Scientific and Techological Challenges*. Washington, DC: National Academy Press.

Elvins, T. (1998). Augmented Reality: The Future's So Bright, I Gotta Wear (See-through) Shades. *Computer Graphics*, *2*, 11–13. doi:10.1145/279389.279411

Fitzmaurice, G., Ishii, H., & Buxton, W. (1995). Bricks: Laying the foundations for graspable user interfaces. Proceedings of the ACM Conference on Human Factors in Computing Systems (CHI'95). pp. 432-449. New York: ACM.

Gee, J. P. (2003). *What video games have to teach us about learning and literacy*. New York: Palgrave Macmillan.

Haller, M., Billinghurst, M., & Thomas, B. (Eds.]. (2007). *Emerging technologies of augmented reality: Interfaces and design*. Hershey, PA: Idea Group.

Hannes, K., & Schmalstieg, D. (2002). *2002 Abstracts and Applications* (pp. 37–41). Mathematics and Geometry Education with Collaborative Augmented Reality. In SIGGRAPH.

Hedley, N. R., Billinghurst, M., Postner, L., May, R., & Kato, H. (2002). Explorations in the use of augmented reality for geographic visualization. *Presence (Cambridge, Mass.)*, *11*(2), 119–133. doi:10.1162/1054746021470577

Höllerer, T., Feiner, S., Terauchi, T., Rashid, G., & Hallaway, D. (1999). Exploring MARS: Developing indoor and outdoor user interfaces to a mobile augmented reality system. *Computers & Graphics*, *23*(6), 779–785. doi:10.1016/S0097-8493(99)00103-X

Horizon Report. (2009) NMC Emerging Technology Initiative. Retrieved January 6, 2010 from: http://www.nmc.org/pdf/2009_Horizon_Report.pdf

Kerawalla, L., Luckin, R., Seljeflot, S., & Woolard, A. (2006). Making it real: exploring the potential of Augmented Reality for teaching primary school science. *Virtual Reality (Waltham Cross)*, *10*(3-4), 163–174. doi:10.1007/s10055-006-0036-4

Klopfer, E., & Squire, K. (2008). Environmental detectives: The development of an augmented reality platform for environmental simulations. *Educational Technology Research and Development*, *56*(2), 203–228. doi:10.1007/s11423-007-9037-6

Liarokapis, F., Petridis, P., Lister, P., & White, M. (2002). Multimedia augmented reality interface for E-learning (MARIE). In *World Transactions on Engineering and Technology Education. UICEE Vol. 1*, No.2.

McDonald's Japan places QR Codes on Big Macs (2007, September 17). Retrieved April 10, 2010 from: http://www.esato.com/news/article.php/id=1094

Mountain, D., & MacFarlane, A. (2007). Geographic information retrieval in a mobile environment: evaluating the needs of mobile individuals. *Journal of Information Science*, *33*(5), 515–530. doi:10.1177/0165551506075333

Norman, D. A. (1988). *The design of everyday things*. New York: Basic Books.

Reality only better. The Economist. Technology Quarterly. Dec 6, 2007. Retrieved November 20, 2010 from: http://www.economist.com/search/displaystory.cfm?story_id=102026233

Resnick, M. (2006). Computer as paintbrush: Technology, play, and the creative society. In D. Singer, R. M. Golinkoff & K. Hirsh-Pasek, K. (Eds.) *Play = Learning*, (pp. 1-16). Oxford: Oxford University Press.

Rogers, Y., Scaife, M., Gabrielli, S., Smith, H., & Harris, E. (2002). A Conceptual Framework for Mixed Reality Environments: Designing Novel Learning Activities for Young Children. *Presence (Cambridge, Mass.)*, *11*(6), 677–686. doi:10.1162/105474602321050776

Rogers, Y., Scaife, M., Gabrielli, S., Smith, I.-I., & Harris, E. A. (2002). Conceptual framework for mixed reality environments: Designing novel learning activities for young children. *Presence (Cambridge, Mass.)*, *11*(6), 677–686. doi:10.1162/105474602321050776

Squire, K. D., & Jan, M. (2007). Mad city mystery: Developing scientific argumentation skills with a place-based augmented reality game on handheld computers. *Journal of Science Education and Technology*, *16*(1), 5–29. doi:10.1007/s10956-006-9037-z

Starner, T., Mann, S., Rhodes, B., Levine, J., Healey, J., & Kirsch, D. (1997). Augmented Reality Through Wearable Computing. *Presence (Cambridge, Mass.)*, *6*(4), 386–398.

Todorov, E., Shadmehr, R., & Bizzi, E. (1997). Augmented Feedback Presented in a Virtual Environment Accelerated Learning of a Difficult Motor Task. *Journal of Motor Behavior*, *29*(2), 147–158. doi:10.1080/00222899709600829

Vandamme, F. (1997). Virtual and Augmented Reality Technology (VAR): The Tool for Realising Education Targets? CC-AI; *The Journal for the Integrated Study of Artificial Intelligence. Cognitive Science and Applied Epistemology*, *14*(1), 73–80.

Wikitude. (2010). Wikitude World Browser. Retrieved April 2, 2010 from http://www.wikitude.org/category/02_wikitude/world-browser

Winn, W. (2003). Learning in artificial environments: embodiment, embeddedness and dynamic adaptation. *Technology, Instruction. Cognition and Learning*, *1*, 87–114.

KEY TERMS AND DEFINITIONS

Augmented Reality (AR): A computer generated information system allowing for the superimposing of text, image, and multimedia layers over video streams of direct or indirect views of a physical real-world environment. It is related to the general concept of mediated reality (MR) in which a view of reality is modified by a computer.

Global Positioning System (GPS): A network system utilizing orbiting satellites to transmit signals to receivers which calculate relative location, speed, direction, and time. GPS is widely used as a navigation tool in a range of consumer electronics products.

Mediated Reality (Med-r): The process of subtracting, adding, or otherwise manipulate a person's perception of their immediate environment through the use of a wearable computer or mobile device.

Mixed Reality Environments (MRE): Refers to systems that blend real and virtual experiences to produce new environments and visualizations allowing for the representation of both physical and computer generated objects co-exist and interact in real time.

Mobile Device: Any computing or communications device intended to frequently move location while maintaining function and operation.

Virtual Reality (VR): Virtual reality is computer generated environment presented in such a way that a user is immersed in this environment with partial or complete exclusion of all other stimuli. Virtual reality offered on personal computers are typically experienced through two of the five **Senses:** sight and sound. The simplest form of virtual reality is a 3-D image that can be explored at a personal computer by using input devices to manipulate the simulated environment and user perspectives on this environment. More sophisticated virtual reality projects involve such approaches as rooms featuring large format displays and wearable computers quick response codes (QR codes): Barcode technology allowing for complex encoding of data through patters and color.

Web 2.0: A term commonly associated with online applications that facilitate interactive sharing of information featuring data interoperability, intuitive user interface designs, and collaboration features. A online application typically labeled as Web 2.0 allows its users to interact and co-author content. This is in contrast to earlier iterations of online content and services that limited users to passive viewing of content provided.

Chapter 13
Student Perceptions and Uses of Wireless Handheld Devices:
Implications for Implementing Blended and Mobile Learning in an Australian University

Raj Gururajan
University of Southern Queensland, Australia

Abdul Hafeez-Baig
University of Southern Queensland, Australia

P. A. Danaher
University of Southern Queensland, Australia

Linda De George-Walker
University of Southern Queensland, Australia

ABSTRACT

If the implementation of blended and mobile learning across disciplines is to be maximized, it is important for researchers to understand how students perceive and use wireless handheld devices, in order to build on those current practices and help to facilitate the next level of adoption. To enhance that understanding, this chapter analyzes data from 228 survey questionnaires conducted in October and November 2008. Students were enrolled in two faculties at the authors' institution. Questions explored students' views and opinions about the uses of wireless handheld devices, such as personal digital assistants, handheld PCs, and smart phones, for teaching and learning activities.

The chapter draws on a case study method using factor and regression analysis to interpret the questionnaire responses about the uses of wireless handheld devices in higher education. The principal findings included that behavior and attitude contribute strongly to the perceived performance of using such devices in the chosen context, and that facilitating conditions have a more complex and mediated

DOI: 10.4018/978-1-60960-511-7.ch013

relationship with behavior and attitude on the one hand and perceived performance on the other. The authors elaborate the implications of those findings for increasing alignment across several different interfaces related to blended and mobile learning in the early 21st century.

INTRODUCTION

Universities are faced with a large number of contemporary challenges. One such challenge is matching the opportunities presented by leading edge technologies with the educational requirements related to curriculum, teaching and learning, and assessment (see also Hansson, 2008). If these two key elements of a university's enterprise are out of alignment, the outcome is likely to be ineffective education and inefficient deployment of resources, resulting in loss of student engagement, community support, and potentially financial viability.

An increasingly popular technological resource used for a progressively diverse variety of professional and personal purposes is the wireless handheld devices such as cell and smart phones, laptop computers, personal digital assistants, and tablet PCs (Dieterle, Dede, & Schrier, 2007). Heightened technical convergence has not only blurred the distinction between public and private in using these devices but also created new possibilities for maximizing the implementation of blended and mobile learning within and across disciplines in all formal educational sectors.

While this is an exciting technological development, it is important to recognize that technical advances do not automatically generate educational outcomes. That is, if wireless handheld devices are to contribute to enhancing the provision of blended and mobile learning in the university sector, considerable attention must be paid to their respective configurations and affordances if the technological capacities are to be matched closely with the educational needs of students and teachers.

One potentially useful strategy for interrogating this prospective matching is to identify and analyze the ways in which students currently perceive and use wireless handheld devices (see also Patten, Armedillo Sánchez, & Tangney, 2006). If those students see those devices as helping to blend their synchronous and asynchronous learning without being tied to predetermined physical locations, and are already using the devices in that way, universities have a strong foundation for building on student ownership and technological capability. If by contrast students regard the devices as being predominantly for private use and as disconnected from their studies, universities might wish to explore whether such an attitudinal gulf can and should be bridged.

This chapter applies this strategy in relation to the technological and educational challenges and opportunities at a single Australian university. Working in faculties of business and education, the authors are actively engaged in researching (Danaher, Gururajan, & Hafeez-Baig, 2009; De George-Walker, Hafeez-Baig, Gururajan, & Danaher, 2010; Hafeez-Baig & Danaher, 2007a, 2007b, accepted for publication) and promoting blended and mobile learning for undergraduate and postgraduate students enrolled in both on- and off-campus modes, the latter including students from around Australia and several other countries (see also Hafeez-Baig, Gururajan, Nazemi, De George-Walker, & Danaher, 2010). They regard such learning as crossing disciplinary boundaries and as potentially harnessing the technical capabilities of contemporary technologies for educational purposes.

The chapter analyzes data from 228 survey questionnaires administered in late 2008 to students enrolled in two faculties at the authors' institution. The questionnaire explored students' views and opinions about the uses of wireless handheld devices for teaching and learning activities. The authors use the findings from the questionnaire to frame and inform a set of implications for implementing blended and mobile learning at their university and across the university sector more broadly. In particular, they argue that any such implementation must be directed comprehensively and simultaneously at several diverse interfaces: between students and the university; between the public and private arenas of students' lives; between the educational and technical dimensions of technologies; and between separate academic disciplines. If those interfaces can be broached and secured, blended and mobile learning outcomes can be maximized and sustained, to the benefit of students, teachers, administrators, and community members alike (Dieterle & Dede, 2007).

The chapter's objectives are therefore to: (a) report the results of the questionnaire; (b) analyze those results in terms of the influences most likely to promote wider scale adoption of wireless handheld devices; and (c) link that analysis to broader issues pertaining to bridging the gaps between blended and mobile learning in contemporary university teaching and learning. The chapter begins by outlining selected themes distilled from relevant literature, and the national and institutional settings in which the study took place.

BACKGROUND

Blended and mobile learning constitutes a rapidly growing area of serious scholarship, both separately and in combination. Blended learning, understood as the convergence of previously separate modes of delivering education (such as distance, face-to-face, and online) in a single course or program, has been extensively researched in higher education contexts (Garrison & Kanuka, 2004; Garrison & Vaughan, 2008). This research has been examined from the perspective of different disciplines such as business (Akyol, Garrison, & Ozden, 2009; Arbaugh, Godfrey, Johnson, Leisen Pollack, Niendorf, & Wresch, 2009; De George-Walker, Hafeez-Baig, Gururajan, & Danaher, 2010) and education (De George-Walker, Hafeez-Baig, Gururajan, & Danaher, 2010). Similarly varied conceptual frameworks have been drawn on in this research, ranging from cognitive presence (Vaughan & Garrison, 2005) and community of inquiry (Arbaugh, Cleveland-Innes, Diaz, Garrison, Ice, Richardson, & Swan, 2008; Shea & Bidjerano, 2009; Swan & Ice, 2010) to quality (Ginns & Ellis, 2007) and collaborative learning (Su & Brush, 2008), including a dissension from the term "blended learning" on both philosophical and practical grounds (Oliver & Trigwell, 2005).

Likewise mobile learning, conceptualized here as the delivery of learning via technologies that are not fixed in permanent place (such as the wireless handheld devices that are the focus of this chapter), is the subject of increasing scholarship in the published literature. Like blended learning, mobile learning has been conceptualized from multiple perspectives, including activity theory (Uden, 2007; Wali, Winters, & Oliver, 2008), learning communities (Danaher, Moriarty, & Danaher, 2009), and learning theory (Sharples, Jaylor, & Vavoula, 2007). Also like blended learning, mobile learning has been evaluated with regard to its utility across several fronts, including in relation to promoting cooperative and collaborative learning (Motiwalla, 2007), enhancing environmental awareness (Uzunboylu, Cavus, & Ercag, 2009), and maximizing the attainment of critical thinking skills (Cavus & Uzunboylu, 2009). Mobile learning has been used equally effectively in engaging with children (Druin, 2009; Kim, Miranda, & Olaciregui, 2008; Kurti, Spilol, & Milrad, 2008) and in higher education settings (Alexander, 2004; Herrington & Herrington, 2007).

The literature analyzing the convergence between blended and mobile learning is necessarily more selective and specialized. Among other themes, that literature has comprised an examination of the impact of such an environment on students' learning behaviors and performance (Wang, Shen, Novak, & Pan, 2008), issues in designing appropriate instruction for digital natives (Wu, 2009), one approach to structuring the combination of blended and mobile learning (Diamantini & Pieri, 2008), research-based guidelines for maximizing the pedagogical benefits of that combination (Sims, Burke, Metcalf, & Salas, 2008), and interrogating the extent to which that combination constitutes a new educational paradigm (Zawacki-Richter, Brown, & Delport, 2007). As with most educational technology literature, the research into the convergence between blended and mobile learning exhibits sometimes excessively optimistic claims about the technical possibilities, leavened by more cautionary tales emphasizing that such possibilities do not always or easily transform into positive pedagogical change.

Finally, there is growing evidence in the research literature of the technological affordances and the educational possibilities of wireless handheld devices. This evidence entails both "straightforward and deep effects" (Dieterle & Dede, p. 1), or surface and deep learning, as well as the distinctive affordances of the devices (Churchill & Churchill, 2008), their status as "effective just-in-time educational tools" (Choi, 2005, p. 825), and their capacity to support secondary schools students' outdoor educational activities (Churchill, Kennedy, Flint, & Cotton, 2010). At the same time, some of the literature highlights limitations of the published research and identifies areas that need to be explored further, such as the costs and potential limitations of using handheld devices for an extended period (Cheung & Hew, 2009).

This inevitably partial review of the research literature demonstrates on the one hand that scholarship is already well-advanced in investigating wireless handheld devices as agents of facilitating effective blended and mobile learning and on the other hand that much remains to be explored in this field. In particular, this chapter and the broader project of which it forms a part contribute a deeper understanding of how students perceive and use these devices in particular contexts, thereby providing a more rigorous basis for making decisions about whether it is worthwhile to invest more time and funding in enhancing the educational applications of these devices and if so which applications are most likely to repay such an investment.

STUDENT PERCEPTIONS AND USES OF WIRELESS HANDHELD DEVICES

Methodology

As noted above, the research presented here is part of a wider study that included three separate phases (N.B.: This section of the chapter is based on the equivalent section of Hafeez-Baig, Gururajan, Danaher and De George-Walker (in press), although that paper reports on a different data selection from that presented here.). The first phase was an exploratory study to identify through focus group discussions the initial views and opinions of users toward the use of wireless technologies in the higher education environment. This phase resulted in the identification of themes for the use of wireless technologies in the higher education environment. The second phase was a confirmatory study based on the findings of phase 1. The third phase used the survey approach to gather students' views about the uses of wireless technologies in the higher education environment.

While the first two phases involved qualitative techniques, the third phase involved a quantitative technique. The findings of the first two phases have already been published; this chapter presents the findings of the third phase. Once the themes had been ascertained from the first two phases, a survey instrument was developed based on these findings to run a university wide survey to quantify the determinants. This study adopted a survey approach to explore the views and opinions of the participating Australian university students toward wireless handheld devices. The instrument was derived from focus group discussions from an earlier study conducted with students from the same university. The instrument was reviewed by academics from the Faculty of Education whose research interest was e-learning/m-learning. This instrument was also pilot tested with students from the Faculties of Business and Education at the same institution.

The mixed method approach was adopted to address the study's aforementioned focus on the influences most likely to promote wider scale adoption of wireless handheld devices. Remenyi, Williams, Money, and Swartz (1998) demonstrated that qualitative approaches can be effective in studying human and psychological factors, and that quantitative approaches can be useful for investigating the prevalence or otherwise of those factors in a wider population. In this research, through focus group techniques, initial information was gathered from the users about their views and opinions of using wireless technologies in a university environment (Remenyi et al., 1998). As noted above, through the findings of the qualitative data analysis, a survey instrument for the quantitative approach was developed to acquire the views and opinions of a large number of students.

Participants in this study were selected randomly from the Faculties of Business and Education for undergraduate and postgraduate courses. A list of courses was identified from each of those faculties with exposure to wireless and ICT technologies. The lecturers in the selected courses were contacted to gauge their willingness to provide an opportunity for one of the authors to talk to the students to explain the research and the role of their participation in the study. The survey instrument clearly mentioned that their participation was voluntary, and the instrument was distributed during the lectures.

The institution where the study took place is an Australian distance education intensive university, having attained that status in the early 1990s after functioning for more than two decades as a college of advanced education. While it operates three campuses in its base region, most of its students enroll as distance and/or online learners and live throughout Australia and several other countries, particularly in Asia, Europe, and North America. Like most Australian universities, it must compete for students in the context of a deregulated higher education market and relatively scarce government funding. If its students are to be attracted and retained, it is crucial to understand how they perceive and use devices that can potentially facilitate their learning outcomes.

A total of 600 survey copies were distributed and 228 usable surveys were coded in the Microsoft Excel spreadsheet. The majority of the participants were undergraduate students and belonged to the Faculties of Business and Education, as was explained above. Participation was almost equally divided by gender, and most participants had one to three years of experience in using educational technologies. The perceived performance (PP) of wireless handheld devices was measured in the context of attitude (A), behavior (B), and facilitating conditions (FC) toward using such devices. Before the higher level statistical analysis was conducted, an initial framework was formulated for the study and is reproduced in Figure 1.

Analyzing the Questionnaire Results

The process of analyzing the data in this part of the study was descriptive, statistical, and inferential. Descriptive analysis helps to summarize and simplify the data, so that large numbers of data can be described in a meaningful manner, such as being able to see how the data are dispersed (Graziano & Raulin, 2006). For example, to understand the characteristics of the data collected, frequency analysis can help the researcher to explore the data in relation to demographic information.

Through inferential analysis, a researcher tries to interpret the findings of the descriptive and other

Figure 1. Initial research framework for the study

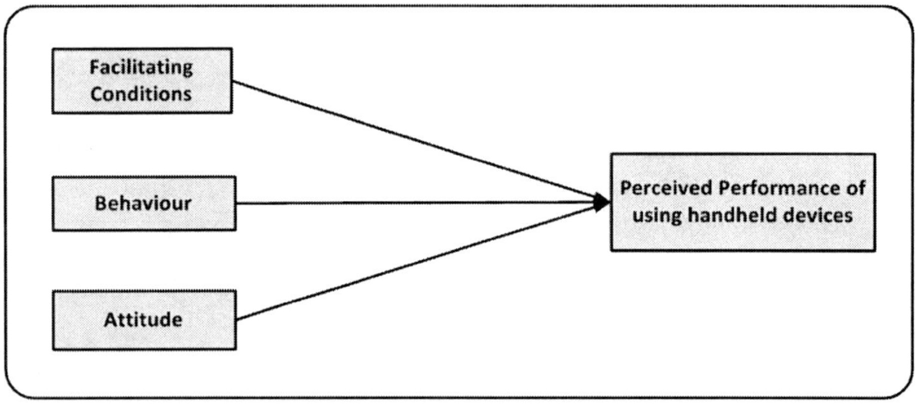

statistical techniques in order to analyze the data and comprehend their meanings and implications. Examples include tests of statistical significance such as the *t*-test, the chi-squared test, and regression analysis (De Vaus, 2002; Graziano & Raulin, 2006; Zikmund, Babin, Carr, & Griffin, 2009).

In this chapter we have used statistical techniques such as descriptive analysis, reliability analysis, correlation analysis, and regression analysis. Zikmund et al. (2009) define reliability as the degree to which a measure is free of error and provides consistent results, and validity as the ability of the scale to measure what is intended to be measured. In the case of multiple regression analysis, the data need to be normally distributed, and there need to be 20 observations for each independent variable (Hair, Black, Babin, & Anderson, 2009).

A descriptive analysis through SPSS was conducted to ensure the data were error free. Descriptive statistics analysis may consist of mean, variance, standard deviation, median, and missing value analysis. In this research, a cross tabulation procedure was used to summarize the data through mean, mode, median, standard deviation, variance, and frequency count, as illustrated in Table 1.

There were four composite variables in the study's initial research framework, as shown in Figure 1. In order to evaluate if there were a significant relationship between them, a Pearson product–moment correlation (abbreviated as Pearson *r*) was conducted. Pearson *r* was suitable as all the variables were measures on the ratio scale. As can be seen from Table 2, there was a positive correlation among all the independent variables *FC*, *B*, and *A* to the dependent variable *PP* ($p < 0.05$ for all, and the *r* values were positive). The value of the correlation ranged from 0.4 to 0.6, as all the correlations were below 0.5 and most of them were quite low, except that the correlation between *PP* and *A* was slightly above the 0.5 value (the actual value was 0.621). To have significant correlation we needed the r value to exceed 0.8 (Hair et al., 2009). Hence, we can assume that all the composite variables were contributing uniquely to the dependent variable *PP*.

Cronbach's alpha was used to ascertain the reliability of the instrument and the composite variables *FC*, *B*, *A*, and *PP*. Cronbach's alpha was calculated from SPSS procedures. Generally, a value of 0.70 is considered an acceptable level of reliability in social science research. Hair et al. (2009) suggested that an acceptable limit can be reduced to 0.60 in exploratory research. In this research, the reliability of the instrument and the various item used to calculate the composite variables are summarized in Table 3.

Table 1. Summary of the descriptive analysis (based on Hafeez-Baig, Gururajan, Danaher, & De George-Walker, in press, p. 23)

Category	Descriptions	Percentage
Participants	Undergraduate	78.1%
	Postgraduate	20.6%
Gender	Male	46.9%
	Female	51.8%
Age		
	Below 20	11%
	20-25	39.5%
	26-30	24.1%
	31-35	12.7%
	36-40	3.9%
	41-45	6.6%
	Above 45	2,2%
Technological experience		
	None	16.2%
	Less than 1 year	19.7%
	Less than 2 years	26.8%
	Between 3 and 5 years	9.2%
	Between 6 and 8 years	9.6%
	Between 9 and 11 years	4.8%
	Between 12 and 14 years	11.8%
	More than 14 years	1.8%

For all the independent composite variables included for this test, Cronbach's alpha was 0.86, which is considered a very high result.

Linear regression analysis was conducted individually for all the independent variables (*FC*, *B*, and *A*) against the dependent variable PP, through the "enter" procedure of SPSS, in order to test the relationship between the dependent variable and the independent variables. In this study, all variables were considered as being at the metric level, with one dependent variable and multiple independent variables as predictors, but entered in the analysis separately. In linear regressions, *R* is used to measure the strength of the relation between the criteria and the predictors. A summary of this analysis is shown in Table 4.

Implications of the Data Analysis

The summary of the reliability analysis in Table 3 provided evidence that the items used to measure the composite independent and dependent variables were the true representation of the composite variables. Moreover, this relationship was further tested by means of a Pearson correlation. The correlation analysis in Table 2 presented confirmation that there was a correlation among the variables,

Table 2. Correlational analysis for the composite variables

IDV (*FC*, *B*, and *A*) and *DV* (*PP*) Correlations				
Variables	Perceived Perform-ance	Facilitating Conditions	Behavior	Attitude
Perceived Performance	1	.422**	.537**	.621**
		.000	.000	.000
Facilitating Conditions	.422**	1	.543**	.451**
	.000		.000	.000
Behavior	.537**	.543**	1	.476**
	.000	.000		.000
Attitude	.621**	.451**	.476**	1
	.000	.000	.000	
**. Correlation is significant at the 0.01 level (2-tailed).				

Table 3. Item descriptions and their reliability for the development of composite variables

No.	Composite variable	Questions included	Cronbach's alpha
1	Perceived Perception	4 items	0.829
2	Facilitating Conditions	3 items	0.654
3	Attitude	3 items	0.847
4	Behavior	3 items	0.862
5	All independent composite variables	9 items	0.859
6	All composite variables	4 items	0.805

and that the three independent variables *FC*, *B*, and *A* are a true measure of the dependent variable *PP*.

We used multiple regression analysis to explore further the relationship between the independent and dependent variables. Table 4 provided a summary of various regression analyses conducted in this study. From the top sections of this table, it was found that the variables *A* and *B* were strong determinants of *PP* ($B = 0.31$, $t = 4.7$, $p < 0.05$) and ($B = 0.46$, $t = 8.0$, $p < 0.05$) respectively. However, it was found that the variable *FC* did not contribute significantly to determining *PP* ($B = 0.07$, $t = 0.97$, $p > 0.05$).

Further regression analysis was conducted to explore the relationship between the variables *A* and *B* and the variable *FC*. From the middle section of Table 4, it was found that there was a strong relationship between *FC* and *A* ($B = 0.21$, $t = 4.1$, $p < 0.05$), as well as with *B* ($B = 0.38$, $t = 6.9$, $p < 0.05$). There was also a direct relationship between *FC* and *PP* ($B = 0.51$, $t = 7.0$, $p < 0.05$).

On the basis of this regression analysis, the study's initial research framework can be modified as represented in Figure 2.

FUTURE RESEARCH DIRECTIONS

The difference between Figures 1 and 2 can be summarized as relating to the rejection of a simplified, linear relationship between the three independent variables and the dependent variable in favor of

Table 4. Summary of the linear regression analysis of the composite variables FC, B, and A in relation to the DV PP

Description of composite variable	*R* value	Adjusted R^2	Degree of freedom	*F*-value	Sig value	Beta value	*t*-value	Sig level
Facilitating Conditions, Behavior, and Attitude are independent variables and Perceived Performance is the dependent variable								
Facilitating Conditions	.681	.456	3, 224	64.505	0.000	.071	0.977	.330
Attitude						.310	4.697	.000
Behavior						.462	7.934	.000
Behavior, and Attitude are independent variables and Facilitating Conditions is the dependent variable								
Behavior	.586	.337	2, 225	58.707	0.000	.378	6.903	0.000
Attitude						.209	4.061	0.000
Facilitating Conditions as independent variables and Perceived Performance is the dependent variable								
Facilitating Conditions	.422	.178	1, 226	48.868	0.000	.509	6.991	0.000

Figure 2. Revised research framework for the study

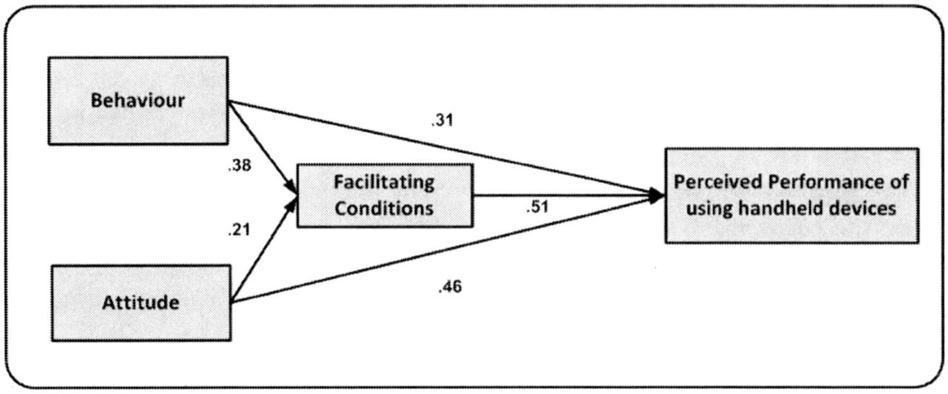

a relationship that is more complex, indirect, and mediated. In particular, while behavior and attitude have been confirmed as having a direct impact on the perceived performance of using wireless handheld devices, facilitating conditions have emerged as something of a mediating or intervening phenomenon. In a sense this should not be a surprising outcome, given that facilitating conditions refer to what educators and administrators do to create conducive environments for learning outcomes, while behavior and attitude constitute both the students' predispositions in relation to such conditions and their responses to those conditions. At the same time, the revised research framework in Figure 2 connotes a more multifaceted and less predictable pedagogical situation than that predicted in the initial research framework in Figure 1.

Although the technological development of wireless handheld devices has expanded rapidly and continues to grow apace, the educational applications of such devices are in many ways still in their infancy – or at least in their early childhood years. The results of this analysis partly explain why that might be so: as with other elements of educational technologies, the pedagogical implications of wireless handheld devices are not easily predicted or planned. The complexity of the relationships among the variables portrayed in Figure 2 is a timely reminder that the transition from the technical to the educational arenas is not necessarily automatic, immediate, or straightforward.

Despite this cautionary note, we consider that Figure 2 constitutes the basis of a model for possible implementation of blended and mobile learning across disciplines. In particular, the centrality of facilitating conditions as providing a potentially integrated link between students' behaviour and attitude on the one hand and their likely take up of wireless handheld devices on the other represents the core of the future research directions that we wish to propose for the project of which this chapter forms a part. More specifically, we have articulated the following questions as a likely guideline for undertaking those directions and thereby for contributing to ongoing and hopefully useful scholarship in blended and mobile learning:

- Which pedagogical conditions are most likely in different contexts to maximize students' educational outcomes from blended and mobile learning environments?
- Which pedagogical conditions are most likely to disrupt the nexus between students' behavior and attitude on the one hand and the effective use of educational technologies on the other?
- In what ways can and should the perceived use of wireless handheld devices feed back

to students' behaviour and attitude on the one hand and specific facilitating conditions on the other?

- How contextualized and changeable and/or fixed and resistant to change are students' behavior and attitude in particular blended and mobile learning environments?
- What does the relationship between students' behavior and attitude on the one hand and facilitating conditions on the other demonstrate about the relationship between blended and mobile learning?

While these questions are complex and widely ranging, we are confident that the chapter's findings as summarized in Figure 2 constitute a rigorous basis for addressing them in subsequent stages of the research project. We are equally confident in turn that addressing those questions, in concert with the contributions to scholarship presented in the other chapters in this book, is an important part of expanding the knowledge, understanding, and practice of blended and mobile learning across disciplines, to the benefit of students, educators, administrators, and other stakeholders alike.

CONCLUSION

Vavoula and Sharples (2009) identified:

...six challenges in evaluating mobile learning: capturing and analysing learning in context and across contexts, measuring mobile learning processes and outcomes, respecting learner/participant privacy, assessing mobile technology utility and usability, considering the wider organisational and socio-cultural context of learning, and assessing in/formality. (p. 54)

We contend that the chapter's findings about the participating students' perceptions and uses of wireless handheld devices synthesizes most if not all of these challenges – certainly in relation to the first, fourth, and fifth of them, and most likely engaging with the second, third, and sixth in some form. Indeed, we assert that a seventh challenge can usefully be added to the list elaborated by Vavoula and Sharples (2009): exploring the extent to which, and the ways in which, mobile learning is facilitated and/or hindered by being aligned with blended learning environments. The mixing of different delivery modes can potentially maximize the educational utility of mobile technologies such as wireless handheld devices; it can also risk adding unnecessary complexity and confusing rather than clarifying the communication of relevant and helpful information and understanding.

More broadly, the foundation for a model outlined in Figure 2 suggests that more research will be needed to explain and frame the complex interrelationships linking students' behavior and attitude, their perceptions of the performance and utility of wireless handheld devices, and the significant but shifting influence of facilitating conditions. Implementing blended and mobile learning, whether across disciplines or delivery modes, remains a crucial challenge for students, educators, administrators, and those seeking to maximize the pedagogical benefits that can and should accrue from technological developments.

ACKNOWLEDGMENT

The authors acknowledge gratefully the participation of the students who completed the survey questionnaire reported here.

REFERENCES

Akyol, Z., Garrison, D. R., & Ozden, M. Y. (2009). Development of a community of inquiry in online and blended learning contexts. *Procedia – Social and Behavioral Sciences, 1*(1), 1834-1838.

Alexander, B. (2004). Going nomadic: Mobile learning in higher education. *EDUCAUSE Review*, *39*(5), 28, 30–32, 34–35.

Arbaugh, J. B., Cleveland-Innes, M., Diaz, S. R., Garrison, D. R., Ice, P., Richardson, J. C., & Swan, K. P. (2008). Developing a community of inquiry instrument: Testing a measure of the community of inquiry framework using a multi-institutional sample. *The Internet and Higher Education*, *11*(3-4), 133–136. doi:10.1016/j.iheduc.2008.06.003

Arbaugh, J. B., Godfrey, M. R., Johnson, M., Leisen Pollack, B., Niendorf, B., & Wresch, W. (2009). Research in online and blended learning in the business disciplines: Key findings and possible future directions. *The Internet and Higher Education*, *12*(2), 71–87. doi:10.1016/j.iheduc.2009.06.006

Cavus, N., & Uzunboylu, H. (2009). Improving critical thinking skills in mobile learning. *Procedia – Social and Behavioral Sciences*, *1*(1), 435-438.

Cheung, W. S., & Hew, K. F. (2009). A review of research methodologies used in studies on mobile handheld devices in K-12 and higher education settings. *Australasian Journal of Educational Technology*, *25*(2), 153–183.

Choi, J. J. C. (2005). Teaching with a personal digital assistant. *Journal of the American College of Radiology*, *2*(10), 825–832. doi:10.1016/j.jacr.2005.02.006

Churchill, D., & Churchill, N. (2008). Educational affordances of PDAs: A study of a teacher's exploration of this technology. *Computers & Education*, *50*(4), 1439–1450. doi:10.1016/j.compedu.2007.01.002

Churchill, D., Kennedy, D., Flint, D., & Cotton, N. (2010). Using handhelds to support students' outdoor educational activities. *International Journal of Continuing Engineering Education and Lifelong Learning*, *20*(1), 51–71.

Danaher, P. A., Gururajan, R., & Hafeez-Baig, A. (2009). Transforming the practice of mobile learning: Promoting pedagogical innovation through educational principles and strategies that work. In Ryu, H., & Parsons, D. P. (Eds.), *Innovative mobile learning: Techniques and technologies* (pp. 21–46). Hershey, PA/ New York, NY: IGI Global/ Information Science Reference.

Danaher, P. A., Moriarty, B. J., & Danaher, G. R. (2009). *Mobile learning communities: Creating new educational futures*. New York, NY: Routledge.

De George-Walker, L. R., Hafeez-Baig, A., Gururajan, R., & Danaher, P. A. (2010). Experiences and perceptions of learner engagement in blended learning environments: The case of an Australian university. In Inoue, Y. (Ed.), *Cases on online and blended learning technologies in higher education: Concepts and practices* (pp. 23–43). Hershey, PA: IGI Global.

De Vaus, D. A. (2002). *Surveys in social research* (5th ed.). Crows Nest, Australia: Allen & Unwin.

Diamantini, D., & Pieri, M. (2008). A blended mobile learning experience: The Nomadis case. *International Journal of Knowledge and Learning*, *4*(2-3), 176–188. doi:10.1504/IJKL.2008.020653

Dierterle, E., & Dede, C. (2006). Straightforward and deep effects of wireless handheld devices for teaching and learning in university settings. Retrieved March 6, 2010, from http://74.125.155.132/scholar?q=cache:nuS6TRgEbaMJ:scholar.google.com/+wireless+handheld+devices&hl=en&as_sdt=2000&as_ylo=2004

Dieterle, E., & Dede, C. (2007). Building university faculty and student capacity to use wireless handheld devices for learning. In van 't Hooft, M., & Swan, K. (Eds.), *Ubiquitous computing in education: Invisible technology, visible impact* (pp. 303–328). Mahwah, NJ: Lawrence Erlbaum Associates.

Dieterle, E., Dede, C., & Schrier, K. (2007). Neomillennial learning styles propagated by wireless handheld devices. In Lytras, M., & Naeve, A. (Eds.), *Ubiquitous and pervasive knowledge and learning management: Semantics, social networking and new media to their full potential* (pp. 32–66). Hershey, PA: Idea Group.

Druin, A. (Ed.). (2009). *Mobile technology for children: Designing for interaction and learning*. Burlington, MA: Elsevier.

Garrison, D. R., & Kanuka, H. (2004). Blended learning: Uncovering its transformative potential in higher education. *The Internet and Higher Education*, *7*(2), 95–105. doi:10.1016/j.iheduc.2004.02.001

Garrison, D. R., & Vaughan, N. D. (2008). *Blended learning in higher education: Framework, principles, and guidelines*. San Francisco, CA: Jossey-Bass.

Ginns, P., & Ellis, R. (2007). Quality in blended learning: Exploring the relationships between online and face-to-face teaching and learning. *The Internet and Higher Education*, *10*(1), 53–64. doi:10.1016/j.iheduc.2006.10.003

Graziano, A. M., & Raulin, M. L. (2006). *Research methods: A process of inquiry* (6th ed.). Boston, MA: Allyn & Bacon.

Hafeez-Baig, A., & Danaher, P. A. (2007a). Future possibilities for mobile learning technologies and applications at the University of Southern Queensland, Australia: Lessons from an academic focus group. In D. P. Parsons & H. Ryu (Eds.), *Mobile learning technologies and applications (MoLTA) 2007: Proceedings of the conference on mobile learning technologies and applications (MoLTA) Massey University, Auckland, New Zealand February 19th 2007* (pp. 16-22). Auckland, New Zealand: Institute of Information and Mathematical Sciences, Massey University.

Hafeez-Baig, A., & Danaher, P. A. (2007b). Challenges and opportunities in facilitating student engagement and empowerment: Perspectives from Information Systems and education courses at the University of Southern Queensland, Australia. In C. Montgomerie & J. Seale (Eds.), *Proceedings of World Conference on Education multimedia, Hypermedia and Telecommunications 2007* (pp. 459-468). Chesapeake, VA: Association for the Advancement of Computing in Education.

Hafeez-Baig, A., & Danaher, P. A. (accepted for publication). Academics' attitudes towards mobile technologies in an Australian university: Implications for enhancing flexibility in cooperative and symbolic communities of Web based educational practice. In Kennedy, D. M., & Kommers, P. A. M. (Eds.), *Mobility: Adding flexibility to Web-based communities. Theme issue of the International Journal of Continuing Engineering Education and Life-long Learning*.

Hafeez-Baig, A., Gururajan, R., Danaher, P. A., & De George-Walker, L. R. (in press). Principles and pressures in managing student attitudes to innovative mobile learning: A view from an Australian distance education intensive university. In Wei, J., Ozok, A. A., & Yang, J. (Eds.), *Managing innovative distance learning for higher education. Theme issue of the.* International Journal of Management in Education.

Hafeez-Baig, A., Gururajan, R., Nazemi, S., De George-Walker, L. R., & Danaher, P. A. (2010). *Selected influences on students' uses of wireless handheld devices: An Iranian exploratory case study.* Paper presented at the Global Learn Asia Pacific conference, Shangri La's Rasa Sayang and Golden Sands Resorts, Penang, Malaysia.

Hair, J. F., Black, W. C., Babin, B. J., & Anderson, R. E. (2009). *Multivariate data analysis* (7th ed.). Upper Saddle River, NJ: Prentice Hall.

Hansson, T. (Ed.). (2008). *Handbook of research on digital Information Technologies: Innovations, methods, and ethical issues*. Hershey, PA: Information Science Reference/IGI Global.

Herrington, A., & Herrington, J. (2007). *Authentic mobile learning in higher education.* Paper presented at the annual conference of the Australian Association for Research in Education. Retrieved March 6, 2010, from https://www.aare.edu.au/07pap/her07131.pdf

Kim, P., Miranda, T., & Olaciregui, C. (2008). Pocket school: Exploring mobile technology as a sustainable literacy education option for underserved indigenous children in Latin America. *International Journal of Educational Development*, *28*(4), 435–445. doi:10.1016/j.ijedudev.2007.11.002

Kurti, A., Spikol, D., & Milrad, M. (2008). Bridging outdoors and indoors educational activities in schools with the support of mobile and positioning technologies. *International Journal of Mobile Learning and Organisation*, *2*(2), 166–186. doi:10.1504/IJMLO.2008.019767

Motiwalla, L. F. (2007). Mobile learning: A framework and evaluation. *Computers & Education*, *49*(3), 581–596. doi:10.1016/j.compedu.2005.10.011

Oliver, M., & Trigwell, K. (2005). Can blended learning be redeemed? *E-Learning and Digital Media*, *2*(1), 17–26.

Patten, B., Armedillo Sánchez, I., & Tangney, B. (2006). Designing collaborative, constructionist and contextual applications for handheld devices. *Computers & Education*, *46*(3), 292–308. doi:10.1016/j.compedu.2005.11.011

Remenyi, D. S. J., Williams, B., Money, A., & Swartz, E. (1998). *Doing research in business and management: An introduction to process and method.* London, UK: Sage Publications.

Sharples, M., Taylor, J., & Vavoula, G. (2007). A theory of learning for the mobile age. In Andrews, R., & Haythornthwaite, C. (Eds.), *The Sage handbook of e-learning research* (pp. 221–247). London, UK: Sage Publications.

Shea, P., & Bidjerano, T. (2009). Community of inquiry as a theoretical framework to foster epistemic engagement and cognitive presence in online education. *Computers & Education*, *52*(3), 543–553. doi:10.1016/j.compedu.2008.10.007

Sims, D. E., Burke, C. S., Metcalf, D. S., & Salas, E. (2008). Research-based guidelines for designing blended learning. *Ergonomics in Design: The Quarterly of Human Factors Applications*, *16*(1), 23–29. doi:10.1518/106480408X282764

So, H.-J., & Brush, T. A. (2008, August). Student perceptions of collaborative learning, social presence and satisfaction in a blended learning environment: Relationships and critical factors. *Computers & Education*, *51*(1), 318–336. doi:10.1016/j.compedu.2007.05.009

Swan, K., & Ice, P. (2010). The community of inquiry framework ten years later: Introduction to the special issue. *The Internet and Higher Education*, *13*(1-2), 1–4. doi:10.1016/j.iheduc.2009.11.003

Uden, L. (2007). Activity theory for designing mobile learning. *International Journal of Mobile Learning and Organisation*, *1*(1), 81–102. doi:10.1504/IJMLO.2007.011190

Uzunboylu, H., Cavus, N., & Ercag, E. (2009). Using mobile learning to increase environmental awareness. *Computers & Education*, *52*(2), 381–389. doi:10.1016/j.compedu.2008.09.008

Vaughan, N. D., & Garrison, D. R. (2005). Creating cognitive presence in a blended faculty development community. *The Internet and Higher Education*, *8*(1), 1–12. doi:10.1016/j.iheduc.2004.11.001

Vavoula, G., & Sharples, M. (2009). Meeting the challenges in evaluating mobile learning: A 3-level evaluation framework. *International Journal of Mobile and Blended Learning, 1*(2), 54–75. doi:10.4018/jmbl.2009040104

Wali, E., Winters, N., & Oliver, M. (2008). Maintaining, changing and crossing contexts: An activity theoretic reinterpretation of mobile learning. *ALT-J: Research in Learning Technology, 16*(1), 41–57.

Wang, M., Shen, R., Novak, D., & Pan, X. (2008). The impact of mobile learning on students' learning behaviours and performance: Report from a large blended classroom. *British Journal of Educational Technology, 40*(4), 673–695. doi:10.1111/j.1467-8535.2008.00846.x

Wu, T. (2009). Flexible learning: Designing blended language instruction for digital natives. In Gibson, I. (Eds.), *Proceedings of Society for Information Technology & Teacher Education international conference 2009* (pp. 3352–3355). Chesapeake, VA: Association for the Advancement of Computers in Education.

Zawacki-Richter, O., Brown, T., & Delport, R. (2008). *Mobile learning–a new paradigm shift in distance education?* Retrieved March 6, 2010, from http://auspace.athabascau.ca:8080/dspace/bitstream/2149/1258/1/Zawacki-Richter_Brown_Delport%20mlearn06.pdf

Zikmund, W. G., Babin, B. J., Carr, J. C., & Griffin, M. (2009). *Business research methods* (8th ed.). Florence, KY: South-Western College Publishing.

ADDITIONAL READING

Ally, M. (2009). *Mobile learning: Transforming the delivery of education and training*. Edmonton, AB: AU Press, Athabasca University.

Bonk, C. J., & Graham, C. R. (Eds.). (2006). *The handbook of blended learning: Global perspectives, local designs*. San Francisco: John Wiley & Sons.

Bruner, G. C., & Kumar, A. (2005, May). Explaining consumer acceptance of handheld Internet devices. *Journal of Business Research, 58*(5), 553–558. doi:10.1016/j.jbusres.2003.08.002

Chinnery, G. M. (2006, January). Emerging technologies: Going to the MALL: Mobile Assisted Language Learning. *Language Learning & Technology, 10*(1), 9–16.

Choi, Y. G., Abbott, T. A., Arthur, M. A., & Hill, D. N. (2007). Toward a future wireless classroom paradigm. *International Journal of Innovation and Learning, 4*(3), 14–25. doi:10.1504/IJIL.2007.011472

Churchill, D., & Hedberg, J. (2008, April). Learning object design considerations for small-screen handheld devices. *Computers & Education, 50*(3), 881–893. doi:10.1016/j.compedu.2006.09.004

Ciussi, M., Rosner, G., & Augier, M. (2009). Engaging students with mobile technologies to support their formal and informal learning. *International Journal of Mobile and Blended Learning, 1*(4), 84–98. doi:10.4018/jmbl.2009090805

Cobcroft, R. S., Towers, S. J., Smith, J. E., & Bruns, A. (2006, September 26). *Mobile learning in review: Opportunities and challenges for learners, teachers, and institutions.* Paper presented at the online learning and teaching conference, Queensland University of Technology, Brisbane, Qld. Retrieved March 6, 2010, from http://eprints.qut.edu.au/5399/

Fang, X., Chan, S., Brzezinski, J., & Xu, S. (2005-2006, Winter). Moderating effects of task type on wireless technology acceptance. *Journal of Management Information Systems, 22*(3), 123–157. doi:10.2753/MIS0742-1222220305

Hastie, M., Hung, I.-C., & Chen, N.-S., & kinshuk. (2010, February). A blended synchronous learning model for educational international collaboration. *Innovations in Education and Teaching International*, *47*(1), 9–24. doi:10.1080/14703290903525812

Jones, G., Edwards, G., & Reid, A. (2009). How can mobile SMS communication support and enhance a first year undergraduate learning environment? *ALT-J Research in Learning Technology*, *17*(3), 201–218.

Kukulska-Hulme, A. (2007, June). Mobile usability in educational contexts: What have we learnt? *International Review of Research in Open and Distance Learning, 8*(2), 1-16. Retrieved February 21, 2010, from http://www.irrodl.org/index.php/irrodl/article/view/356/907

Liu, C.-C., & Kao, L.-C. (2007). Do handheld devices facilitate face-to-face collaboration? Handheld devices with large shared display groupware to facilitate group interactions. *Journal of Computer Assisted Learning*, *23*, 285–299. doi:10.1111/j.1365-2729.2007.00234.x

Myers, B. A. (2005, May). Using handhelds for wireless remote control of PCs and appliances. *Interacting with Computers*, *17*(3), 251–264. doi:10.1016/j.intcom.2004.06.010

Parsons, D. P., & Ryu, H. (Eds.). (2007). *Mobile learning technologies and applications (MoLTA) 2007: Proceedings of the conference on mobile learning technologies and applications (MoLTA) Massey University, Auckland, New Zealand February 197th 2007.* Auckland, New Zealand: Institute of Information and Mathematical Sciences, Massey University.

Percival, J., & Percival, N. (2009). A case of a laptop learning campus: How do technology choices affect perceptions? *ALT-J Research in Learning Technology*, *17*(3), 173–186.

Pfeiffer, V. D. I., Gemballa, S., Jarodzka, H., Scheiter, K., & Garjets, P. (2009, November). Situated learning in the mobile age: Mobile devices on a field trip to the sea. *ALT-J: Research in Learning Technology*, *17*(3), 187–199.

Roschelle, J., Patton, C., & Tatar, D. (2007). Designing networked handheld devices to enhance school learning. In Zelkowitz, M. V. (Ed.), *Advances in computers* (*Vol. 70*, pp. 1–61). London: Academic Press.

Ryu, H., & Parsons, D. P. (Eds.). (2009). *Innovative mobile learning: Techniques and technologies*. Hershey, PA: Information Science Reference/IGI Global.

Shen, R., Wang, M., & Pan, X. (2007). Increasing interactivity in blended classrooms through a cutting-edge mobile learning system. *British Journal of Educational Technology*, *39*(6), 1073–1086. doi:10.1111/j.1467-8535.2007.00778.x

Singh, D., & Abu Bakar, Z. (2007). Wireless implementation of a mobile learning environment based on students' expectations. *International Journal of Mobile Learning and Organisation*, *1*(2), 198–215. doi:10.1504/IJMLO.2007.012678

Traxler, J. (2007, June). Defining, discussing, and evaluating mobile learning: The moving finger writes and having writ…. *International Review of Research in Open and Distance Learning, 8*(2), 1-12. Retrieved February 21, 2010, from http://www.irrodl.org/index.php/irrodl/article/view/346/882

Traxler, J. (2008). Learning in a mobile age. *International Journal of Mobile and Blended Learning*, *1*(1), 1–12. doi:10.4018/jmbl.2009010101

Wang, M., Novak, D., & Shen, R. (2008). Assessing the effectiveness of mobile learning in large hybrid/blended classrooms. In J. Fong, R., Kwan & F. L. Wang (Eds.), *Hybrid learning and education: First international conference, ICHL 2008 Hong Kong, China, August 13-15, 2008 proceedings* (pp. 304-315). Berlin, Germany: Springer.

Zhang, D. (2007). Web content adaptation for mobile handheld devices. *Communications of the ACM*, *50*(2), 75–79. doi:10.1145/1216016.1216024

KEY TERMS AND DEFINITIONS

Attitude: Assumption held by individual learners and groups of learners about particular technologies.

Behavior: Actions initiated by individuals and groups of learners in relation to particular technologies.

Blended Learning: The mixing of multiple modes of delivering learning (for example, distance, face-to-face, and online) in a single course or program.

Facilitating Conditions: Circumstances that encourage and support the alignment of particular technologies with specific groups of learners to achieve their desired outcomes.

Mobile Learning: The delivery of learning via technologies that are not fixed in permanent place (for example, wireless handheld devices).

Perceived Performance: Learners' attitudes toward the utility of a particular technology in a specific context in helping to achieve their desired outcomes.

Wireless Handheld Devices: Technologies such as personal digital assistants, handheld and tablet PCs, and cell and smart phones that are portable and can be used for communication and learning.

Chapter 14
Using Students' Own Mobile Technologies to Support Clinical Competency Development in Speech Pathology

Trish Andrews
The University of Queensland, Australia

Bronwyn Davidson
The University of Queensland, Australia

Anne Hill
The University of Queensland, Australia

Danielle Sloane
The University of Queensland, Australia

Lynn Woodhouse
The University of Queensland, Australia

ABSTRACT

The need to adequately prepare students for the workplace competencies of a health professional in the 21st century demands exploration of alternative learning opportunities. Two such examples are the appropriation of mobile technologies and the use of standardised patients to support clinical learning. This chapter will discuss the appropriation of students' own mobile devices to support the development of clinical competency for speech pathology students in a standardised patient clinic. The chapter includes descriptions of a project that focussed on the role of mobile technologies in supporting learning across different contexts. The results indicated that the use of mobile technologies in a clinical practice setting can make a positive contribution to clinical competency development. Issues for future integration of mobile technologies in clinical practice are raised.

DOI: 10.4018/978-1-60960-511-7.ch014

INTRODUCTION

This chapter describes and reports on a project that explored the use of students' own mobile technologies to support clinical learning and competency development within a clinical module. This module uses standardised patients who are actors or real patients, carefully trained by professional staff to enact the role of a patient according to prescribed educational goals and specific skill development requirements (Barrows, 1971). In particular the chapter focuses on:

- The importance of competency development in clinical practice including an outline of the challenges and barriers faced in developing clinical competencies
- An outline of standardised patient clinics and their role in workplace learning and practice education
- A discussion of 'mobility' in learning and how this can be enabled through the use of mobile devices
- An outline of the use of mobile learning to support learning across different contexts
- A discussion of applications of students' own mobile technologies to support teaching and learning activities
- A detailed description of the project phases
- A discussion of the project outcomes and implications for mobile learning in relation to practice and workplace learning
- Implications for future developments in the appropriation of mobile technologies in clinical practice.

BACKGROUND

Tertiary education programs in the health sciences rely on the inclusion of clinical practice and work-integrated learning components in order to graduate professionals competent to work in their chosen discipline. University programs report increasing difficulty in obtaining sufficient traditional clinical practicum placements for students within the workplace. In response to this, allied health professionals are challenged to be innovative and to embrace new technologies within the development and implementation of a clinical education curriculum. Consequently, in 2008, the Division of Speech Pathology within the School of Health and Rehabilitation Sciences at The University of Queensland introduced a clinical education module utilising standardised patients in a third year undergraduate student voice clinic.

While the standardised patient clinic was deemed successful in providing opportunities for clinical practice, it was identified that more needed to be done to support the development of clinical competencies, in particular to encourage and support transfer of learning across different learning contexts. Accordingly, in 2009, mobile technologies were introduced into the program to create a blended learning environment that linked different learning contexts and modes of learning including clinical practice, mobile learning, eLearning and face-to-face activities.

COMPETENCY DEVELOPMENT

Developing competency for professional practice is critical for students in the health sciences. Professional competency is the ability to perform the required tasks of a designated profession to an appropriate standard as determined by that profession. Competency and the capacity to integrate effective reasoning within a professional domain develops gradually. Higgs and Titchen (2000) suggest it is the result of interaction of a complex mixture of propositional knowledge, professional skills or wisdom and personal knowledge and attitudes. Undergraduate and professional Masters tertiary programs provide opportunities for students to develop competency, in particular knowledge and skills, through academic curricula with a strong foundation in case-based learning.

Knowledge can be tangibly measured through assessment in academic courses, while professional skills and attributes are less readily assessed in such situations and hence often inferred. It is also acknowledged that knowledge learnt in academic environments is not always easily transferred to competent practice in the workplace. Clinical placements (variously called 'clinical practicum' or 'fieldwork placements') allow for more concrete observation of all aspects of competency, and provide opportunities for students to demonstrate their integration of knowledge, skills and professional attributes in their work with clients. Such placements occur in many health professional programs such as physiotherapy, dietetics, pharmacy, occupational therapy and speech pathology.

Clinical placements in the workplace provide opportunities for students to practise those skills which are significant to their profession. Practice education demands the integration of their acquired knowledge, skills and attitudes and situates their learning within genuine professional experience. The opportunity to integrate theory with practice and to see 'real-life' examples of their theoretical knowledge is invaluable. Placements provide students with the capacity to develop clinical reasoning, "the sum of the thinking and decision-making processes associated with clinical practice" (Higgs & Jones, 2008, p. 4). The process of clinical reasoning is undertaken on a constant basis, is a consistent skill used by health professionals in making decisions about client needs and goals and is a critical professional competency.

The requirements for work-based learning and the management and allocation process of clinical placements are organised in diverse ways by professional programs but all have a shared goal of taking students' learning beyond the classroom into professional domains. Traditionally, placements take the form of a one to one, or two to one, student:educator ratio. Whilst these offer valuable learning opportunities for students, their future viability is not assured. Changing professional, resource and fiscal environments make the provision of an appropriate number of quality placements difficult (Rodger et al., 2008). In addition, suitable placement allocation is jeopardised by the increasing number of health science university programs across the world and the consequent increase in student numbers. A further challenge facing programs is managing the provision of quality placements which reflect the full range of core professional areas of competency. This full range is not always available within traditional workplace clinical experiences for students due to factors such as client variability, service provider priorities and educator experience. It is apparent, then, that development of competency in clinical settings cannot be readily sustained in a traditional format. In order to address this issue simulation has been explored as a viable alternative to traditional models of clinical placements in the health sciences. One approach to simulation is the use of standardised patients.

STANDARDISED PATIENTS

The use of standardized patients has been reported extensively in medical and nursing literature and is becoming more prevalent in allied health education programs to complement more traditional clinical placement experiences for students (Hill, Davidson, & Theodoros, 2010). The benefits of using standardised patients (SPs) are widely acknowledged (Lysaght & Hill, in press). Since each SP engaged in a given clinical activity is trained in the same format according to the same learning objectives, their interaction with students is standardised, thereby creating an equal learning opportunity for each student and equal capacity to meet assessment requirements. Students value the SP learning experience (Lane & Rollnick, 2007) and report that they are able to learn more gradually, safely and with more freedom than with 'real clients'. An additional benefit for student learning is the extensive amount of high quality feedback which is provided by both educators and SPs during and following interactions. This feedback allows students to reflect on their performance immediately and in context. Feedback from the SP is particularly useful in prompting students to acknowledge the personal factors at play in an interview situation and to focus primarily on developing effective communication skills (Bokken et al., 2009).

Recruitment of motivated SPs who are committed to student education is vital to ensure the success of an SP clinical program. Training procedures may differ according to educational need and program goals. All training will, however, share a focus on initially building case scenarios based on real client data and then using these scenarios to provide SPs with a clear concept of their required client portrayal including essential personality and clinical features of the case (Hill et al., 2010). This training seeks to standardise client presentation in order to reduce client variability and increase the focus of the clinical activity on student learning. In addition, giving guidelines to SPs about the provision of feedback in an effective manner is essential.

The use of SPs in the clinical program described below aimed to develop specific clinical competencies for students in an area of professional practice not readily available within the traditional workplace environment, that of management of clients presenting with voice disorders. While students had access to essential case-based classroom learning in this area, the opportunity to situate their specific learning within professional practice contexts was limited. It is important to note that, in this clinical program, the use of SPs significantly reduced the ethical and consent issues traditionally associated with using real clients. As the SPs were actors employed to portray the role described above, they consented to being videoed and audiotaped within the clinic with the use of students' mobile technologies. The ethical considerations of using real clients in any further studies relating to students' mobile technology use warrant particular attention.

While the use of SPs in clinics result in positive outcomes for learners, transferring learning to different contexts remains an issue, with many students struggling to perform required techniques confidently and competently despite learning these skills in a clinical context. Technologies and devices that allow for recording of therapy techniques, practice and review of treatment tasks, discussion with educators and peers, and opportunities for review and reflection, provide valuable tools for the novice clinician.

USING STUDENTS' OWN MOBILE DEVICES FOR TEACHING AND LEARNING

Mobile technologies are those technologies that are generally seen as both portable and personal (Naismith, Lonsdale, Vavoula & Sharples, 2006). This includes such devices as laptop computers, tablets, mobile phones, MP3 players, audience response systems, PDAs, digital cameras, gaming consoles and GPS systems (Dyson, Litchfield, Lawrence, Raban, & Leijdekkers, 2009; Traxler, 2009b).

Widespread ownership of mobile devices by students is undisputed (Traxler, 2009b). In the developing world, many students own at least three mobile devices – a mobile phone, a laptop computer and an MP3 player. In some countries ownership of multiple mobile phones is not uncommon (Mohammed & Mohan, 2010). This broad access to mobile devices means that their integration into teaching and learning activities is becoming increasingly viable. The popularity of data plans and prepaid plans encourages students to use their mobile devices for a range of activities including accessing the internet and social networking sites, recording, storing and viewing videos and photos.

While the widespread ownership of mobile devices and students' willingness to use them for teaching and learning activities creates many opportunities, the diversity of devices owned by students also creates a number of issues and challenges. Operational systems vary widely across devices, limiting the use of specialised applications. Applications of technologies such as videoconferencing can also be challenged by availability and reliability of wireless networks. Notwithstanding these issues, applications such as recording, playing vodcasts, accessing the web and texting are all viable and are not generally impacted by operating system differences.

When considering the integration of mobile technologies into teaching and learning activities, many projects and activities have focussed on the use of particular mobile devices such as 3G phones (Ferry, 2008), smart phones and iPods (Herrington, Herrington, Mantei, Olney, & Ferry, 2009), PDAs (Huffstutler Wyatt, & Wright, 2002; Torre, Simpson, Sebastian, & Elnicki, 2005) and tablet PCs (Andrews, Rayner, & Edwards, 2009; Koile & Singer, 2006). While these studies have yielded useful information about the applications, benefits and limitations of using these devices for selected learning activities, the realty of student ownership of devices is characterised by diversity in the nature of the devices. Some studies (Pettit & Kukulska-Hulme, 2008, for example) show that students are often unwilling to use an additional device to the ones they already own, further raising questions about the sustainability and scalability of using a specific device to support mobile learning in the medium to long term.

MOBILITY

The notion of mobility and its influence on learning in a mobile world is emerging as a key area of research in the 21st century (Kukulska-Hulme & Sharples, 2009; Traxler, 2009a). The mobility provided by mobile devices frees up the learner from location and time dependency in relation to learning, providing opportunities for learning activities and learning support to be accessed in a variety of locations and times, 'anywhere, anytime' (Katz & Aakhus, 2002). Further to this, the capabilities of mobile devices means that students can be mobile within a specific location, both through accessing resources and also through such activities as back channel communications (Maddrell, 2008) that can enable the learner to access immediate feedback and information from peers or educators to support learning. Mobility can enable students to 'escape' the classroom (Naismith et al., 2006) and interact with others in ways which significantly change common concepts of classroom interaction and challenge the ways in which educators think about and plan for educational interactions (Traxler, 2008).

LEARNING ACROSS DIFFERENT CONTEXTS

One of the major challenges of designing effective learning is finding ways to enable learning to be successfully transferred between different contexts. Hull (1993) believed that in order to learn effectively, learners need to be able to utilise new knowledge and information in ways that make sense to them. In addition, learners look for meaning in context. Learning can be designed for numerous environments and contexts to incorporate many different kinds of experiences in order to achieve the intended learning outcome. These learning environments might include classrooms, labs, field excursions and clinical placements, amongst many. Learning in these different environments provides learners with opportunities to make links between the different kinds of learning they experience. It is acknowledged, however, that this is problematic. One of the characteristics of mobile learning is its perceived capacity to support learning across and within a range of learning contexts. Kukulska- Hulme and Sharples (2009) contend that:

Learning when mobile means that context becomes all-important, since even a simple change of location is an invitation to revisit learning, in both a literal sense (to apply it, reflect on it, reinforce it, share it) and metaphorical, to reconsider what constitutes learning or what makes it effective. (2009, p. 159)

Naismith et al. (2006), however, acknowledge the critical importance of mobile devices as tools to enable learners to access resources 'anywhere, anytime', and consequently enable learning in and between different contexts.

MOBILE LEARNING IN A STANDARDISED PATIENT CLINIC

An innovative project was developed to explore the use of mobile technologies to support clinical learning and competency development within a standardised patient clinical module. The clinical module was a compulsory component of the clinical curriculum for third year undergraduate students enrolled in a four year Bachelor of Speech Pathology Program at The University of Queensland in Australia. This clinical module focussed on the development of students' clinical competency in assessment and intervention for adults presenting with an acquired voice disorder. The overall aim of the Standardised Patient (SP) Voice Clinic is to provide students with an opportunity to develop competency in case history-taking and assessment of clients with voice disorders, to gain experience with explanation and demonstration of voice intervention techniques to clients and to develop skills in reporting to referring agencies.

Students were assigned to clinic groups of eight peers facilitated by a clinical educator. The SP Voice Clinic ran over four weeks with students attending three face-to-face clinic sessions (of 4 hours duration) and completing other clinical learning tasks online. These online learning activities were supported by Blackboard and provided opportunities for students to participate in asynchronous online discussions, reflection tasks, and access to recorded podcasts and vodcasts of demonstrations of voice assessment and intervention techniques. In weeks 1 and 4 of the module, students interviewed, assessed and provided therapy to SPs who presented with a clinically authentic voice disorder. Support from a small Teaching and Learning Grant allowed the authors to examine students' use of mobile technologies within this clinical module.

The aim of the mobile technologies in clinical education project was to investigate and evaluate the appropriation of students' mobile technologies within the clinical practice learning environment of the SP Voice Clinic. The project team examined how mobile technologies were used by students to maximise situated and personalised approaches to clinical learning. Students' competency development was facilitated through students' utilising and accessing course materials on their mobile devices and examining how these devices might be used for the recording of events of interest or importance to students.

The specific objectives of the "mobile learning in clinical practice" project were:

- For students to utilise mobile technologies to support development of reflective practice and clinical reasoning
- To develop peer learning skills through sharing and discussion of items of interest / importance in clinical learning and practice
- To investigate and evaluate the use of mobile technologies to facilitate the development of clinical competencies
- To explore the concept of 'mobility' in clinical practice education.

The project had two major phases. Phase 1 involved the development, distribution, and

Table 1. Speech pathology students' responses to question "What type of technology do you currently use and how often?"

	Daily	2/3 times per week	Fortnightly	Monthly	Never
Laptop Computer	168	28	5	6	23
Tablet PC	35	16	6	12	102
Mobile Phone	225	0	0	0	6
Smart Phone or 3G Enabled Phone	56	3	4	2	143
MP3 Player	99	66	19	10	20
Gaming Device (e.g. Nintendo)	4	19	14	37	114
PDA	2	0	0	1	172

completion of a survey to determine speech pathology students' access to and current use of mobile technologies. This survey was completed by students in years 2-4 of the four year Bachelor of Speech Pathology program and Graduate Entry Masters students (GEMS) enrolled in the Masters of Speech Pathology Studies. Key findings from the survey of 234 students enrolled in speech pathology courses in 2009 are detailed below. This student cohort consisted of 228 females and 6 males. Table 1 records the students' responses to the question "What type of technology do you currently use and how often?" Not surprisingly 225 students reported daily use of mobile phones and a large proportion also reported daily use of a laptop computer. In addition, 165 students reported frequent use (daily or 2/3 times per week) of MP3 players.

The survey canvassed students' willingness to use their mobile phone plans for learning purposes (see Figure 1). A greater proportion of second year students indicated a willingness to use their own plans, while for the other groups of students the decision for and against using their plan was divided.

Figure 2 highlights students' responses to the question "For what reasons do you currently use your mobile technology?" The major reported use relates to social communication including texting, emailing and social networking. The use of mobile

Figure 1. Students' willingness to use their plan for learning activities

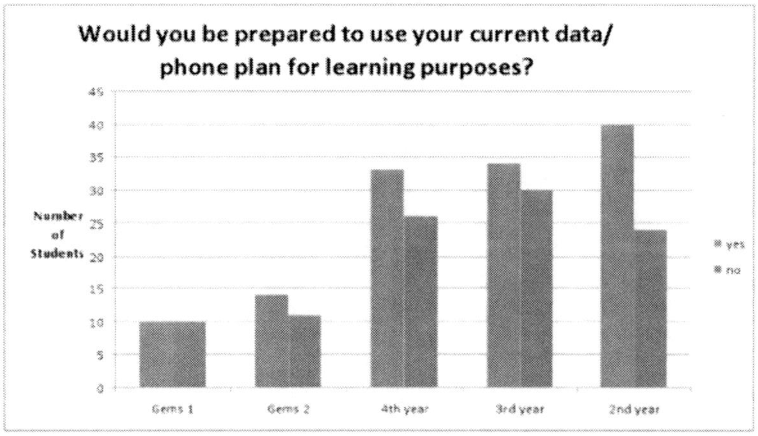

Figure 2. Graph recording students' current use of mobile technologies

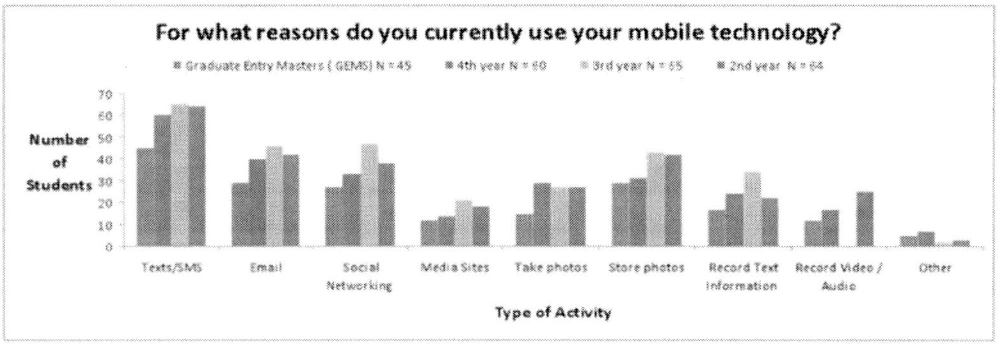

devises for storing photos, recording written information and also audio and video files and accessing media sites was also noted.

There was a general willingness by a portion of the student groups to use mobile technologies across a range of learning activities including recording information, use of blogs and wikis and particularly for accessing course information (See Figure 3). Again, the use of mobile technologies for informal peer interaction in learning activities was rated positively by half to two-thirds of the students.

Survey findings informed planning for Phase 2 of the project which had a focus on examining clinical learning in the voice standardised patient clinical module, and in exploring students' use of their own mobile technologies within specific learning activities.

Phase 2 required the development of teaching and learning activities for the four week SP Voice Clinic Module which was undertaken by the third year undergraduate Speech Pathology students. Fifty-nine (59) students consented to participate in the research project. Phase 2 included development and implementation of the clinical module (a total of 16 hours formal clinic time, consisting of 4x4 hour clinical learning sessions) and evaluation of outcomes. Technical support was

Figure 3. Graph recording students' willingness to use mobile technologies for learning activities

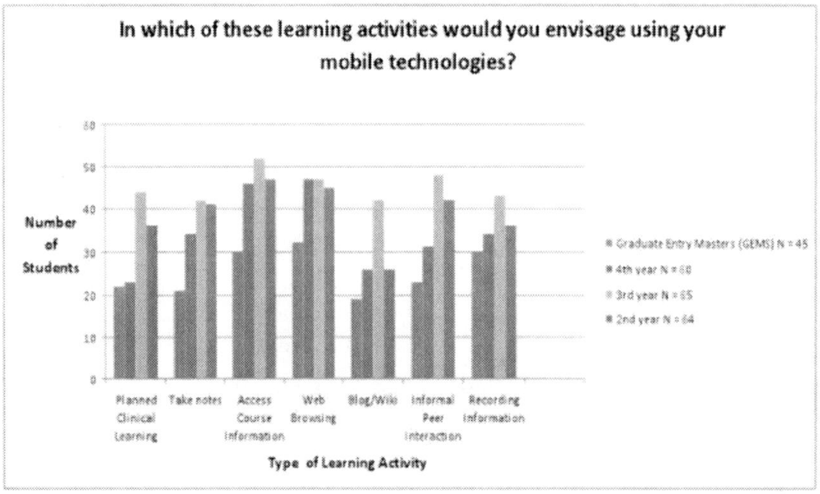

sought and provided in order to assist students in uploading files from their own mobile devices and to help students to problem solve technical dilemmas. Preparation for the SP Voice Clinic included collation of a student workbook and the setting up on online activities on the Course Blackboard site. Teaching and learning activities in relation to mobile technology use are detailed further in Table 2.

Project Outcomes

At the completion of the four week module, 59 students completed a questionnaire designed to provide feedback on their use of mobile technologies in this clinical module. Open and closed questions were included. It is of note that 50 of the 59 students reported that mobile technology enhanced their learning. However, when asked to comment on the overall effectiveness of mobile technology use for facilitation of clinical learning, students were equivocal in their response with about half indicating that it was effective or highly effective (See Figure 4). This finding raises issues for further investigation including seeking specific feedback from students on those aspects that may have been barriers to the effective use of mobile technologies in this particular program. In focus groups, students identified the critical role of technical support in accessing materials and that lack of such support can be a barrier to utilising mobile technologies.

Students were also asked to rate the effectiveness of the various eLearning activities (e.g. podcasts) in facilitating clinical learning. The graphs in Figures 5 and 6 particularly focus on the students' feedback on the effectiveness of the Wimba vodcasts to assist them in clinical skill development. The following graphs (Figures 5 and 6) record that Wimba podcasts that demonstrated vocal techniques for both assessment and therapy were perceived as effective or highly effective by a majority of the students.

STUDENTS' PERCEPTIONS OF MOBILE TECHNOLOGIES IN CLINICAL LEARNING

Students were asked questions such as "How has mobile technology enhanced your clinical learning in the Voice clinic?" The following quotes highlight key learning for the students.

- *"Listening to the Wimba podcast was so helpful in understanding different ways to introduce techniques and explain concepts to the client"*
- *"Wimba podcasts were fantastic"*
- *"It was good to be able to access information across various mobile technologies whenever I wanted to and the information can be kept in future to assist me with my learning"*
- *"Gives us an opportunity to learn and discuss with others from various locations. Podcasts great help to us to use and explain techniques. Download to use anywhere was good"*
- *"It is useful to access resources wherever we are, especially when you have to meet with a partner"*
- *"It enabled me to access information, anywhere, anytime"*

Clearly, students were enthusiastic about the Wimba podcasts providing a flexible means to practise voice techniques 'anywhere, anytime' (Katz & Aakhus, 2002) prior to seeing a client. Podcasts can be made by an experienced clinician and serve as a role modeling of "good practice" and "specific skill development". The fact that they can be downloaded and listened to in different environments, ranging from on a bus, or in a park, or when meeting informally with their student pair, was seen as a distinct advantage. The 'mobility' factor (Naismith et al., 2006; Traxler, 2008), was further highlighted by students who valued being able to learn and discuss clinic

Table 2. Teaching and learning activities for Standardised Patient (SP) Voice Clinic and the use of Mobile Technology

When?	Type of Activity	Use of Mobile Technology
Preclinic classroom Tutorial	Classroom introduction to the SP Voice Clinic Module and the use of mobile technologies to facilitate clinical learning by academic staff member and learning designer.	Lap top Blackboard demonstration
Web-based preparation	Students view the content of an online "Helpful hints" folder to learn about on-line discussion etiquette, giving feedback, and accessing Blogs, Wimba and Online Discussion Boards	Students access Blackboard site and resources through Laptop, Smart phone or PDA
Pre Week 1	Students undertake the following learning tasks: (a) Listen to demonstration Wimba Podcasts. These podcasts were designed to demonstrate aspects of voice assessment and reflect the content of previous voice lectures. These podcasts served as a trigger to encourage students to consider how they would elicit an accurate representation of the client's vocal capabilities (e.g. What language will they use? How would they demonstrate each task?) (b) Develop and role play presentation of assessment tasks with partner in preparation for SP Voice Assessment Interview (Week 1) (c) Self rating/ Discussion. This activity requires the student to reflect on their presentation of assessment techniques with their partner. (e.g. What assessments tasks were difficult? Why? Did students feel comfortable to demonstrate the tasks? Why? Why not? What aspects of the assessment may be confronting or difficult for the client?)	Access Wimba podcasts by various technology-i.e. laptop, mobile phone, other Watch or listen to Wimba podcasts Consider use of MP3, Laptop, Skype or other Wiki– allocated groups on Blackboard (refer to 'Guidelines on giving feedback; CRC- Commendation, Recommendation, Commendation.')
Week 1	Students (in pairs) complete SP Interview - Case History and Voice Assessment Students receive feedback on performance from peers, clinical educator and SP	Interviews recorded onto DVD for review by students and/ or clinical educator MP3 Mobile phone Other
Week 2	Students encouraged to listen to recorded activities Students complete an online, delayed reflection task in the form of a private Blog Students listen to Vodcasts - "on-line recordings of four clients with a voice disorder"	Students encouraged to review recordings of interviews made using their mobile technology and/or the DVD recordings made in the clinic.
Week 3 (Workshop with Clinical Educator)	Students rate recordings of four clients with a voice disorder and discuss findings and plan for intervention options Student groups construct a letter to the referring ENT for each client. Students use online Discussion Board to collaboratively plan and review drafts of the letter	Each student pair posts letter on Blackboard discussion board. Group discussion re each pairs' draft letter. Subsequent discussion and modification. Email to CE for comments prior to Week 4.
Week 4 (CE and SP)	SP feedback and therapy session	Interviews recorded onto DVD and students' own MP3, Mobile Phone, or other devices

Figure 4. Students' rating of effectiveness of mobile technology in facilitating clinical learning

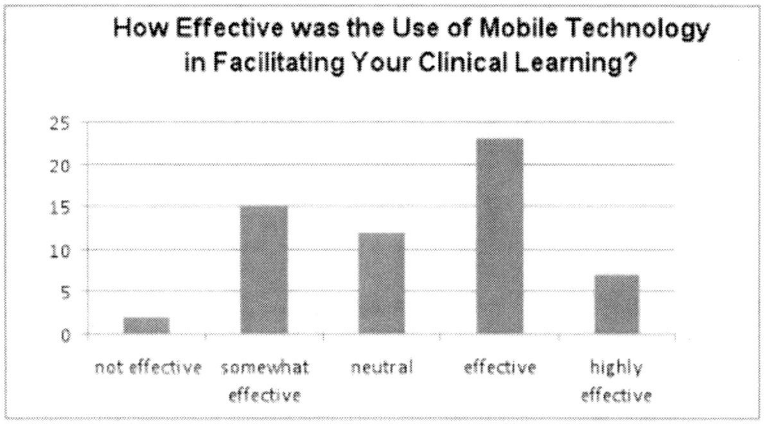

preparation or review their clinical performance with peers at different times and in different locations. Additionally, students found the ability for 'just in time' learning enabled and supported by the mobile devices beneficial to their learning.

Students used an array of devices when recording their assessment of the standardized patient. Students predominantly used MP3 players (49%), followed by recording onto a DVD utilizing the overhead cameras in the clinic room. Considerable diversity in use of mobile technology was indicated when students were required to record their own attempts at introducing therapy techniques. Survey results were as follows: MP3 players (39%), digital voice recordings (27%), mobile phones (10%), Skype on laptops (10%), Dictaphones and minicam recorders (6%) and 8% of the students chose simply to practice with a peer. The use of recording devices was particularly valuable in extending opportunities for reflective practice.

Students responded that the barriers to the use of mobile technologies in the SP clinic predominantly revolved around technical issues. File size was identified as a problem, and issues of internet speed and wireless access were also noted. Students were observed to learn from each other through discussion on the Blackboard Discussion Group site and through sharing of resources (including articles, web-sites, and suggested texts). Reflective practice was facilitated by students'

Figure 5. Students' rating of effectiveness of Podcasts for development of client assessment skills

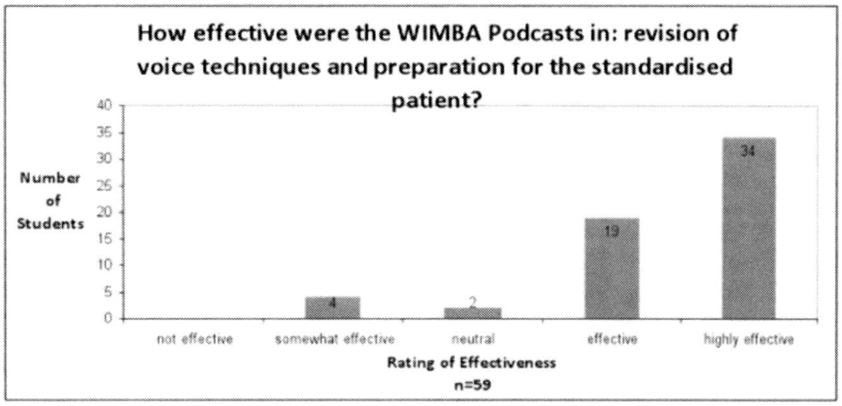

Figure 6. Students' rating of effectiveness of Podcasts for development of therapy skills

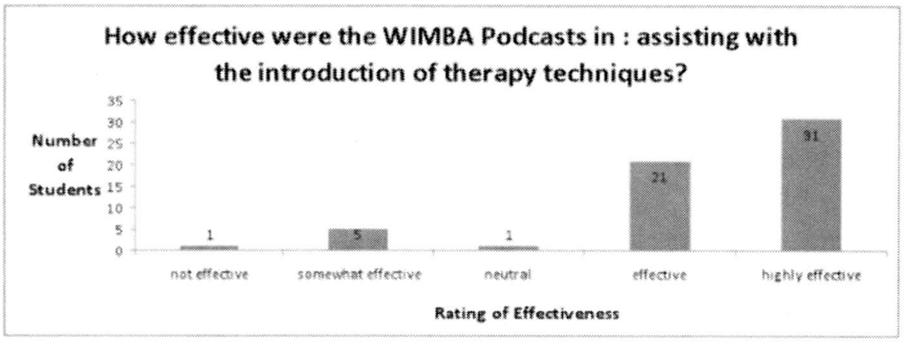

completion of a delayed reflection in a digital blog on the Blackboard site.

Another barrier to the use of mobile technology for clinical learning is equity of access in terms of the level and type of devices owned by studies. Some devices may lack the tools such as cameras, note taking ability and audio recording that are useful in the clinical practice setting and this needs to be taken into consideration in the planning process. Differences in digital literacy can also be considered a barrier, with many students lacking the skills to make effective use of their technologies.

Students also responded to a question relating to ways in which they could envisage using mobile technologies in future clinical practice. Students identified that mobile technologies could be utilized for both communication with colleagues and for improving clients' access to services. In particular students described possibilities for working with clients in rural and remote communities and also for linking with specialist clinicians through mobile devices. The following quotes illustrate the breadth of application of new technologies in clinical practice and the vision that students hold for their future professional work. When asked to describe ways they may use mobile technologies in clinical practice, students stated:

- *"Communication with fellow speech pathologists or other allied health professionals will be enhanced by using these mobile technologies"*
- *"Depending on where I am working, technologies such as file sharing, podcasting and video calling via the phone or internet would be useful for those clients who cannot regularly access a clinic"*
- *"I would use clinical discussion boards, videos of techniques and explanations and use of Skype and podcasts"*
- *"To make clinical resources more readily available"*
- *"To contact colleagues and share thoughts and suggestions etc about best practice"*

LIMITATIONS OF THE STUDY

While the study explored the ways that mobile technologies might support competency development and found that staff and students perceived their use to be valuable in this regard, it did not measure improvements in competency as a direct result of utilizing the mobile technologies.

FUTURE DIRECTIONS FOR MOBILE LEARNING IN CLINICAL PRACTICE

An action learning approach (Zuber-Skerritt, 1993), incorporating planning, action, analysis and reflection, has informed our learning from this initial project. The process of project planning drew on the complementary expertise of a range of professionals including clinical educators, academic expertise and educational design, and technical support. Input from the various professionals encouraged reflection on the process and refinement of project activities and goals at all stages of the project cycle. The process included evaluating and reflecting on the outcomes of this initiative, which have facilitated changes in practice within the clinical curriculum and illuminated areas for further development. The following paragraphs suggest ways in which mobile technologies might be further integrated into practice education.

Looking to the future, we envisage two major areas of expansion in our use of mobile technologies in a health and rehabilitation sciences curriculum. The first is in integrating mobile technology applications throughout the curriculum by introducing students to peer learning and eLearning applications that harness the use of students' mobile technologies early in their undergraduate and masters programs. Within the curriculum we have a particular interest in further integration of eLearning and mLearning and in exploring both creative and systematic ways in which student learning can be enhanced. One step in this process is facilitating the preparedness of staff and also of students to utilize new technologies and optimize the use of their own mobile devices. As discussed by Kukulska-Hulme and Traxler (2005) and Lefoe, Olney, Wright, and Herrington (2009), comprehensive staff development and support are key aspects of ensuring effective use of new technologies within the curriculum. The recent increased availability of Smart phones makes new applications for learning and clinical skill development a distinct possibility. The potential for mobile technologies to be an integral component of practices that support life-long learning is both a challenge and a reality.

A second step is to explore new developments in other areas of clinical practice (not just in the standardised patient arena). Student exposure to service delivery that offers clients therapy programs that can be accessed electronically and through mobile devices holds great promise for the future. As highlighted by one of our students *"I think that including mobile technologies in this course helps us to think about what we could use in our real clinics"*. The introduction of mobile technologies in the standardised patient voice clinic assisted students to think about applications in clinical practice in the workplace. Certainly, further consideration is required to problem-solve ethical issues related to applications in real-life clinical practice: for example, recording videos of clients on mobile phones. Issues of "privacy" and "confidentiality" arise. Whilst this is a complex issue, the ethical debate needs to happen, the rights of the public need to be protected and protocols established.

Further to these two initiatives another future direction for mobile learning could include the development and integration of tasks to develop a wider range of communication skills for clinical practice. Additionally, exploring the inclusion of tasks to facilitate reflection would be valuable in a range of practice education settings.

Areas for future research include the conduct of a controlled study to measure the influence of mobile learning on competency development and issues relating to digital literacy and the digital divide.

CONCLUSION

The use of mobile technologies to support the development of clinical competencies was deemed successful by both staff and students. Students enjoyed the flexibility for learning 'on the move', 'anywhere, anytime' and 'just in time' enabled by the use of mobile technologies and felt that it contributed to both their confidence and their competence in the standardised patient clinic placement activities. The use of vodcasts were particularly useful for students in this regard. Some students also found benefit in recording events for later reflection. However, access to wireless networks and lack of understanding of technical issues relating to using mobile devices still create barriers for integrating mobile technologies into clinical practice activities. The importance of providing technical support for these kinds of innovations cannot be understated. Additionally students appear to want direction into how to effectively use their mobile devices for teaching and learning activities. Notwithstanding these issues it can be considered that mobile technologies create many opportunities for teaching and learning and provide benefits for students in supporting teaching and learning across contexts. Integral to clinical education is evaluation of "what" and "how" students are learning and the creation of opportunities to extend learning beyond the scheduled clinical session (Cruice, 2005). Further understanding of the use of students' own mobile technologies to support their clinical learning and competency development across a range of clinical practice contexts is warranted. This project has implications for a wide range of practice education settings. For example, the use of mobile technologies to support competency development would have application in physiotherapy clinical practice settings in ways similar to that used in the speech pathology clinic. Practice teaching is another area where mobile technologies could be effectively used to support competency development including the development of classroom management techniques. The increasingly ubiquitous ownership of mobile devices and access to reliable wireless networks will provide many more opportunities to explore the ways in which mobile learning can support practice education.

ACKNOWLEDGMENT

This project was supported by funding from a Strategic Teaching and Learning Grant at The University of Queensland.

REFERENCES

Andrews, T., Rayner, A., & Edwards, J. (2009). Tablet PCs, DyKnow and collaborative learning: Developing mathematical problem solving skills in large undergraduate classes. In *Proceedings of the MLearn 2009 Conference*, 24-28 Oct 2009, Orlando, Florida.

Barrows, H. S. (1971). *Simulated patients (programmed patients): The development and use of a new technique in medical education*. Springfield, IL: Charles C. Thomas.

Bokken, L., Rethans, J. J., van Heurn, L., Duvivier, R., Scherpbier, A., & van der Vleuten, C. (2009). Students' views on the use of real patients and simulated patients in undergraduate medical education. *Academic Medicine*, *84*, 958–963. doi:10.1097/ACM.0b013e3181a814a3

Cruice, M. (2005). Common issues but alternative solutions and innovations. *Advances in Speech Language Pathology*, *7*, 162–166. doi:10.1080/14417040500181429

Dyson, L., Litchfield, A., Lawrence, E., Raban, R., & Leijdekkers, P. (2009). Advancing the m-learning research agenda for active, experiential learning: Four case studies. *Australasian Journal of Educational Technology*, *25*(2), 250–267.

Ferry, B. (2008). Using mobile phones to augment teacher learning in environmental education. In *Proceedings, ASCILITE 2008*, Melbourne.

Herrington, J., Herrington, A., Mantei, J., Olney, I., & Ferry, B. (2009). Using mobile technologies to develop new ways of teaching and learning. In J. Herrington, A. Herrington, J. Mantei, I. Olney, & B. Ferry (Eds.), *New technologies, new pedagogies: Mobile learning in higher education.* (pp.1-14). Faculty of Education, University of Wollongong. Retrieved from http://nla.gov.au/nla.arc-101041

Higgs, J., & Jones, M. A. (2008). Clinical decision making and multiple problem spaces. In Higgs, J., Jones, M. A., Loftus, S., & Christensen, N. (Eds.), *Clinical reasoning in the health professions* (3rd ed., pp. 3–17). Oxford, UK: Elsevier Butterworth-Heinemann.

Higgs, J., & Titchen, A. (2000). Knowledge and reasoning. In Higgs, J., & Jones, M. (Eds.), *Clinical reasoning in the health professions* (2nd ed., pp. 23–32). Oxford, UK: Butterworth-Heinemann.

Hill, A. E., Davidson, B. J., & Theodoros, D. G. (2010). A review of standardised patients in clinical education: Implications for speech-language pathology programs. *International Journal of Speech-Language Pathology, Early Online,* 1-12.

Huffstutler, S., Wyatt, T., & Wright, C. (2002). The use of handheld technology in nursing education. *Nurse Educator*, *27*(6), 271–275. doi:10.1097/00006223-200211000-00008

Hull, D. (1993). *Opening minds, opening doors: The rebirth of American education.* Waco, TX: Center for Occupational Research and Development.

Katz, J., & Aakhus, M. (2002). *Perpetual contact: Mobile communication, private talk, public performance.* Cambridge, UK: Cambridge University Press. doi:10.1017/CBO9780511489471

Koile, S., & Singer, D. (2006). Development of a tablet–PC based system to increase instructor-student classroom interactions and student learning. In Berque, D., Prey, J., & Reed, R. (Eds.), *The impact of pen-based technologies on education: Vingettes, evaluations and future directions*. Purdue University Press.

Kukulska-Hulme, A., & Sharples, M. (2009). Mobile and contextual learning. *ALT-J: Research in Learning Technology*, *17*(3), 159–160.

Kukulska-Hulme, A., & Traxler, J. (Eds.). (2005). *Mobile learning: A handbook for educators and trainers*. Oxon, UK: Routledge.

Lane, C., & Rollnick, S. (2007). The use of simulated patients and role-play in communication skills training: A review of the literature to August 2005. *Patient Education and Counseling*, *67*, 13–20. doi:10.1016/j.pec.2007.02.011

Lefoe, G., Olney, I., Wright, R., & Herrington, A. (2009). Faculty development for new technologies: Putting mobile learning in the hands of the teachers. In J. Herrington, A. Herrington, J. Mantei, I. Olney, & B. Ferry (Eds.), *New technologies, new pedagogies: Mobile learning in higher education* (pp. 15-27). Wollongong: University of Wollongong. Retrieved from http://nla.gov.au/nla.arc-101041

Lysaght, R., & Hill, A. (in press). The use of standardized patients as an alternative to fieldwork. In McAllister, L., Paterson, M., Higgs, J., & Bithell, C. (Eds.), *Innovations in allied health fieldwork education: A critical appraisal*. The Netherlands: Sense Publishers.

Maddrell, J. (2008). *The effect of backchannel interactions on cognitive load.* Retrieved July 2, 2009, from http://www.scribd.com/doc/3727549/BackchannelResearch-Paper-Jennifer-Maddrell

Mohammed, P., & Mohan, P. (2010). Combining digital games with culture: A novel approach towards boosting student interest and skill development in computer science programming. In A. Gadomski, B. Kramer, C. Lester, & M. Popescu (Eds.), *Proceedings of the Second International Conference on Mobile, Hybrid and On-line Learning, eL & mL, 2010,* St Maarten, Netherlands, Antilles.

Naismith, L., Lonsdale, P., Vavoula, G., & Sharples, M. (2006). *Literature review in mobile technologies and learning. Future lab Series, Report 11*. University of Birmingham.

Pettit, J., & Kukulska-Hulme, A. (2008). Do smart devices make smart learners? In *Proceedings of the MLearn 2008 Conference*, 7-10 Oct 2008, Ironbridge Gorge, Shropshire, UK.

Rodger, S., Webb, G., Gilbert, J., Wrightson, P., McKeeken, J., & Devitt, L. (2008). Clinical education & practice placements in the allied health professions: An international perspective. *Journal of Allied Health*, *37*, 53–62.

Torre, D., Simpson, D., Sebastian, J., & Elnicki, D. (2005). Learning/feedback activities and high-quality teaching: Perceptions of third year medical students during and inpatient rotation. *Academic Medicine*, *80*(10), 950–954. doi:10.1097/00001888-200510000-00016

Traxler, J. (2008). *Mobility, modernity and development.* Paper presented at first m4d Karlstadt, Sweden.

Traxler, J. (2009a). Learning in a mobile age. *International Journal of Mobile and Blended Learning*, *1*(1), 1–12. doi:10.4018/jmbl.2009010101

Traxler, J. (2009b). Students and mobile devices: Choosing which dream. In *Proceedings: ALT-C 2009 "In dreams begins responsibility" - choice, evidence and change*, 8-10 September 2009, Manchester.

Zuber-Skerritt, O. (1993). Improving learning and teaching through action learning and action research. *Higher Education and Development*, *12*(1), 45–58. doi:10.1080/0729436930120105

ADDITIONAL READING

Ally, M. (2007). Guest Editorial: Mobile Learning. *International Review of Research in Open and Distance Learning*, *8*(2), 1–4.

Biggs, J. (1999). What the student does: Teaching for enhanced learning. *Higher Education Research & Development*, *18*(1), 57–75. doi:10.1080/0729436990180105

Caladine, R., Andrews, T., Tynan, B., Smyth, R., & Vale, D. (2010). New Communications Options: a renaissance in IP use. In G. Veletsianos, G. (Ed.), *Using Emerging Technologies in Distance Education.* Athabasca: Athabasca University Press.

Cochrane, T., & Bateman, R. (2010). Smartphones give you wings: pedagogical affordances of mobile Web 2.0. *Australasian Journal of Educational Technology*, *26*(1), 1–14.

Delany, C., & Molloy, E. (Eds.). (2009). *Clinical education in the health professions*. Sydney: Churchill Livingston Elsevier.

Garrett, B. M., & Jackson, C. (2006). A mobile clinical e-portfolio for nursing and medical students, using wireless personal digital assistants (PDAs). *Nurse Education Today*, *26*, 647–654. doi:10.1016/j.nedt.2006.07.020

Herrington, J., Herrington, A., Mantei, J., Olney, I., & Ferry, B. (Eds.). (2009). *New technologies, new pedagogies: Mobile learning in higher education.* Faculty of Education, University of Wollongong. http://nla.gov.au/nla.arc-101041.

Higgs, J., Jones, M. A., Loftus, S., & Christensen, N. (Eds.). (2008). *Clinical reasoning in the health professions* (3rd ed.). Oxford: Elsevier Butterworth-Heinemann.

John, P. (2002). *Teaching and Learning with ICT: New Technology, New Pedagogy?* Paper presented at the BERA conference, 12-14 September, University of Exeter.

Kervin, L., & Mantei, J. (2009). Collaborative gathering, evaluating and communicating 'wisdom' using iPods. In J. Herrington, A. Herrington, J. Mantei, I. Olney, & B. Ferry (Eds.), *New technologies, new pedagogies: Mobile learning in higher education* (pp. 99-109). Wollongong: University of Wollongong. http://nla.gov.au/nla.arc-101041.

Kim, S. H., Mims, C., & Holmes, K. P. (2006). An introduction to current trends and benefits of mobile wireless technology use in higher education. *AACE Journal*, *14*(1), 77–100.

Kukulska-Hulme, A. (2007). Mobile Usability in Educational Contexts: What have we learnt? *International Review of Research in Open and Distance Learning*, *8*(2), 1–16.

Kukulska-Hulme, A. (2009). Conclusions: Future Directions in Researching Mobile Learning. In G. Vavoula, N. Pachler, & A. Kukulska-Hulme, A. (Eds.), *Researching Mobile Learning: Frameworks, tools and research designs* (pp. 353–365). Oxford, UK: Peter Lang Verlag.

Kukulska-Hulme, A., & Pettit, J. (2009). Practitioners as innovators: Emergent practice in personal mobile teaching, learning, work and leisure. In M. Ally (Ed.), *Mobile Learning: transforming the delivery of education and training. Issues in Distance Education* (pp. 135–155). Athabasca: Athabasca University Press

Kukulska-Hulme, A., Sharples, M., Milrad, M., Arnedillo-Sánchez, I., & Vavoula, G. (2009). Innovation in Mobile Learning: a European Perspective. *International Journal of Mobile and Blended Learning*, *1*(1), 13–35. doi:10.4018/jmbl.2009010102

Kukulska-Hulme, A., & Traxler, J. (Eds.). (2005). *Mobile Learning: A Handbook for Educators and Trainers*. London: Routledge.

Kukulska-Hulme, A., & Traxler, J. (2007). Designing for mobile and wireless learning. In Beetham, H., & Sharpe, R. (Eds.), *Rethinking pedagogy for a digital age: Designing and delivering e-learning*. London: Routledge.

Ladyshewsky, R. K., & Gardner, P. (2008). Peer assisted learning and blogging: A strategy to promote reflective practice during clinical fieldwork. *Australasian Journal of Educational Technology*, *24*(3), 241–257.

McAllister, L., Paterson, M., Higgs, J., & Bithell, C. (Eds.)(in press). *Innovations in Allied Health Fieldwork Education: A Critical Appraisal.* The Netherlands; Sense Publishers.

Palmer, S., & Holt, D. (2009). Staff and student perceptions of an online learning environment: Difference and development. *Australasian Journal of Educational Technology*, *25*(3), 366–381.

Rose, M., & Best, D. (Eds.). (2005). *Transforming Practice through Clinical Education, Professional Supervision and Mentoring*. Oxford: Elsevier Chruchill Livingstone.

Sharples, M. (2002). Disruptive devices: mobile technology for conversational learning. *International Journal of Continuing Engineering Education and Lifelong Learning*, *12*(5/6), 504–520. doi:10.1504/IJCEELL.2002.002148

Sharples, M. (2005). Learning as Conversation: Transforming Education in the Mobile Age. In *Proceedings of Conference on Seeing, Understanding, Learning in the Mobile Age,* (pp. 147-152). Budapest, Hungary.

Syder, D. (1996). The use of simulated clients to develop the clinical skills of Speech and Language Therapy students. *European Journal of Disorders of Communication, 31*, 181–192. doi:10.3109/13682829609042220

KEY TERMS AND DEFINITIONS

Mobile Devices: Handheld and portable technologies such as mobile phones, MP3 players, laptop computers, gaming consoles, digital cameras, audience response systems, etc.

Mobility: The ability to learn anywhere, anytime supported by access to resources regardless of time and location.

Contextual Learning: Learning in a variety of contexts that support the development of knowledge and understanding.

Standardised Patients: Actors who are trained to portray a variety of clinical conditions to facilitate students' participation in clinical practice activities.

Clinical Education: Learning and teaching and in clinical settings that supports the development of clinical competency.

Chapter 15
The New Age "Information Dowser" and Mobile Learning Opportunities:
The Use of Library Classification and Subject Headings in K-20 Education - Today and Tomorrow

Tom Adamich
Muskingum University, USA

ABSTRACT

This treatise will profile a case study exploration involving the possible use of library classification and subject headings as an element of contextual identification – evidence that a particular material's subject matter and content (as a whole or a portion of it) may satisfy an informational need. One of the challenges of this task is to develop a materials evaluation process using library classification that is both user-friendly and technologically savvy. In this case, mobile technology has been selected as the possible mode of information delivery.

The other challenge is to determine if the resulting use of contextual identification (using library classification and subject headings) accessible via a mobile device is appropriate for a particular institution's information/material retrieval needs, user population, and budget (in this case, a small academic library). The result is the development of the "RMU Information Dowser" project by the Robert Morris University (RMU) Libraries. This project, also designed to possibly satisfy the RMU Libraries mandate to assist in university-wide application of the ACRL Information Literacy Competency Standards for Higher Education in the future, will profile how the university has been exploring use of a combination of mobile technology and reference processes to create a tool to promote rapid library catalog information retrieval and materials access in a student-centered, socially-friendly context.

DOI: 10.4018/978-1-60960-511-7.ch015

INTRODUCTION

In today's information age, the "desire" to retrieve information quickly and easily is increasing at an exponential rate. Appropriately, the "desire" to identify, review, and use the resources (i.e. steps in the materials selection and information literacy processes) in order to satisfy an information need quickly and easily is also rising at a rapid pace. In unison with these factors is the proliferation of technology tools designed to comfortably and successfully locate and use these items. Furthermore, the ability of young people (identified arbitrarily as "Millennials" ages 12 to 24) to use technology (cell phones, Palm computers, and the like) has also grown exponentially and can be attributed to a broad awareness of technology tools information provided by popular media outlets as well as convenient access to the equipment, peer association with the items, and widespread use of technology tools in the daily lives of teens – both inside and outside the classroom.

Conversely, while the widespread use of technology tools and the "desire" to retrieve information and information materials is rapid, the mastery of information literacy skills – the ability to identify specific information and the sources used to obtain that information -- in the post K-20 environment has declined significantly over the last quarter century. Furthermore, the "psychological" effects of information literacy skills decline have lead to an increase in the number of young adults entering the academic environment with the "fear" of research and its associated negative connotations.

One key aspect of information retrieval involves the use of terms and classification associated with a specific item (i.e. subject headings and library classification). In the past, library patrons used printed bibliographies or "finding aids" to locate library materials corresponding to specific subject headings (typically the Library of Congress Subject Headings) and classification (mainly the Library of Congress Classification System or the Dewey Decimal Classification System). These location tools also included institution-specific information on where the items were located in the library and how they were obtained for use.

Today, the same identification tools exist (i.e. library classification and subject headings) but are primarily housed and delivered electronically. However, the electronic mode of information delivery, in most instances, still remains very "traditional" (i.e. use of a computer terminal or laptop-type computer in a "fixed" location). As will be explored in this discussion, the challenge is to determine how the same useful identification tools can be used in world that increasingly relies on "mobile technology" (i.e. hand-held devices) as information delivery tools.

BACKGROUND

As was mentioned earlier, locating library materials using library classification (i.e. an item's classification-derived "home on the shelf") is, obviously, not a new concept. Likewise, the use of the descriptive labels associated with a particular classification number to validate item selection also has its origins in basic library theory and practice. According to Mills (2004), indexing and searching function as the two fundamental operations of retrieval. The classification number itself – using a combination of numbers, letters, and, in some cases words -- works to establish the placement context for a single unit or multiple units with respect to the tangible placement of that entity in a prescribed physical area. The classification's descriptive labels function as "access points" in creating an intellectual link between the subjective labels and the objective number. Library catalogers create the conceptual bond between the classification number and its descriptive labels to develop a tool users may manipulate to encourage information retrieval, since such retrieval is the "final and, therefore, the most obvious of the processes that contribute to an information system" (Jacob, 2004, p. 515).

A significant contribution to analyzing the issue of the relationship between a classification number, its descriptive labels, and materials access was conducted by Chandler and LeBlanc at Cornell University. Using Columbia University's Hierarchical Interface to Library of Congress Classification (HILCC), Chandler and LeBlanc conducted a classification-to-subject analysis (via extractions of both subject headings and classifications obtained from representative metadata). The HILCC was developed by Stephen Paul Davis of Columbia University to take advantage of newly-developed library automation system user interfaces which allow users more direct and accurate subject access to all library collections. In Columbia's case, development of the HILCC allows Columbia to obtain greater subject access to its electronic collections, hereby creating "virtual catalog" access to Columbia's electronic resources. (Davis, 2001, p. 20)

Using HILCC's breakdown of key subject categories as the primary subject framework, Chandler and LeBlanc proceeded to write a Perl script which would link the LC call numbers in Cornell's undergraduate catalog with the developed HILCC subject map. Chandler and LeBlanc then proceeded to examine the resulting delimited file, with the goal of obtaining valid links between subject categories and items in Cornell's catalog. However, unlike the HILCC project at Columbia (with its "electronic resources" focus), Chandler and LeBlanc's main objective at Cornell was to use the link between print items in the Cornell University undergraduate library catalog and the HILCC classification/subject classes to create a "virtual catalog" of print resources (some of which are located in off-campus storage) as a means of addressing library collections space issues (Chandler and LeBlanc, 2006, p. 158).

Unfortunately, the Cornell-to-HILCC correlated mapping produced disappointing results with respect to the desire to obtain a working "retrieval set"; several subject categories contained more than 1,000 titles, with other sets yielding 2,500 and 10,000 titles respectively. (Chandler and LeBlanc, 2006) In order to test the validity of their processes, Chandler and LeBlanc examined data acquired from several ARL (Association of Research Libraries) and, upon analysis, obtained a Pearson correlation of the results. Chandler and LeBlanc then proceeded to modify the existing HILCC table to yield a more "universal" subject structure, rather than one tailored specifically to reflect Columbia's undergraduate catalog (Chandler and LeBlanc, 2006, p. 159).

Another important reason for examining links between library classification and subject descriptive access is the ability of such a relationship to contribute favorably to demonstrating how a particular educational material's content matter may contain parts and portions of "teachable units" (which will be described in greater detail shortly when the concept of the "learning object" is discussed). These units are used as the basis for developing test questions that might be associated with both formative and summative assessment. The RMU Information Dowser concept project illustrates how, by using library classification and subject matter to create contextual mapping accessible via mobile technology, the tool has the potential to contribute favorably to a user's ability to locate the "teachable unit" within a particular educational item, extract that item, and use the concept to support elements of formative and summative assessment (the aforementioned test question development as well as selection of item content for use in course design and selection of teaching strategy). The RMU Information Dowser project may also have the potential to contribute to successful information and resource discovery that will promote incorporating important psychometric models (e.g., Bayesian inference) into the instructional process.

FORMATIVE ASSESSMENT

According to Schwarz and Sykes' (2004)*Psychometric foundations of formative assessment,* formative assessments are frequent and evaluate a "small number of data points (items)." Other favorable characteristics of formative assessment include providing information to accomplish the following (Schwarz & Sykes, 2004, p. 2):

- Identify current student knowledge levels at a "highly granular level"
- Support prescriptive teaching information at multiple levels (student, class, school)
- Align content and materials to state standards*
- Promote instructional effectiveness
- Predict how well students are prepared to meet state standards and annual summative state assessments

Please note that the third bullet in the list is starred to further emphasize the goal of content-to-materials alignment in applying formative assessment techniques. However, as will be profiled later in this article, the methodologies and thought processes associated with content-to-materials alignment differ widely with respect to the body of assessment strategies used, the alignment procedures utilized, and the selection of modes of instructional delivery (i.e. virtual, asynchronous, live, etc.).

Another key feature of formative assessment is the "diagnostic" nature of the basic concept. While summative assessments like standardized testing incorporate a "comprehensive" approach to testing (i.e. using single, paper-based tests populated with a preponderance of concrete, single-response test questions covering a wide array of topics and often administered on a "one-time-only" basis), Schwarz and Sykes (2004) illustrate the formative assessments' more specific and fundamental differences:

- Content contains high levels of technical rigor
- Assessments are deep and broad, often incorporating constructed-response items
- Assessments are given at multiple times for progression tracking
- Both traditional and non-traditional (i.e. electronic) modes of assessment are available
- Local educators can "customize" some aspects of the formative assessment (FA) system
- Fine-grain alignment to content standards
- FA system is dynamic and improves over time (Schwarz & Sykes, p. 3)

Once again, the strong relationship of content-to-materials alignment to formative assessment is clearly evident. According to Schwarz and Sykes (2004), alignment is a form of "validation" in the formative assessment model and involves an examination of the "items that comprise the assessment and an alignment of that test content to the goal(s) of the assessment, obtaining item and test statistical characteristics and may even involve steps such as ascertaining the usability of score reports.

One may ask the question, "How does formative assessment's "diagnostic characteristics" relate to the RMU Information Dowser project and to the learning object/smallest teachable concept discussed earlier?" Simply stated, since formative assessment involves frequent testing of concepts – ideas which are often highly-technical, granular and directly aligned to assessment goals – the learning object (particularly the electronic-based interactive variety, to be discussed shortly) and the smallest teachable concept are, in essence, the "core elements" of formative assessment. The RMU Information Dowser project, in concept, facilitates the contextual and location-based "link" between the material used to teach the learning object/smallest teachable concept and its corresponding assessment-based element (i.e.

the test question) as it exists in the formative assessment tool.

The frequent testing and learning object/smallest teachable concept link issues are further illustrated in formative assessment's basic testing structure, which may include the use of "benchmark" tests and "growth strands". Schwarz and Sykes (2004) identify benchmark testing as small tests (historically called "pre-tests" and "post-tests") that may "closely match the scope and sequence of instruction that match to a state's content standards [and] can provide information that can more effectively impact instruction and learning." Furthermore, growth strands can be defined as subsets of benchmark tests and include student-initiated diagnostic references to concept comprehension and educator-based references to a particular content standard and appropriate instructional interventions which may be recommended to the student (Schwarz & Sykes, 2004, p. 7). As with formative assessment in general, the learning object /smallest teachable concept are the core elements to any "effective FA-based" material and instructional outputs developed by educators and educational materials suppliers. Once again, the RMU Information Dowser's possible core mission (to use a combination of library classification and subject description metadata – accessible via mobile technology - to place the item in context and facilitate rapid retrieval) would contribute greatly to the success of this process.

Now, the question, "Why do the learning object/smallest teachable concept (and the RMU Information Dowser's "contextual locator-based" information inquiry design) play such a pivotal role in improving student understanding of individual, test-related concepts in either a formative or summative testing/standardized testing situation? The discussion of the connections that may exist between frequent testing, learning objects/smallest teachable concept, and rapid (as well as meaningful) information retrieval (possibly via the mobile technology associated with the RMU Information Dowser project) may help to answer that student understanding-based question.

BAYESIAN INFERENCE MODELS AND THE "PSYCHOMETRIC APPROACH" IN FORMATIVE ASSESSMENT

Earlier, the role of formative assessment's "diagnostic characteristics" was defined to include the "frequent testing of concepts, ideas which are often highly-technical, granular and directly aligned to assessment goals." As a result, educational theorists developed several psychometric approaches that may be used to promote the frequent testing (i.e. benchmark tests and growth strands) and instructional interventions which are the key elements of formative assessment. Additionally, these psychometric models incorporate "item response theory" or IRT, which directly supports another key element of formative assessment – the alignment of test questions (and also, with respect to this study, educational materials) to content assessment goals – in this case, state academic standards).

One model, identified by Schwarz and Sykes (2004), is the Unified (Fusion) Model, which combines knowledge of a particular rule or theory in a "discrete/deterministic" context in unison with a stochastic IRT-based element. In simple terms, the model mathematically predicts how a student's knowledge of a particular rule will determine her/his ability to answer a test question that indirectly incorporates the rule in successfully answering that question.

Another model, which relates to this study, is the use of Bayesian belief networks, also known as adaptive probability networks. By definition, a Bayesian model features the use of probability in creating a "probabilistic relationship" between variables in an historical context. For example, when using content from an educational resource to develop testing materials to use when incorporating a "benchmark testing's pre-test and post-test" structure into a course's testing structure, if both a pre-test and post-test question are analyzed (called "pre-cursor" and "post-cursor" by Schwarz and Sykes, 2004, 18), an educator can make an inference as to the "probability" that a student understands either part or all of a concept. Furthermore, intervention strategies may be identified to promote remediation by the student if necessary.

As a result of this analysis, the learning object/smallest teachable concept, once again, becomes, clearly, an important element in the achievement of standards-based, individual test-related concepts. Both the size of the learning object/smallest teachable concept and the diagnostic opportunities for prediction and remediation are identified advantages for use of that element. Some educational theorists venture to label the learning object/smallest teachable concept only as a "learning object. What, then, by definition, is a learning object?

LEARNING OBJECT

According to the Learning Technology Standards Committee of the Institute of Electrical and Electronics Engineers (IEEE), a *learning object* is any entity, digital or non-digital, that may be used for learning, education, or training (IEEE 2002). An additional definition identifies the learning object as "any digital resource that can be reused to support learning." (Wiley 2001). Yet another definition talks about the learning object as "web-based interactive chunks" of e-learning" designed to explain a stand-alone learning objective (CETL 2005). In every case, these definitions are a direct link to both an information query's individual questions and groups of question categories respectively. Learning objects also form the basis for the discussion on both formative assessment/Bayesian Inference Models theory profiled earlier in this article.

The instructional goal behind the use of the learning object is to attempt to place the "smallest teachable concept" in a tangible framework to better represent its connection to parts/portions of library materials (i.e. chapters, graphs, charts, figures, visual clips, sound bytes, etc.). By defining a learning object in "tangible terms", educators can better understand the "smallest teachable concept" – both individually and in context with other "smallest teachable concepts" contained within a particular library resource. The additional derived benefit of using a learning object is to promote the "diagnostic and probability-based" benefits of teaching smaller, more specific pieces of information (i.e. parts of a library resource) as part of a school's comprehensive formative assessment utilization program – one which also effectively incorporates Bayesian concepts into teaching. Thus, the decision has been made to refer to the learning object/smallest teachable concept in tandem to best reflect both their intangible (i.e. the virtual contextual access to/references of the learning object) and tangible characteristics (i.e. with respect to the actual parts/portions of a resource to which they refer – either in traditional "hard copy" form or electronically).

FORMATIVE ASSESSMENT, BAYESIAN PSYCHOMETRIC APPROACHES, AND LEARNING OBJECTS IN EDUCATION: SUCCESSFUL APPLICATIONS AND IMPLICATIONS FOR ALIGNMENT OF INSTRUCTIONAL MATERIALS

Now that we have a fundamental understanding of the use of formative assessment and how it incorporates Bayesian psychometric probabilities and learning objects/smallest teachable part as the "chunks" and "crumbs" of information retrieval and evaluation, let's look at several examples of successful utilization of these concepts in the education context.

Fisher, Grant, Frey, and Johnson (2007) discuss how the effectiveness of using oral questioning, individual concept testing, and writing prompts – all formative assessment strategies – act to encourage precision in teaching and discourage student misunderstanding of concepts in K-12 education. The authors developed a four-step process which was applied school-wide at a high school in San Diego, California (USA). The four steps include the following: (1) develop pacing guidelines, (2) design common assessments, (3) conduct item analysis, and (4) engage in instructional conversation.

Many of these strategies rely heavily on the identification and use of instructional resources that promote precision teaching and reduce concept misunderstanding. In one example, Fisher, Grant, Frey, and Johnson take the results of an item level analysis and determine that most students answered the question being analyzed incorrectly because they confused one type of democracy with another. Thus, the importance of locating and using resources that promote effective item-level mastery becomes critical.

How were the educators able to address the democracy comprehension problem illustrated above with such precision? By generating an item level analysis report, teachers used statistical algorithms to generate item level statistics and examined these statistics to determine student mastery. In cases where mastery levels were low for a particular item, teachers worked together to identify instructional materials and resources that had the potential to increase student mastery and reduce confusion. Such identification involved mainly informal methods such as word-of-mouth recognition of resources and best practices/experience.

Another K-12-based school-wide application of formative assessment is profiled by Priestly and Sime (cited in Fisher, Grant, Frey, & Johnson, 2007). In *Formative assessment for all: a whole-school approach to pedagogic change,* the authors present data associated with the "Assessment is for learning" initiative being taught in all schools in Scotland. Once again, the emphasis is placed on educators working together in groups to develop related item-level assessments and alleviate student-based concept confusion and reduce misunderstanding levels. Where the Priestly and Sime study differs from the Fisher, Grant, Frey, and Johnson study is in its focus on educator dynamics and school culture and how the resistance to using such methodology as item-level assessment may impede the potential for instructional success.

Also, as with the Fisher, Grant, Frey, and Johnson study (2007), Priestly and Sime identify four elements of the Scottish formative assessment model which incorporate a number of instructional resources utilization-based strategies: (1) questioning (using 'wait time' during oral questioning to allow students to process questions and generate answers); (2) feedback through marking (using 'feedforward' – feedback targeted at improvement rather than mere reporting); (3) peer assessment and self assessment (promoting two-way communication of understanding – e.g., the red light/green light approach of 'traffic lighting'); and (4) formative use of summative tests (permitting students to redraft work after research to validate choices and to set/mark summative questions (Fisher, Grant, Frey, and Johnson, p. 65).

Thus, in the Fisher, Grant, Frey, and Johnson study, there is implied evidence that the ability to effectively identify and use appropriate instructional material in formative assessment-based instruction is dynamic and unscientific. However, the conceptual existence of appropriate instructional resource identification models and technologies does exist and the conceptualization process and possible utilization of such a model will be discussed in greater detail relative to the RMU Information Dowser project's goals and objectives (discussed shortly). In this discussion, the creation/delivery of metadata to reflect materials-to-assessment goals alignment is intended to answer the identification/use question.

FORMATIVE ASSESSMENT, BAYESIAN PSYCHOMETRIC APPROACHES, AND LEARNING OBJECTS IN BUSINESS: IMPLICATIONS FOR CONTEXTUAL ALIGNMENT OF INSTRUCTIONAL MATERIALS

Even though the RMU Information Dowser projects' possible positive impact and implementation apply primarily to the use of contextual identification (using library classification and subject headings) of learning object/smallest teachable object-based library materials (which accessible via a mobile device) in an academic setting, other quasi-education and training areas (some of which are directly related to the workforce preparation skill goals of education) also are exploring use of contextual information retrieval/materials retrieval technology to support their formative assessment testing goals and formative/summative evaluation success.

Remember, the core definition of a learning object/smallest teachable concept is that it is an entity that may be used for "learning, education or training". In a business context, the ability to connect mastery of a particular concept to the assessment used to determine mastery (and to identify and locate relevant materials to teach that concept) is often as simple as whether or not a piece of equipment produced via a manufacturing process (or processes) passes final inspection. In this example, locating the bulletin or manual that contains a section or portion that discusses that concept is critical to the process.

As in the educational context, business often incorporates formative assessment's "one concept focus", benchmark testing, and growth strands in training to determine whether a worker can demonstrate mastery. Also, the business-based use of the predictability of mastery associated with assessing a worker's ability to understand a topic or process taught in a training session is related conceptually to incorporating Bayesian psychometric concepts into the business training assessment process. However, while the methodologies and approaches to using formative assessment in business and education are similar, the goals and end products differ slightly.

Berstene, in his article *Hawthorne's twice-taught topics* (Berstene 2006), defines the use of learning objects in a business context by telling the story of a business meeting that features a discussion of the need to "chunk" the company's training curriculum into one-hour segments (to compensate for the need to change the program as the organization's needs change). He explains that by using the chunking of tasks (or the "learning object"), the process allows, "…the most novice of employees to function at the same level of performance as the most qualified employee." (Berestene, 2006, p. 31).

Berstene (2006) later argues that the idea of making all of the learning "chunks" one hour in length is impractical because some concepts require a few minutes to master while others "might require two hours to learn proficiently with practice. He makes reference to Hamerly's *The great chunky debate*, which talks about identifying the "terminal learning objective" and taking each element of the training program down to the "smallest teachable concept". Later, Berstene refers to Edgar Allan Poe's famous analysis of Nathaniel Hawthorne's "twice-told tales", which was a commentary on the newly-developed (at the time) genre called the short story. Poe makes several references to a "poem too brief may produce a vivid, but never an intense or enduring impression". Berstene expands on Poe's analogy to chunking and use of the learning object/smallest teachable part by saying:

Today, the same may be said for too small a chunk or 'infonugget'. Learning cannot take place if we do not provide the opportunity to allow for practice and failure, or to tell, show, do, and tell the student again what it is that he/she needs to succeed in the task. (Berstene, 2006, p. 30).

Another business-based similarity to the use of learning objects in the education context mentioned earlier is the "one concept focus" that suggests developing a training module that "delivers one and only one skill or knowledge point. All of the text, exercises and feedback, skill and knowledge checks should add only to acquisition of that one item"(Berstene, 2006, p. 31). However, unlike remedial interventions in education (which may be the result of data obtained from benchmark testing and growth strands assessment), Berstene feels that students in a business context should "...be able to complete the module in one sitting, whether that [module] is fifteen minutes or three hours."(Berstene, 2006, p. 31). This cost/benefit-related element for business remediation differs from that of education remediation in that the end product of business remediation has an immediate end product value; conversely, education's cost/benefit end product value may be deferred and recognized many years later when the student completes graduation requirements (either K-12 or higher education- based) and enters the workforce.

An additional business-based reference to the learning object comes from Autodesk Media & Entertainment and its attempt to establish guidelines for learning associated with "Maya" – the company's digital image creation, 3-D effects, and other special effects tool. In *Maya Fundamentals Standards* (Autodesk Media & Entertainment, 2007), the company outlines several discussion points arranged by topic, sub-topic, and content. According to the documentation, the "content "area listing reflects the "smallest teachable piece of content." Once again, the use of the learning object/smallest teachable concept in business training encourages the use of formative assessment techniques and psychometric approaches to remediation that may include Bayesian belief networks/adaptive probability networks; however, once again, the cost/benefit model for business requires evidence of mastery which is directly related to digital image/3-D effects/special effects as they incorporate directly Maya commands and processes, whereas the education model may require demonstration of mastery only at the general level – i.e. understanding digital image/3-D effects/special effects concepts.

At this point, the theoretical relationship between the use of formative assessment and Bayesian psychometric approaches in both a business and educational context has been established. Additionally, business-based examples of both theories are featured. However, as with education's use of formative assessment-based teaching elements, the question arises, "How does one effectively and quickly identify and locate instruction materials which teach specific concepts – supporting the use of formative assessment techniques and learning objects in tandem to improve employee understanding of individual, test-related concepts in a summative testing/standardized testing situation (i.e. in the case of a business application, the successful application of a training concept in either a specific production or business output-related application)? It is fortunate that the mass utilization and accessibility of the Internet has allowed for the potential to mass distribute materials-to-information goal/objective aligned instructional materials identification and location information. With respect to the RMU Information Dowser project and mobile technology, the possibilities of business replication of the project's goals (i.e. to locate appropriate training manuals and learning objects/smallest teachable concepts within those manuals using mobile technology) is tangible.

REVISITING ACADEMIC ISSUES: MOBILE TECHNOLOGY AND THE "MILLENNIAL FACTOR"

Again the "materials-to-information goal/objective " focus of most information seeking and evaluating endeavors (as evidenced by its role in promoting formative assessment and the use of psychometric models (including Bayesian psychometrics), Chandler and LeBlanc's adaptation of

Columbia University's Hierarchical Interface to Library of Congress Classification reflects clearly the efforts of both business and library researchers to develop useful links between classification/subject categories and materials (either as a whole or in part) in today's Internet-based technology landscape. To return, briefly, to the academic side of the equation, as with the Columbia project to create a "virtual catalog" of electronic resources, Cornell's "virtual catalog" of print resources represents a commitment by Cornell University Libraries to use tools and techniques (in this case, universal Internet access and subject-based, keyword search strategies) familiar to today's "technology-savvy" populace – particularly academic library patrons ages 18-24. Using computer and, specifically, library automation system technology, the goal is for the library patron (in the case of the academic library, the student or faculty member) to quickly obtain materials relevant to her/his research needs.

Sometimes, however, the ability of the library patron to achieve a point of retrieval (either information or materials-based) is interrupted by two factors – a sufficient level of understanding relative to a particular subject area and an inability to contextually locate a particular resource or information element in an information retrieval area (i.e. library shelf, online database, traditional personal computer station, etc.). Thus, the effective use of the classification element (its objective core number and subjective descriptive labels functioning as a unit) is dependent upon the searcher's pedagogical relationship to mastering the information retrieval landscape (often viewed in the abstract by educators in all curricular areas) and the physical reality (either tactile or virtual) of the placement of a resource in the retrieval mechanism provided.

With respect to the Robert Morris University (RMU) Library's "Information Dowser" project, the awareness that today's typical RMU library patron – "Millennials" those persons between the ages of 18-24 who are technologically "savvy" but often lacking the pedagogical understanding of classification/subject analysis as well as the time to achieve such understanding – was a key factor in seeking a way to, conceptually, facilitate quick, accurate materials retrieval (which satisfied research needs) using mobile technology readily available and familiar to the library patron (in this case, a Blackberry or similar device). This project was effective labeled the "information dowser" by Christopher Devine, Head, Public Services, RMU Libraries. The term "information dowser" was actually coined as a sort of analogy to the dowsing or divining rod that is still sometimes used to detect groundwater. http://en.wikipedia.org/wiki/Dowsing .

In the case of effective classification element and contextual understanding of its role in locating library information and resources, the Robert Morris University (RMU) Libraries identified, using student inquiry statistics and informal faculty surveys, a perceived lack of understanding relative to the Libraries' classification system and its use (in this case, the Dewey Decimal Classification System). Details of these measures will be discussed later in this treatise. In addition, Robert Morris University instituted a university-wide initiative to integrate components of the *ACRL Information Literacy Competency Standards for Higher Education* in 2004. Through the use of bibliographic instructional programs developed and presented by Jacqueline Klentzin, Bibliographic Instruction Professor, RMU Libraries, several areas of the *Standards* are addressed (in particular, Standards 2, 3, and 5)(C. Devine, personal communication, May 18, 2010). Student and faculty evaluations indicated a need to analyze the level of student understanding with respect to the Dewey Decimal Classification System. This measure is in direct correlation to the following elements example of the *ACRL Information Literacy Competency Standards for Higher Education (*ALA, 2006):

Standard Two

The information literate student accesses needed information effectively and efficiently.

1. The information literate student retrieves information online or in person using a variety of methods.

   Outcomes Include:

   a. Uses various search systems to retrieve information in a variety of formats
   b. Uses various classification schemes and other systems (e.g., call number systems or indexes) to locate information resources within the library or to identify specific sites for physical exploration
   c. Uses specialized online or in person services available at the institution to retrieve information needed (e.g., interlibrary loan/document delivery, professional associations, institutional research offices, community resources, experts and practitioners)

The combination of information retrieval theory, an analysis of query results from actual reference inquiries, and location/retrieval of materials using technology forms the basis of the information dowser project from a *conceptual* standpoint. The emphasis on the fact that the project is conceptual places the idea of utilizing mobile technology as a searching/location/tracking tool – containing data directly linked to the RMU Libraries Online Catalog – in a realistic context. Knowing that several key components of the concept would require the acquisition and testing of the interaction of bibliographic and directional terminology by the user prior to translating that interaction to technology means that the RMU Information Dowser project can be classified as a "phase project" – with theory, product development, and application components.

Basically, the use of the RMU "information dowser" involves the library patron's interaction with information need, time factors, final information product requirements, etc. using technology specifically related to the library's key information components – the library collection and the library's public services, or reference, staff. Each portion of the information dowser model is dependent upon the other for functionality (See Figure 1).

For example, if a student were searching for information on "geographic landforms", she or he, using the information dowser, would use the core term, structured as a Library of Congress Subject Heading, as the basis for forming an information "tree" with several branches including the following:

- LCSH "See" or "See also" references for related terms
- Dewey classification associated with the term
- Location of library collection resources (print)
- Options for searching in library's electronic databases
- Options for search assistance (public services staff, self-service opportunities)

Here are the results associated with the term "landforms" and the first branch of the information tree as searched in the "LC Authorities" database (Library of Congress, 2009; see Table 1).

While this is by no means a comprehensive list, the see-also references for "landforms" originate from an accepted thesaurus (i.e., LCSH) and would be accessible for inclusion in the information dowser product.

In the second branch of the "landforms" tree, listed below is the result of a Dewey classification search (the classification scheme used at Robert Morris University) using the subject "landforms" in the RMU ROBCAT catalogue (Robert Morris University, 2009; see Table 2).

Figure 1. Information Dowser Model ©2010 Tom Adamich

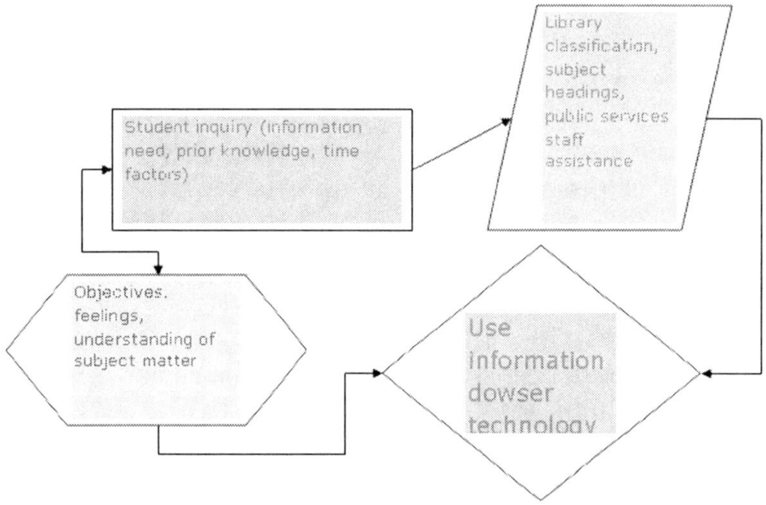

The third branch of the "information dowser tree" involves showing where this particular print resource would be located in the RMU Libraries facility and the status of that item for use; ideally, a location/tracking device (which could interact with a RFID—Radio Frequency Identification Device—electronic security strip or similar device) would indicate on a screen where the item actually resided on the shelf (Robert Morris University, 2009; see Table 3).

Using the fourth branch of the "information dowser tree", the patron would then be directed to a listing of electronic databases like those listed below for additional searching (Robert Morris University, 2009, see Table 4).

Finally, at the same time, the student would also receive information associated with the fifth branch of the "information dowser tree" – how to receive assistance from the RMU Libraries Public Services staff at one of the two RMU Libraries locations listed below (phone and e-mail details not in accordance with the institution's information access policies)(Robert Morris University, 2009, see Table 5).

Information similar to the example above would be programmed into the mobile device and made available for student and faculty use at all Robert Morris University campuses. Each component of the "information dowser tree" would interact simultaneously – prompting the user to select the information access path corresponding to her/his level of need.

In January, 2006, Dr. Valerie Powell, Professor of Computer and Information System, Robert Morris University initiated preliminary discussions with members of the RMU Libraries staff to discuss information dowser theory, project logistics, and access to the relevant assessment data discussed earlier in this paper. An ad-hoc committee – composed of Dr. Powell, RMU Libraries staff members, Dr. Fran Caplan, Libraries Director; Chris Devine, Head of Public Services; Don Luisi, Head, Acquisitions Department; Tom Adamich, Head, Cataloging Department; and Emmet Devine, Department of Information Services -- was assembled to discuss the concept.

During the early stages of the discussion process, references were made to Dr. Powell's past research involving the use of library classification and thesauri to create finding aids for patron use. In 1977, Dr. Powell was involved in the development of the "Resource Handbook/Manual Para

Table 1. Landforms search in LC Authorities database ©2009 Library of Congress)

Search	Term
Search Also Under (5XX)	Alluvial fans
Search Also Under (5XX)	Bars (Geomorphology)
Search Also Under (5XX)	Beaches
Search Also Under (5XX)	Bogs
Search Also Under (5XX)	Caves
Search Also Under (5XX)	Coasts
Search Also Under (5XX)	Deltas
Search Also Under (5XX)	Deserts
Search Also Under (5XX)	Duricrusts
Search Also Under (5XX)	Earth pyramids
Search Also Under (5XX)	Fens
Search Also Under (5XX)	Glacial landforms
Search Also Under (5XX)	Islands
Search Also Under (5XX)	Karst
Search Also Under (5XX)	Lava tubes
Search Also Under (5XX)	Mounds
Search Also Under (5XX)	Natural bridges
Search Also Under (5XX)	Pans (Geomorphology)
Search Also Under (5XX)	Pediments (Geology)
Search Also Under (5XX)	Piedmonts (Geology)
Search Also Under (5XX)	Plains
Search Also Under (5XX)	Potholes
Search Also Under (5XX)	Reefs
Search Also Under (5XX)	Sand dunes
Search Also Under (5XX)	Seashore
Search Also Under (5XX)	Slopes (Physical geography)
Search Also Under (5XX)	Swamps
Search Also Under (5XX)	Terraces (Geology)
Search Also Under (5XX)	Tundras
Search Also Under (5XX)	Valleys
Search Also Under (5XX)	Volcanoes
Search Also Under (5XX)	Watersheds
Search Also Under (5XX)	Wetlands
Search Also Under (5XX)	Cliffs
Search Also Under (5XX)	Benches (Geomorphology)
Search Also Under (5XX)	Uplands
Search Also Under (5XX)	Kettle holes
Search Also Under (5XX)	Potholes (Lakes) [proposed update]
Search Also Under (5XX)	Hoodoos (Geomorphology)

continued on following page

Table 1. Continued

Search Also Under (5XX)	Badlands
Search Also Under (5XX)	Planetary landforms
Search Also Under (5XX)	Geomorphology
Used For/See From (4XX)	Mayer, Cassie. Landforms

Table 2. ROBCAT search results, second branch ©2009 Robert Morris University (Moon Township, PA)

150	__ |**a** Landforms
450	__ |**a** Land forms
551	__ |**w** g |**a** Earth |**x** Surface
550	__ |**a** Geomorphology

Table 3. ROBCAT search results, third branch ©2009 Robert Morris University (Moon Township, PA)

092	__ |**a** 551.4 |**b** D96
100	1_ |**a** Dury, G. H. |**q** (George Harry), |**d** 1916-
245	10 |**a** Essays in geomorphology.
Location:	Moon Campus - General Collection
Call Number:	551.4 D96
Number of Items:	1
Status:	Not Charged

Table 4. ROBCAT databases, fourth branch ©2006 Robert Morris University (Moon Township, PA)

Science & Math Databases	
Conference Papers Index Conference Papers Index: Conference Papers Index indexes papers and poster sessions presented at major scientific conferences in the fields of the life sciences, enviornmental sciences, and aqauatic sciences. Environmental Science Collection Environmental Science Collection:The Environmental Subject Collection is an eJournal collection of over 70 environmental science serials published by Elsevier. This collection uses the ScienceDirect interface for content access. Full text coverage extends from 2001 - present. IngentaConnect Ingenta provides a search interface for over 4500 journals in all fields. On-campus users can register for the journal alert and saved searches features. MathSciNet Ingenta provides a search interface for over 4500 journals in all fields. On-campus users can register for the journal alert and saved searches features.	Project Muse The full text to over 100 scholarly journals in the humanities and social sciences. Scopus Scopus: Scopus is a new interdisciplinary database published by Elsevier. Scopus contains 25 million abstracts from over 14,000 titles across 4,000 publishers in the fields of agricultural science, biological science, chemistry, economics, engineering, environmental science, general science, health science, life science, mathematics, psychology, physics, and social science. Toxline TOXLINE: TOXLINE is a information clearinghouse for toxicology data that is mainiatined by U.S. National Library of Medicine. TOXLINE utilizes a "rolling 5-year" backfile, consisting of records from the last 5 years plus the current year.

Table 5. RMU Libraries contact information, fifth branch ©2006 Robert Morris University (Moon Township, PA)

For More Information,Contact: Library Main Campus Patrick Henry Center
Center for Adult and Continuing Education

Recursos", published by the Federation Bilingual Training Resources Center at East Texas State University in Denton, Texas. This publication uses selected portions of the LC Classification and subjects/descriptors from the LC Subject Headings Thesaurus to identify bilingual education materials that were housed in the Gee Library on the Commerce Campus of ETSU.

In the document, the bilingual education classification areas and subject headings are coupled with specific library circulation, reference, and general instructional information on using the LC Classification System and LC Subject Headings. Additionally, information on the use of the ERIC (Educational Resources Information Center) research service, including a list of ERIC keywords is featured.

While the Information Dowser Committee considered the "Resource Handbook/Manual Para Recursos" a "finding aid from the past", they also felt it reflects the core concepts of the information dowser concept–combining information on the Robert Morris University Libraries' materials and materials acquisitions procedures in a searchable, electronic structure using the Libraries' chosen classification and thesaurus as well as access to electronic resources and library reference services. The key to the success of the "Resource Handbook/Manual Para Recursos" was the fact that the three components necessary for a successful finding aid–identification of terminology, information location, and information access procedures – was present.

In April, 2006, the Information Dowser Committee discussed ways to obtain information on reference queries and analyze those queries for content, structure, and function. Thus, in June, 2006, the RMU Libraries compiled public services reference inquires for the month at the Campus Libraries' reference desk, located in the Patrick Henry Center on the RMU Main Campus in Moon Township. Listed below is a sample of some of the inquiries that were recorded (Robert Morris University, 2009):

- Subject heading query log ©2007 Robert Morris University (Moon Township, PA)
- The impact of communicating via email on Emotional Intelligence
- The decision to drop the atomic bomb
- The history of Wishing (yes, that is not a typo)
- African Americans as a target market
- RFID technology in healthcare
- Female American spies during World War II
- Technology in the classroom
- Technological innovations
- Mandated volunteer programs
- History, evolution and future of manufacturing in Western Pennsylvania
- Home schooling of children
- Communication problems between 911 centers and police departments

The Information Dowser Committee analyzed the results for content and form to determine how the programming needed to facilitate the information dowser's dynamic interaction between multiple layers of terminology relationships (i.e. main headings, derivatives, See references, See also references, etc.), library classification, and reference assistance opportunities would function when applied to RMU-specific information inquiries.

Another area of analysis involved segmenting a portion of the RMU Libraries' collection and conducting a resources identification and mapping project by determining what Dewey (and LC Classification) numbers as well as Library of Congress Subject Headings corresponded to a particular collection area. Using a bibliography of "Information security" resources developed by the late Henrietta Angus, Head, Acquisitions Department, RMU Libraries, the Information Dowser Committee extracted keywords from titles listed in the bibliography, identified corresponding LC Subject Headings for the extracted words, and mapped those LCSH Thesaurus terms to the corresponding Dewey/LC Classification numbers [Author's note: At the time, the RMU Libraries had also embarked on research to determine the feasibility of switching from the Dewey Decimal Classification System to the LC Classification System, thus the reasoning motivating the inclusion of LCC numbers in the analysis.] The results of that study are listed below (Robert Morris University, 2009):

- Course Keyword:
- Information security
- LCSH Equivalent(s) and LCC/Dewey:
- Computer security. QA76.9/658.478
- Data protection. R855.3/651.5
- Computer networks—Security measures. TK5105.59/005.8
- Computer networks—Access control. TK5105.59/005.8
- Course Keyword:
- Information assurance
- LCSH Equivalent(s) and LCC/Dewey:
- Computer networks—Security measures. TK5105.59/005.8
- Computer networks—Access control. TK5105.59/005.8
- Course Keyword:
- Sarbanes Oxley
- LCSH Equivalent(s) and LCC/Dewey:
- United States. Sarbanes-Oxley Act of 2002 KF1357/346.73
- Auditing, Internal—Law and legislation—United States. KF1357/346.73

In late June, 2006, the Information Dowser Committee met to discuss the future direction of product development and determined that there is great potential for the development of an information dowser tool. While the RMU Libraries would continue to analyze the feasibility of using mobile technology to facilitate both information and materials retrieval (relative to the aforementioned identification of learning objects/smallest teachable part; supporting RMU's formative assessment initiatives; promoting the use of Bayesian psychometric models as well as information literacy initiatives being pursued university-wide at Robert Morris University), the application of the RMU Information Dowser project was viewed in 2010, according to Devine, in the following way:

I can say that if such a utility were available (i.e already developed) [which was not the case with the RMU Information Dowser project concept], we would probably conduct a pilot study to investigate whether or not it would be helpful to our patrons in simply navigating and locating materials physically housed in our stacks. Frankly, I would suspect that a large library with a complex and arbitrary organization that was housed in several adjoined buildings (the Pattee Library at Penn State comes readily to mind) might find this application of the Dowser more useful than a small, simple library such as ours. Nevertheless, assuming that the price of implementation was not prohibitive, we would, as I have said, probably be open to testing it. Beyond that, we would not have an interest in any broader application of the concept and/or device at this time. (C. Devine, personal communication, May 18, 2010)

FUTURE TRENDS AND CONCEPTS

The "youth information explosion" – prompted by the widespread use of information-based technology tools by young people (defined arbitrarily as "Millennials" ages 12-24) and the desire to successfully use those tools in information-rich environments such as the World Wide Web – will certainly lead to the proliferation and continued development of tools like the "information dowser". The demand for rapid retrieval of relevant information continues to grow. According to a recent study by the Artificial Intelligence Laboratory at the Massachusetts Institute of Technology, only three out of every thirteen Web searches involve information the searcher has seen before; a corresponding number of these searches are general, non-specific types of searches (i.e. browsing, contact information searches), which lend themselves to using a multifaceted combination of specific and general information to obtain the desired search result – similar to the Information Dowser concept (Ackerman, 2002). Since that survey was conducted, the proliferation of multifaceted combination searches has increased significantly.

Whether or not the information community can effectively develop computer searching software and related technology using the RMU Information Dowser's multifaceted combination of general and specific search elements remains to be seen. Yet, there is growing evidence that such "liberal control" of descriptive terminology is not only necessary but essential to the success of information access and retrieval in the future.

REFERENCES

Ackerman, M. (2002). *Surviving the information explosion* (in *CSAIL Publications*). Cambridge, MA: MIT Computer Science and Artificial Intelligence Laboratory. (pp. 349-351). Retrieved October 6, 2007, from http://www.ai.mit.edu/research/abstracts/abstracts2002/user-interfaces/01ackerman.pdf

American Library Association. Association of College & Research Libraries. (2002). *Information literacy competency standards for higher education.* Chicago, IL: ALA. Retrieved October 9, 2007 from http://www.ala.org/ala/acrl/acrlstandards/informationliteracycompetency.htm

American Library Association. (2006). *ACRL information literacy competency standards for higher education.* Retrieved July 9, 2009, from http://www.ala.org/ala/mgrps/divs/acrl/standards/informationliteracycompetency.cfm

Berstene, T. 2006. Hawthorne's twice-taught topics. *The Journal for Quality and Participation, 29*(3), 29-32. Retrieved April 9, 2009 from http://web.ebscohost.com/ehost/pdf?vid=3&hid=120&sid=f3140458-726e-48f4-a6d4-03b33f8b755a%40sessionmgr108

Centre for Excellence in Teaching and Learning in Reusable Learning Objects. (2005). *CETL FAQs.* Retrieved October 5, 2007, from http://www.rlo-cetl.ac.uk/faqs.htm

Chandler, A., & LeBlanc, J. (2006). Exploring the potential of a virtual undergraduate library collection based on the hierarchical interface to LC classification. *Library Resources & Technical Services, 50*(3), 157–165.

Davis, S. P. (2001). HILCC, a hierarchical interface to Library of Congress classification: A project report on the development of a operational prototype LCC-based subject interface to electronic resources at Columbia University Libraries. *Journal of Internet Cataloging, 5*(4), 19–49. doi:10.1300/J141v05n04_03

Devine, C. 2010. *Electronic communication, May 18, 2010.* Retrieved June 4, 2010, from Muskingum University Microsoft Outlook Electronic Mail.

Fisher, D., Grant, M., Frey, N., & Johnson, C. (2007). Taking formative assessment schoolwide. *Educational Assessment,* 64-68. Retrieved June 15, 2009, from http://web.ebscohost.com/ehost/pdfviewer/pdfviewer?vid=4&hid=112&sid=0d9b84bd-3a6b-498b-9528-422d3ecc28d7%40sessionmgr110

Jacob, E. K. (2004). Classification and categorization: A difference that makes a difference. *Library Trends*, *52*(3), 515–541.

Learning Technology Standards Committee. (2002). *IEEE standard for learning object metadata. (IEEE Standard 1484.12.1).* New York, NY: IEEE.

Library of Congress. (2006). *Library of Congress Authorities database*. Washington, D.C.: Library of Congress. Retrieved September 11, 2006, from http://authorities.loc.gov

Library of Congress. (2009) *LC authorities search of term: Landforms*. Retrieved July 9, 2009, from http://authorities.loc.gov

Mills, J. (2004). Faceted classification and logical division in information retrieval. *Library Trends*, *52*(3), 541–571.

Powell, V. J. (1977). *Resource handbook*. Commerce, TX: Federation Bilingual Training Resources Center, East Texas State University.

Robert Morris University. (2007). *Subject heading query log – internal memo*.

Robert Morris University. (2007). *Information security term and Library of Congress/Dewey Decimal System equivalencies document – internal memo*.

Robert Morris University. (2009). *ROBCAT catalog search*. Retrieved July 9, 2009, from http://robcat.rmu.edu/cgi-bin/Pwebrecon.cgi?DB=local&PAGE=First

Robert Morris University. (2009). *Database listing*. Retrieved July 9, 2009, from http://sentry.rmu.edu/web/cms/academics/library/research-assistance/Pages/databases.aspx

Schwarz, R. D., & Sykes, R. (2004). *Psychometric foundations of formative assessment.* Conference paper presented at the National Educational Computing Conference, New Orleans, 2004 June. Retrieved October 5, 2007, from http://www.iste.org/Content/NavigationMenu/Research/NECC_Research_Paper_Archives/NECC_2004/Schwarz-Richard-NECC04.pdf

Wiley, D. A. (Ed.). (2001). Connecting learning objects to instructional design theory: A definition, a metaphor, and a taxonomy. *The Instructional Use of Learning Objects: Online version*. Retrieved October 5, 2007, from http://reusability.org/read/chapters/wiley.doc

ADDITIONAL READING

Arroyo-Vázquez, N. (2009). Web móvil y bibliotecas. (Spanish). *El Profesional de la Información*, *18*(2), 129–136. doi:10.3145/epi.2009.mar.02

Bell, A. (2007). *Handheld computers in schools and media centers*. Worthington, OH: Linworth.

Buczynski, J. (2008). Libraries Begin to Engage Their Menacing Mobile Phone Hordes Without Shhhhh! *Internet Reference Services Quarterly*, *13*(2/3), 261–269. doi:10.1080/10875300802103916

Corley, P. (2004). Planning, Promoting, and Conducting a PDA Technology Fair. *Journal of Electronic Resources in Medical Libraries*, *1*(3), 73–82. doi:10.1300/J383v01n03_07

Cuddy, C. (2007). RSS Feeds for PDAs. *Journal of Electronic Resources in Medical Libraries*, *4*(3), 123–129. doi:10.1300/J383v04n03_09

Cuddy, C. (2008). The iPhone in Medical Libraries. *Journal of Electronic Resources in Medical Libraries*, *5*(3), 287–292. doi:10.1080/15424060802222513

Cuddy, C. (2009a). Twittering in Health Sciences Libraries. *Journal of Electronic Resources in Medical Libraries*, *6*(2), 169–173. doi:10.1080/15424060902932284

Cuddy, C. (2009b). Mobile Computing. *Journal of Electronic Resources in Medical Libraries*, *6*(1), 64–68. doi:10.1080/15424060802705301

David, R., Trevor, M., & Taeho, K. (2009). Integrating NAND Flash Devices onto Servers. *Communications of the ACM*, *52*(4), 98–106. doi:10.1145/1498765.1498791

Ellero, N., Looney, R., & Ragon, B. (2007). P.O.D. Principles. *Medical Reference Services Quarterly*, *26*, 69–90. doi:10.1300/J115v26S01_05

Forster, S. (2006). Using instant messaging for online reference service. *The Australian Library Journal*, *55*(2), 147–158.

Gelfand, J. (2004). The Ubiquity of Grey Literature in a Connected Content Context. *Publishing Research Quarterly*, *20*(1), 54–61. doi:10.1007/BF02910860

Greene, C. (2010). *The anywhere library: a primer for the mobile web*. Chicago: Association of College and Research Libraries.

Griffey, J. (2010). *Mobile technology and libraries*. New York: Neal-Schuman Publishers.

Honeybourne, C., Sutton, S., & Ward, L. (2006). Knowledge in the Palm of your hands: PDAs in the clinical setting. *Health Information and Libraries Journal*, *23*(1), 51–59. doi:10.1111/j.1471-1842.2006.00621.x

Iroka, L. (2007). Use of Mobile Phones Among Library Users in the Medical Library at the College of Medicine, University of Nigeria, Enugu Campus, Enugu, Nigeria. *Journal of Hospital Librarianship*, *7*(1), 17–28. doi:10.1300/J186v07n01_02

Le Ber, J., Lombardo, N., Weber, A., & Bramble, J. (2004). Portable Classroom Leads to Partnership. *Medical Reference Services Quarterly*, *23*(2), 41–48. doi:10.1300/J115v23n02_04

Library Mobile Initiatives. (2008)... *Library Technology Reports*, *44*(5), 33–38.

Marcoux, E. (2009). the phone: a tool for learning. *Teacher Librarian*, *36*(4), 73–74.

Marshall, C., Golovchinsky, G., & Price, M. (2001). DIGITAL LIBRARIES AND MOBILITY. *Communications of the ACM*, *44*(5), 55–56. doi:10.1145/374308.374340

Mi, M., Suresh, S., & Kamat, D. (2008). Assessing PDA Use and Needs Among Pediatric Residents. *Journal of Electronic Resources in Medical Libraries*, *5*(1), 1–15. doi:10.1080/15424060802093336

Mostéfaoui, S. K. Zakaria, M., and Giaglis, G. (eds.). (2008). Advances in ubiquitous computing: future paradigms and directions. Hershey PA: IGI Pub.

Needham, M., & Ally, M. (2009). M-libraries: libraries on the move to provide virtual access. *Journal of Web Librarianship*, *3*(4), 378–379. doi:10.1080/19322900903294470

Robinson, K. (2005). Point-of-Need Customer Service: Mobile Service Using Wireless Technology. *Journal of Access Services*, *3*(2), 1–14. doi:10.1300/J204v03n02_01

Satoh, I. (2006). Location-based services in ubiquitous computing environments. *International Journal on Digital Libraries*, *6*(3), 280–291. doi:10.1007/s00799-006-0006-1

Starkweather, W., & Stowers, E. (2009). Smartphones: A Potential Discovery Tool. *Information Technology & Libraries*, *28*(4), 187–188.

Walton, G., Childs, S., & Blenkinsopp, E. (2005). Using mobile technologies to give health students access to learning resources in the UK community setting. *Health Information and Libraries Journal*, *22*, 51–65. doi:10.1111/j.1470-3327.2005.00615.x

Wei-Po, L., & Cheng-Che, L. (2003). Customising WAP-based information services on mobile networks. *Personal and Ubiquitous Computing*, *7*(6), 321–330. doi:10.1007/s00779-003-0247-6

West, M., Hafner, A., & Faust, B. (2006). Expanding Access to Library Collections and Services Using Small-Screen Devices. *Information Technology & Libraries*, *25*(2), 103–107.

KEY TERMS AND DEFINITIONS

Formative Assessment: Evaluation of content understanding conducted at strategic intervals in the learning process.

Summative Assessment: Evaluation of content understanding conducted normally at the end of the learning process.

Psychometric Models: Measuring knowledge, abilities, attitudes, and personality traits in order to assign a measure of predictability to an area of study or process.

Subject Heading: A specific word or phrase, chosen from a selected list of preferred terms, that best describes the content of the work it describes.

Controlled Vocabulary: A list of preferred terms whose entry conventions have been documented to encourage consistent and uniform use across sources and platforms.

Taxonomy: The practice and/or technique of describing, identifying, naming, and classifying.

Library Classification Systems: The coding and organizing of library materials according to subject in which a particular combination of numbers (or, in the case of some library classification systems like the Library of Congress, a combination of numbers and letters) is assigned to each material.

Dowsing: The act of discovering/finding. In ancient times, dowsing was associated with discovering water supplies. In the context of this article, dowsing is related to discovering/finding information.

Compilation of References

Abas, Z. W., Chng, L. P., & Mansor, N. (2009). A study on learner readiness for mobile learning at Open University Malaysia. In *Proceedings of IADIS International Conference on Mobile Learning* (pp. 151-57).

Abt, G., & Barry, T. (2007). The quantitative effect of students using podcasts in a first year undergraduate exercise physiology module. *BioScience Education Journal*, 10. Retrieved June 1, 2008, from http://www.bioscience.heacademy.ac.uk/ journal/vol10/beej-10-8.aspx

Ackerman, M. (2002). *Surviving the information explosion* (in *CSAIL Publications*). Cambridge, MA: MIT Computer Science and Artificial Intelligence Laboratory. (pp. 349-351). Retrieved October 6, 2007, from http://www.ai.mit.edu/research/abstracts/abstracts2002/user-interfaces/01ackerman.pdf

Adams, P. (2006). Exploring social constructivism: Theories and practicalities. *Education, 34*(3), 243–257.

Adams, N. B., DeVaney, T. A., & Sawyer, S. G. (2009). *Measuring conditions conducive to knowledge development in virtual learning environments: Initial development of a model-based survey*. (pp. 1-23). Retrieved July 2, 2010, from www.jtla.org

Adatia, F., & Bedard, P. (2003). Palm reading: 2. Handheld software for physicians. *Canadian Medical Association Journal, 168*(6), 727–734.

Aguera, B. (2007). *Arcas demos Photosynth*. Retrieved April 10, 2010, from http://www.ted.com/talks/blaise_aguera_y_arcas_demos_photosynth.html

Akyol, Z., Garrison, D. R., & Ozden, M. Y. (2009). Development of a community of inquiry in online and blended learning contexts. *Procedia – Social and Behavioral Sciences, 1*(1), 1834-1838.

Alexander, B. (2004). Going nomadic: Mobile learning in higher education. *EDUCAUSE Review, 39*(5), 28, 30–32, 34–35.

Alexander, B. (2005). Podcasting and the liberal arts. *The Newsletter of the National Institute for Technology and Liberal Education*, *4*(3). Retrieved on November 1, 2006 from http://newsletter.nitle.org/v4_n3_summer 2005/podcasting.php

Allen, I., & Seaman, J. (2006). *Making the grade: Online education in the United States, 2006. Sloan Consortium Group*. Needham, MA: Sloan-C.

Allen, I., & Seaman, J. (2009). *Learning on demand: Online education in the United States, 2006. Sloan Consortium Group*. Needham, MA: Sloan-C.

Allen, K. (2005). Online learning: Constructivism and conversation as an approach to learning. *Innovations in Education and Teaching International, 42*(3), 247–256. doi:10.1080/01587910500167985

Ally, M. (Ed.). (2009). *Mobile learning: Transforming the delivery of education & training*. Athabasca, Canada: Athabasca University.

Al-Ubaydli, M. (2006). *The doctor's PDA and smartphone handbook: A guide to handheld healthcare*. London, UK: Royal Society of Medicine Press.

American Library Association. Association of College & Research Libraries. (2002). *Information literacy competency standards for higher education*. Chicago, IL: ALA. Retrieved October 9, 2007 from http://www.ala.org/ala/acrl/acrlstandards/informationliteracycompetency.htm

American Library Association. (2006). *ACRL information literacy competency standards for higher education.* Retrieved July 9, 2009, from http://www.ala.org/ala/mgrps/divs/acrl/standards/informationliteracycompetency.cfm

Amin, A. H. M., Mahmud, A. K., Abidin, A. I. Z., Rahman, M. A., Iskandar, B. S., & Ridzuan, P. D. (2006). M-learning management tool development in campus-wide environment. *Issues in Informing Science and Information Technology, 3*, 423–434.

Anderson, M. C. (2009). The medium is the messenger: Using podcasting to deliver consumer health information. *Journal of Consumer Health on the Internet, 13*(2), 119–128. doi:10.1080/15398280902896428

Andrews, T., Rayner, A., & Edwards, J. (2009). Tablet PCs, DyKnow and collaborative learning: Developing mathematical problem solving skills in large undergraduate classes. In *Proceedings of the MLearn 2009 Conference*, 24-28 Oct 2009, Orlando, Florida.

Apple. (2007). *100 million iPods sold.* April 9, 2007. Retrieved January 3, 2010, from http://www.apple.com/pr/library/2007/ 04/09ipod.html

Arbaugh, J. B., Cleveland-Innes, M., Diaz, S. R., Garrison, D. R., Ice, P., Richardson, J. C., & Swan, K. P. (2008). Developing a community of inquiry instrument: Testing a measure of the community of inquiry framework using a multi-institutional sample. *The Internet and Higher Education, 11*(3-4), 133–136. doi:10.1016/j.iheduc.2008.06.003

Arbaugh, J. B., Godfrey, M. R., Johnson, M., Leisen Pollack, B., Niendorf, B., & Wresch, W. (2009). Research in online and blended learning in the business disciplines: Key findings and possible future directions. *The Internet and Higher Education, 12*(2), 71–87. doi:10.1016/j.iheduc.2009.06.006

Arrington, M. (2010). The unauthorized TechCrunch iPad review. *TechCrunch,* Retrieved July 2, 2010, from http://techcrunch.com/2010/04/02/the- unauthorized-techcrunch-review/

Artino, A. R. (2008). Promoting academic motivation and self-regulation: Practical guidelines for online instructors. *TechTrends, 52*(3), 37–45. doi:10.1007/s11528-008-0153-x

Association of Faculties of Medicine of Canada. (2006). *Results of a survey of distributed medical education activities at Canadian Faculties of Medicine.* Retrieved from http://afmc.ca/education-distributed-med-edu-e.php

Attewell, J. (2004). *Mobile technologies and learning.* Learning and Skills Development Agency. Retrieved from http://www.mlearning.org/docs/The%20m- learning%20project%20%20technology% 20update%20and%20project%20 summary.pdf

Attewell, J., & Savill-Smith, C. (2004). Mobile learning and social inclusion: Focus on learners and learning, (pp. 3-11). *Proceedings of MLEARN 2003 Conference.* London, UK: LSDA.

Aukstakalnis, S., & Blatner, D. (1992). *Silicon mirage-the art and science of virtual reality.* Berkeley, CA: Peachpit Press.

Azuma, R. T. (1997). A survey of augmented reality. *Presence (Cambridge, Mass.), 6*(4), 355–385.

Azzarito, L., & Ennis, C. D. (2003). A sense of connection: Toward social constructivist physical education. *Sport Education and Society, 8*(2), 179–198. doi:10.1080/13573320309255

Bach, S., Haynes, P., & Lewis Smith, J. (2007). *Online learning and teaching in higher education.* Maidenhead, UK: McGraw Hill, Open University Press.

Baddeley, A. D. (1986). *Working memory.* New York, NY: Oxford University Press.

Baird, D. E., & Fisher, M. (2006). Neomillennial user experience design strategies: Utilizing social networking media to support always on learning styles. *Journal of Educational Technology Systems, 34*(1), 5–32. doi:10.2190/6WMW-47L0-M81Q-12G1

Balacheff, N. (2006). 10 issues to think about the future of research on TEL. *Les Cahiers Leibniz, Kaleidoscope Research Report, 147.*

Baldwin, A. (1999). Taking the laptop home. *Appalachia, 32*(1), 10–15.

Banas, J. (2009). Generating motivation to learn via tailored lesson introductions. *International Journal of Information and Communication Technology Education, 5*(4), 1–17. doi:10.4018/jicte.2009041001

Banas, J. (2008). Behavioral construct tailing: A motivational design technique for electronic learning environments. In M. Simonson (Ed.), *31st Annual Proceedings: Selected Research and Development Papers at the Annual Convention of the Association for Educational Communications and Technology* (pp. 23-38). Orlando, FL: Nova Southeastern University.

Barnard, L., Yi, J. S., Jacko, J. A., & Sears, A. (2007). Capturing the effects of context on human performance in mobile computing systems. *Personal and Ubiquitous Computing, 11*(2), 81–96. doi:10.1007/s00779-006-0063-x

Barnes, C., & Tynan, B. (2007). The adventures of Miranda in the brave new world: Learning in a Web 2.0 millennium. *ALT-J Research in Learning Technology, 15*(3), 189–200.

Barrios, (2004). *Laptops for learning: Final report and recommendations of the Laptops for Learning Task Force.* Retrieved November 18, 2006, from www.etc.usf.edu/L4L/Cover.pdf

Barrows, H. S. (1971). *Simulated patients (programmed patients): The development and use of a new technique in medical education.* Springfield, IL: Charles C. Thomas.

Bartlett, J. F. (1995). *Experience with wireless World-Wide Web client.* (Technical report, note TN-46). Digital Western Research Lab. Retrieved July 2, 2010, from www.research.digital.com/wrl/techreports /abstracts/TN-46.html

Baumgart, D. C. (2005). Personal digital assistants in healthcare: Experienced clinicians in the palm of your hand? *Lancet, 366*, 1210–1222. doi:.doi:10.1016/S0140-6736(05)67484-3

BC Ministry of Education. (2005). *$2.1 million for computer pilot projects in schools.* Retrieved March 12, 2006, from http://www.bced.gov.bc.ca/onetoone/

BC Ministry of Education. (2010). *Student statistics-2009/10 92-Nisga'a.* Retrieved July 12, 2010, from http://www.bced.gov.bc.ca/reports /pdfs/student_stats/092.pdf

Bebell, D. (2005). *Technology promoting student excellence: An investigation of the first year of 1:1 computing in New Hampshire Middle Schools.* Boston, MA: Boston College, Technology and Assessment Study Collaborative.

Bebell, D., & O'Dwyer, L. M. (2010). Educational outcomes and research from 1:1 computing settings. *Journal of Technology, Learning, and Assessment, 9*(1). Retrieved June 26, 2010, from http://www.jtla.org

Bedny, G. Z., & Harris, S. R. (2008). Working sphere/engagement and the concept of task in activity theory. *Interacting with Computers, 20*(2), 251–255. doi:10.1016/j.intcom.2007.07.005

Bee, H. L., & Bjorkland, B. R. (2004). *The journey to adulthood* (5th ed.). Englewood Cliffs, NJ: Prentice Hall.

Berstene, T. 2006. Hawthorne's twice-taught topics. *The Journal for Quality and Participation, 29*(3), 29-32. Retrieved April 9, 2009 from http://web.ebscohost.com/ehost/pdf?vid=3&hid=120&sid=f3140458-726e-48f4-a6d4-03b33f8b755a%40sessionmgr108

Bertelsen, P., Kanstrupand, A. M., & Christiansen, E. (2007). Problem-based learning in medicine: Performance support with m-learning. [nd *International Conference on e-learning.* New York, USA.]. *Proceedings of the ICEL, 2007*, 2.

Billings, D. M. (2005). From teaching to learning in a mobile, wireless world. *The Journal of Nursing Education, 44*(8), 343.

Blaisdell, M. (2006). Academic MP3s: Is it time yet? *Campus Technology*. Retrieved on November 1, 2006, from http://campustechnology.com/article.asp?id=18001

Blake, R. J. (2008). *Brave new digital classroom.* Washington, DC: Georgetown University Press.

Boardman, D., Casalegno, F., McMurray, B., & Pomeroy, S. (2008). Rethinking the mobile social shopping experience. Retrieved February 4, 2010, from http://mobile.mit.edu/research/community/or

Bodrova, E. (2008). Make-believe play versus academic skills: A Vygotskian approach to today's dilemma of early childhood education. *European Early Childhood Education Research Journal, 16*(3), 357–369. doi:10.1080/13502930802291777

Boelen, C. (2000). *Towards unity for health: Challenges and opportunities for partnership in health development, a working paper.* World Health Organization. Retrieved March 21, 2010, from http://www.who.int/hrh/documents/en/TUFH_challenges.pdf

Bokken, L., Rethans, J. J., van Heurn, L., Duvivier, R., Scherpbier, A., & van der Vleuten, C. (2009). Students' views on the use of real patients and simulated patients in undergraduate medical education. *Academic Medicine*, *84*, 958–963. doi:10.1097/ACM.0b013e3181a814a3

Bomsdorf, B. (2005). *Adaptation of learning spaces: Supporting ubiquitous learning in higher distance education.* Paper presented at the Mobile Computing and Ambient Intelligence: The Challenge of Multimedia, Dagstuhl Seminar Proceedings.

Bonds-Raacke, J. M. (2006). Students' attitudes toward the introduction of a course website. *Journal of Instructional Psychology*, *33*(4), 251–255.

Bonds-Raacke, J. M., & Raacke, J. D. (2008). Using tablet PCs in the classroom: An investigation of students' expectations and reactions. *Journal of Instructional Psychology*, *35*(3), 235–239.

Bonk, C. J., & Zhang, K. (2006). Introducing the R2D2 model: Online learning for the diverse learners of this world. *Distance Education*, *27*(2), 249–264. doi:10.1080/01587910600789670

Borau, K., & Ullrich, C. (2009). Mobile learning with open learning environments at Shanghai Jiao Tong University, China. *Learning Technology Newsletter*, *11*(1-2), 7–8.

Bouzeghoub, A., Defude, B., Duitama, F., & Lecocq, C. (2006). A knowledge-based approach to describe and adapt learning objects. *International Journal on E-Learning*, 95–102.

Bouzeghoub, A., & Do, K. (2008). *A situation based metadata for describing pervasive learning objects*. Paper presented at the mLearn 2008: 1st International Conference on Mobile Learning, Wolverhampton, Ironbridge, UK.

Bouzeghoub, A., & Do, K. (2009). *Situation-aware adaptive recommendation to assist mobile users in a campus environment*. Paper presented at the AINA '09: The IEEE 23rd International Conference on Advanced Information Networking and Applications, Bradford, United Kingdom, 2009.

Bouzeghoub, A., & Do, K. (2010). *Active sharing of contextual learning experiences among users in personal learning environments using a peer-to-peer network*. Paper presented at the The 10th IEEE International Conference on Advanced Learning Technologies, ICALT, Sousse, Tunisia (à paraître).

Bouzeghoub, A., Do, K., & Lecocq, C. (2008). *A situation-based delivery of learning resources in pervasive learning*. Paper presented at the Creating new learning experiences on a global scale: Second European Conference on Technology Enhanced Learning, EC-TEL 2007, Crete, Greece.

Brandon Hall Research. (2009). *Mobile learning comes of age: How and why organizations are moving to learning on mobile devices*. Retrieved July 2, 2010, from www.brandon-hall.com/publications/ mobilelearning/mobile-learning- comesofage.shtml

Bransford, J., Brown, A., & Cocking, R. (Eds.). (2000). *How people learn: Brain, mind, experience, and school*. Washington, DC: National Academy Press.

Brézillon, P. (2005). *Task-realization models in contextual graphs*. Paper presented at the International and Interdisciplinary Conference CONTEXT 2005.

Brian, M. (2010). Google Goggles releases an exciting new feature. Lets you translate…the world. Retrieved June 1, 2010, from http://thenextweb.com/mobile/2010/05/06/google-googles-releases-exciting-feature-lets-translatethe-world/

Bridge, P. D., & Ginsburg, K. A. (2001). An integrated approach for evaluating students' achievements of clinical objectives. *Medical Education Online*, *6*(9), 1–6.

Bright, G. W., Waxman, H. C., & Williams, S. E. (1993). *Impact of calculators on mathematics instruction*. Lanham, MD: University Press of America.

Brittain, S., Glowacki, P., Van Ittersum, J., & Johnson, L. (2006). Podcasting lectures: Formative evaluation strategies helped identify a solution to a learning dilemma. *Educause Quarterly, 3*. Retrieved May 28, 2008, from http://connect.educause.edu/Library/ EDUCAUSE+Quarterly/Podcasting Lectures/39987

Brock, T. P., & Smith, S. R. (2007). Using digital videos displayed on personal digital assistants (PDAs) to enhance patient education in clinical settings. *International Journal of Medical Informatics*, *76*(11/12), 829–835. doi:10.1016/j.ijmedinf.2006.09.024

Brodersen, C., Christensen, B. G., Gronboek, K., Dindler, C., & Sundararajah, B. (2005). *eBag: A ubiquitous Web infrastructure for nomadic learning*. Paper presented at the Proceedings of the 14th international conference on World Wide Web.

Brooks-Young, S. (2010). *Teaching with the tools kids really use: Learning with Web and mobile technologies*. Thousand Oaks, CA: Corwin.

Brown, T. (2006). Using instant messenger programs for virtual office hours. *Feedback*, *47*(1).

Brown, T. (2007). Gratifications of handheld computers. *Academic Exchange Quarterly*, *11*(4).

Bruner, J. (1987). *Actual minds, possible worlds*. Cambridge, MA: Harvard University Press.

Bullard, J. (2003). Constructivism: Does your practice match your conceptual framework? *Journal of Early Childhood Teacher Education*, *24*, 157–162. doi:10.1080/1090102030240304

Burdette, S. D., Herchline, T. E., & Oehler, R. (2008). Practicing medicine in a technological age: Using smartphones in clinical practice. *Clinical Infectious Diseases*, *47*, 117–122. doi:10.1086/588788

Burgess, H., & Mayes, A. S. (2003). Tutoring primary trainee teachers through e-conferencing. *Mentoring & Tutoring*, *11*(3), 285–305. doi:10.1080/1361126032000138337

Calbraith, D. (2010). *Towards an effective pedagogy and evaluation framework for learning objects in medical education and healthcare*. Unpublished doctoral thesis, Nottingham University.

Calbraith, D., & Dennick, R. (2009). *Pedagogical strategies for learning objects*. Paper presented at the ICERI international conference 2009, Madrid, Spain, 16-18th November. Full paper published by IATED. Retrieved July 2, 2010, from www.iated.org/iceri2009/index.php?section=19

Camarata, K., Yi-Luen Do, E., Gross, M. D., & Johnson, B. R. (2002). *Navigational blocks: Tangible navigation of digital information*. Minneapolis, MN: ACM Press.

Campbell, G. (2005). Podcasting in education. *EDUCAUSE*, Nov/Dec 05.

Carroll, J. M. (2000). *Making use: Scenario-based design of human-computer interactions*. Retrieved July 2, 2010, from www.mitpress.mit.edu/catalog/item/default.asp?ttype=2&tid=4114

Caudell, T. P., & Mizell, D. W. (1992). Augmented reality: An application of heads-up display technology to manual manufacturing processes. *Proceedings of 1992 IEEE Hawaii International Conference on Systems Sciences*, (pp 659-669).

Caudill, J. G. (2007). The growth of m-learning and the growth of mobile computing: Parallel developments. *International Review of Research in Open and Distance Learning*, *8*(2). Retrieved July 10, 2009, from http://www.irrodl.org/index.php/irrodl/ article/view/348/873

Cavus, N., & Uzunboylu, H. (2009). Improving critical thinking skills in mobile learning. *Procedia – Social and Behavioral Sciences*, *1*(1), 435-438.

Cebeci, Z., & Tekdal, M. (2006). Using podcasts as audio learning objects. *Interdisciplinary Journal of Knowledge and Learning Objects*, *2*, 7–57.

Centre for Excellence in Teaching and Learning in Reusable Learning Objects. (2005). *CETL FAQs*. Retrieved October 5, 2007, from http://www.rlo-cetl.ac.uk/faqs.htm

Chaari, T. (2007). *Adaptation d'applications pervasives dans des environnements multi-contextes*. Institut national des science appliquées de Lyon, Lyon, France.

Chan, A., & Lee, M. (2005). *An MP3 a day keeps the worries away-exploring the use of podcasting to address preconceptions and alleviate pre-class anxiety amongst undergraduate information technology students*. Student Experience Conference, Charles Stuart University. Retrieved on November 1, 2006, from http://www.csu.edu.au/division/studserv/ sec/papers/chan.pdf

Chandler, A., & LeBlanc, J. (2006). Exploring the potential of a virtual undergraduate library collection based on the hierarchical interface to LC classification. *Library Resources & Technical Services*, *50*(3), 157–165.

Chang, B., Lee, S., & Kim, B. (2006). Exploring factors affecting the adoption and continuance of online games among college students in South Korea: Integrating uses and gratification and diffusion of innovation approaches. *New Media & Society*, *8*(2), 295–319. doi:10.1177/1461444806059888

Chapin, A. (2005). *iPods and second language acquisition*.

Chase, W. G., & Simon, H. A. (1973). Perception in chess. *Cognitive Psychology*, *4*, 55–81. doi:10.1016/0010-0285(73)90004-2

Chehimi, F., Coulton, P., & Edwards, R. (2006). Delivering 3D advertising to mobile phone. *Proceedings of the IEEE International conference on consumer electronics*, Las Vegas, January 2006. Retrieved July 2, 2010, from http://eprints.lancs.ac.uk/22928/1/ getPDF17.pdf

Cheung, W. S., & Hew, K. F. (2009). A review of research methodologies used in studies on mobile handheld devices in K-12 and higher education settings. *Australasian Journal of Educational Technology*, *25*(2), 153–183.

Cheverst, K., Mitchell, K., & Davies, N. (1998). Design of an object model for a context sensitive tourist guide. *Computers & Graphics*, *23*, 883–891. doi:10.1016/S0097-8493(99)00119-3

Choi, J. J. C. (2005). Teaching with a personal digital assistant. *Journal of the American College of Radiology*, *2*(10), 825–832. doi:10.1016/j.jacr.2005.02.006

Christensen, C. M., Horn, M. B., & Johnson, C. W. (2008). *Disrupting class: How disruptive innovation will change the way the world learns*. New York, NY: McGraw Hill.

Churchill, D., & Churchill, N. (2008). Educational affordances of PDAs: A study of a teacher's exploration of this technology. *Computers & Education*, *50*(4), 1439–1450. doi:10.1016/j.compedu.2007.01.002

Churchill, D., Kennedy, D., Flint, D., & Cotton, N. (2010). Using handhelds to support students' outdoor educational activities. *International Journal of Continuing Engineering Education and Lifelong Learning*, *20*(1), 51–71.

Cleveland Clinic. (n.d.). *Videos, webcasts & podcasts*. Retrieved from http://my.clevelandclinic.org/health _edge/about_podcast.aspx

Clough, G., Jones, A. C., McAndrew, P., & Scanlon, E. (2007). Informal learning with PDAs and smartphones. *Journal of Computer Assisted Learning*, *24*(5), 359–371. doi:.doi:10.1111/j.1365-2729.2007.00268.x

CNET. (2010). Best smartphones. Retrieved July 2, 2010, from http://reviews.cnet.com/best-smartphones

Cobanoglu, C. (2006). An analysis of blogs as a teaching tool as perceived by hospitality management students. *Journal of Hospitality, Leisure, Sport and Tourism Education*, *5*(2), 83–88. doi:10.3794/johlste.52.138

Cohen, A., Oxford, R., & Chi, J. (2006). Learning style survey: Assessing your learning styles. In *Styles- and Strategies-Based Instruction: A Teacher's Guide*. Center for Advanced Research on Language Acquisition.

Cole, M. (1996). *Cultural psychology: A once and future discipline*. Cambridge, MA/London, UK: Belknap Press of Harvard University Press.

Colevins, H., Bond, D., & Clark, K. (2006). Nurses get a hand from handhelds. *Computers in Libraries*, *26*(4), 6–48.

Collis, B., & Moonen, J. (2001). *Flexible learning in a digital world: Experiences and expectations* (pp. 154–155). London, UK: Kogan.

Cooper, S., Dale, C., & Spencer, S. (2009). A tutor in your back pocket: Reflections on the use of iPods and podcasting in an undergraduate popular music programme. *British Journal of Music Education*, *26*, 85–97. doi:10.1017/S0265051708008280

Cooper, H. (1982). *Scientific guidelines for conducting integrative literature reviews*. Paper presented at the annual meeting of the American Educational Research Association, New York, NY.

Cooper, I., & Shufflebotham, R. (1995). *PDA Web browsers: Implementation issues*. Retrieved July 2, 2010, from http://alethea.ukc.ac.uk/wp?94ihc

Copely, J. (2007). Audio and video podcasts of lectures for campus-based students: Production and evaluation of student use. *Innovations in Education and Teaching International*, *44*(4), 387–399. doi:10.1080/14703290701602805

Cotton, D. R. E., & Gresty, K. A. (2007). The rhetoric and reality of e-learning: Using the think-aloud method to evaluate an online resource. *Assessment & Evaluation in Higher Education*, *32*(5), 583–600. doi:10.1080/02602930601116920

Creswell, J. W. (2005). *Educational research: Planning, conducting, and evaluating quantitative and qualitative research* (2nd ed.). New Jersey: Pearson Education.

Creswell, J. W. (2008). *Educational research: Planning, conducting, and evaluating quantitative and qualitative research* (3rd ed.). Upper Saddle River, NJ: Pearson/ Merrill Prentice Hall.

CRN. (2010). *Mobile learning*. Retrieved from July 2, 2010, from http://www.crn.com.mobile/ 222002587;jsessionid=DTFGA1N1W4DPVQE1GPDA

Cruice, M. (2005). Common issues but alternative solutions and innovations. *Advances in Speech Language Pathology*, *7*, 162–166. doi:10.1080/14417040500181429

Cuddy, L. J., & Jacoby, L. L. (1982). When forgetting helps memory: An analysis of repetition effects. *Journal of Verbal Learning and Verbal Behavior*, *21*(4), 451–467. doi:10.1016/S0022-5371(82)90727-7

CVMD. (2010). *The cardiovascular multimedia information network*. Retrieved March 21, 2010, from http:// www.cvmd.org/

Dale, C. (2007). Strategies for using podcasting to support the learning of tourism students. *Journal of Hospitality, Leisure, Sport and Tourism Education*, *6*(1), 49–57. doi:10.3794/johlste.61.155

Dale, C. (2008). iPods and creativity in learning and teaching: An instructional perspective. *International Journal of Learning and Teaching in Higher Education*, *20*(1), 1–9.

Dale, C., & Hassanien, A. (2008). Podcasting in tourism education. *Journal of Teaching in Travel & Tourism*, *7*(3), 43–60. doi:10.1080/15313220801909361

Dale, C., & Lane, A. M. (2007). A wolf in sheep's clothing? An analysis of student engagement with virtual learning environments. *Journal of Hospitality, Leisure, Sport and Tourism Education*, *6*(2), 100–108. doi:10.3794/ johlste.62.156

Dale, C., & McCarthy, P. (2006). I like your style: The learning approaches of leisure, tourism and hospitality students studying generic modules. *Journal of Hospitality, Leisure, Sport and Tourism Education*, *5*(2), 48–58. doi:10.3794/johlste.52.116

Dale, C., & Povey, G. (2008). Learner generated content and podcasting. *Journal of Hospitality, Leisure, Sport and Tourism Education*, *8*(1), 117–123. doi:10.3794/ johlste.81.214

Dalgarno, B. (2001). Interpretations of constructivism and consequences for computer assisted learning. *British Journal of Educational Technology*, *32*(2), 183–194. doi:10.1111/1467-8535.00189

Danaher, P. A., Moriarty, B. J., & Danaher, G. R. (2009). *Mobile learning communities: Creating new educational futures*. New York, NY: Routledge.

Danaher, P. A., Gururajan, R., & Hafeez-Baig, A. (2009). Transforming the practice of mobile learning: Promoting pedagogical innovation through educational principles and strategies that work. In Ryu, H., & Parsons, D. P. (Eds.), *Innovative mobile learning: Techniques and technologies* (pp. 21–46). Hershey, PA/ New York, NY: IGI Global/ Information Science Reference.

Davis, F. D., & Venkatesh, V. (2004). Toward preprototype user acceptance testing of new Information Systems: Implications for software project management. *IEEE Transactions on Engineering Management*, *51*(1), 31–46. .doi:10.1109/TEM.2003.822468

Davis, J., Chryssafidou, E., Zamora, J., Davies, D., Khan, K., & Coomarasamy, A. (2007). Computer-based teaching is as good as face to face lecture-based teaching of evidence based medicine: A randomized controlled trial. *BMC Medical Education*, *7*(23). doi:.doi:10.1186/1472-6920-7-23

Davis, S. P. (2001). HILCC, a hierarchical interface to Library of Congress classification: A project report on the development of a operational prototype LCC-based subject interface to electronic resources at Columbia University Libraries. *Journal of Internet Cataloging*, *5*(4), 19–49. doi:10.1300/J141v05n04_03

de Freitas, S., & Levene, M. (2003). Evaluating the development of wearable devices, personal data assistants and the use of other mobile devices in further and higher education. Developing the next generation of education. *Proceedings of the EDEN 10th Anniversary Conference*, Stockholm, June 10-13, 2001.

De George-Walker, L. R., Hafeez-Baig, A., Gururajan, R., & Danaher, P. A. (2010). Experiences and perceptions of learner engagement in blended learning environments: The case of an Australian university. In Inoue, Y. (Ed.), *Cases on online and blended learning technologies in higher education: Concepts and practices* (pp. 23–43). Hershey, PA: IGI Global.

De Groot, A. (1965). *Thought and choice in chess*. The Hague, Netherlands: Mouton.

De Vaus, D. A. (2002). *Surveys in social research* (5th ed.). Crows Nest, Australia: Allen & Unwin.

De Ville, K. (2008). The cure is in hand? The brave new world of handheld computers in medicine. *Cambridge Quarterly of Healthcare Ethics*, *17*, 385–400. doi:. doi:10.1017/S0963180108080523

Deal, A. (2007). *A teaching with technology white paper: Podcasting*. (Document ID: CSD5003). Retrieved October 4, 2007, from http://connect.educause.edu/files/CMU_Podcasting_Jun07.pdf

Dearnley, C., Haigh, J., & Fairhall, J. (2008). Using mobile technologies for assessment and learning in practice settings: A case study. *Nurse Education in Practice*, *8*(3), 197–204. doi:.doi:10.1016/j.nepr.2007.07.003

Dearnley, C. A., Taylor, J., Hennessy, S., Parks, M., Coates, C., & Haigh, J. (2009). Using mobile technologies for assessment and learning in practice settings: Outcomes of five case studies. *International Journal on E-Learning*, *8*(2), 193–207.

Dearnley, C. A., & Walker, S. A. (2009). Mobile enabled research. In Vavoula, G. (Ed.), *Researching mobile learning*. Oxford, UK: Peter Lang.

Deltecs InfoTech Pvt Ltd. (2009). *Mobile training or Blackberry training-a new buzz word in corporate training programs*. Mambai, India. Retrieved August 2, 2009, from http://www.pressreleasepoint.com/ mobile-training-or-blackberry-training- new-buzz-word-corporate-training-programs

Derycke, A., Chevrin, V., & Vantroys, T. (2007). *P-learning and e-retail: A case and a flexible software architecture*. Paper presented at the Pervasive Learning 2007, an International Workshop on Pervasive Learning, Toronto, Ontario, Canada.

Devinder, S., & Zaitun, A. B. (2006). Mobile learning in wireless classrooms. *Malaysian Online Journal of Instructional Technology*, *3*(2), 26–42.

Devine, C. 2010. *Electronic communication, May 18, 2010*. Retrieved June 4, 2010, from Muskingum University Microsoft Outlook Electronic Mail.

Dey, A. K. (2001). Understanding and using context. *Personal and Ubiquitous Computing*, *5*(1), 4–7. doi:10.1007/s007790170019

Diamantini, D., & Pieri, M. (2008). A blended mobile learning experience: The Nomadis case. *International Journal of Knowledge and Learning*, *4*(2-3), 176–188. doi:10.1504/IJKL.2008.020653

Diddi, A., & LaRose, R. (2006). Getting hooked on news: Uses and gratifications and the formation of news habits among college students in an internet environment. *Journal of Broadcasting & Electronic Media*, *50*(2), 193–210. .doi:10.1207/s15506878jobem5002_2

Dierterle, E., & Dede, C. (2006). Straightforward and deep effects of wireless handheld devices for teaching and learning in university settings. Retrieved March 6, 2010, from http://74.125.155.132/scholar?q=cache:nuS6TRgEbaMJ:scholar.google.com/+wireless+handheld+devices&hl=en&as_sdt=2000&as_ylo=2004

Dieterle, E., Dede, C., & Schrier, K. (2007). Neomillennial learning styles propagated by wireless handheld devices. In Lytras, M., & Naeve, A. (Eds.), *Ubiquitous and pervasive knowledge and learning management: Semantics, social networking and new media to their full potential* (pp. 32–66). Hershey, PA: Idea Group.

Dieterle, E., & Dede, C. (2007). Building university faculty and student capacity to use wireless handheld devices for learning. In van 't Hooft, M., & Swan, K. (Eds.), *Ubiquitous computing in education: Invisible technology, visible impact* (pp. 303–328). Mahwah, NJ: Lawrence Erlbaum Associates.

Dimmick, J. W., Kline, S., & Stafford, L. (2000). The gratification niches of personal e-mail and the telephone. *Communication Research*, *27*(2), 227–248. doi:10.1177/009365000027002005

Dimmick, J. W., Sikand, J., & Patterson, S. J. (1994). The gratifications of the household telephone: Sociability, instrumentality, reassurance. *Communication Research*, *21*(5), 643–663. doi:10.1177/009365094021005005

Dong, Q. (2008). *Explaining why young adults use MySpace and Facebook through uses & gratification theory*. Conference Papers-National Communication Association.

Doolittle, P., Lusk, D., Byrd, C., & Mariano, G. (2009). iPods as mobile multimedia learning environments. In H. Ryu & D. Parsons (Eds.). *Innovative mobile learning: Techniques and technologies*. (pp. 83-101). Hershey, PA: IGI Global.

Douglas, J., Douglas, A., & Barnes, B. (2006). Measuring student satisfaction at a UK university. *Quality Assurance in Education, 14*(3), 251–267. doi:10.1108/09684880610678568

Dourish, P. (2004). What we talk about when we talk about context. *Personal and Ubiquitous Computing, 8*(1), 19–30. doi:10.1007/s00779-003-0253-8

Dourish, P. (2001). *Where the action is: The foundations of embodied interaction*. Cambridge, MA: MIT Press.

Downing, K. (2001). Information Technology, education and healthcare: Constructivism in the 21st century. *Educational Studies, 27*(3), 229–235. doi:10.1080/03055690120076510

Drascic, D., & Milgram, P. (1996). Perceptual issues in augmented reality. In *Stereoscopic Displays and Virtual Reality Systems III* (pp. 123–134). San Jose, CA: SPIE.

Drucker, P. F. (1999). Knowledge-worker productivity: The biggest challenge. *California Management Review, 41*(2), 41–49.

Druin, A. (Ed.). (2009). *Mobile technology for children: Designing for interaction and learning*. Burlington, MA: Elsevier.

Duke University. (2005). *Duke digital initiative*. Retrieved on November 1, 2006, from http://www.duke.edu/ddi/

Dyson, L., Litchfield, A., Lawrence, E., Raban, R., & Leijdekkers, P. (2009). Advancing the m-learning research agenda for active, experiential learning: Four case studies. *Australasian Journal of Educational Technology, 25*(2), 250–267.

Eaton, K. (2010). The iPad's biggest innovation: Its $500 price. *Fast Company*. Retrieved July 2, 2010, from http://fastcompany.com/article/apples-tablet-introduced?page=0%2C0

Ebersole, S. (2000). Uses and gratifications of the Web among students. *Journal of Computer-Mediated Communication, 6*(1).

EDUCAUSE. (2005). *Seven things you should know about podcasting*.

Edwards, A. (2005). Let's get beyond community and practice: The many meanings of learning by participating. *Curriculum Journal, 16*(1), 49–65. doi:10.1080/0958517042000336809

Ellis, C. (2006). *The ethnographic I: A methodological novel about autoethnography*. Lanham, MD: Rowan & Littefield.

Ellis, R. (1994). *The study of second language acquisition*. Oxford, UK: Oxford University Press.

Esquire Magazine. (2010). *Augmented reality issue*. Retrieved January 6, 2010, from http://www.esquire.com/the-side/augmented-reality

Evans, C. (2008). The effectiveness of m-learning in the form of podcast revision lectures in higher education. *Computers & Education, 50*, 491–498. doi:10.1016/j.compedu.2007.09.016

Ewing, J. (2000). Enhancement of online and offline student learning. *Educational Media International, 37*(4), 205–217. doi:10.1080/09523980050210394

Fadel, C., & Lemke, C. (2006). *Technology in schools: What the research says*. Culver City, CA: Metiri Groups's Technology Solutions that Work/Cisco Systems.

Farmer, J., Lindstaedt, S. N., Droschl, G., & Luttenberger, P. (2004). *Ad-hoc-work-integrated technology-supported teaching and learning*. Paper presented at the 5th conference on Organisational Knowledge, Learning and Capabilities, Innsbruck, Austria.

Farrell, M. J., & Rose, L. (2008). Use of handheld computers in clinical nursing education. *The Journal of Nursing Education, 47*(1), 13–19. doi:.doi:10.3928/01484834-20080101-03

Feiner, S. K. (2002). Augmented reality: A new way of seeing: Computer scientists are developing systems that can enhance and enrich a user's view of the world. *Scientific American*, 34–41.

Felix, U. (2005). E-learning pedagogy in the third millennium: The need for combining social and cognitive constructivist approaches. *ReCALL, 17*(1), 85–100. doi:.doi:10.1017/S0958344005000716

Fernandez, V., Simo, P., & Sallan, J. M. (2009). Podcasting: A new technological tool to facilitate good practice in higher education. *Computers & Education, 53*(2), 385–392. doi:10.1016/j.compedu.2009.02.014

Ferneley, E. (2010). *Feasibility study into MP3 and MP4 assisted learning*. Retrieved July 2, 2010, from www.edu.salford.ac.uk/scd/tlqis/ abstracts/ferneley0506.rtf

Ferry, B. (2008). Using mobile phones to augment teacher learning in environmental education. In *Proceedings, ASCILITE 2008*, Melbourne.

Fischer, S., Stewart, T. E., Mehta, S., Wax, R., & Lapinsky, S. E. (2003). Handheld computing in medicine. *Journal of the American Medical Informatics Association, 10*(2), 139–149. doi:10.1197/jamia.M1180

Fisher, D., Grant, M., Frey, N., & Johnson, C. (2007). Taking formative assessment schoolwide. *Educational Assessment,* 64-68. Retrieved June 15, 2009, from http://web.ebscohost.com/ehost/pdfviewer/pdfviewer?vid=4&hid=112&sid=0d9b84bd-3a6b-498b-9528-422d3ecc28d7%40sessionmgr110

Fitzmaurice, G. M., Laird, N. M., & Ware, J. H. (2004). *Applied longitudinal analysis*. Hoboken, NJ: Wiley-Interscience.

Fjeld, M., Lauche, K., Bichsel, M., Voorhorst, F., Krueger, H., & Rauterberg, M. (2002). Physical and virtual tools: Activity theory applied to the design of groupware. *Computer Supported Cooperative Work, 11*, 153–180. doi:10.1023/A:1015269228596

Florida, R. (2005). *The world is spiky*. Retrieved July 13, 2009, from http://www.creativeclass.com/rfcgdb/articles /other-2005-The World is Spiky.pdf

Frontline Group. (2001). *Building a culture that supports and encourages electronic learning*. Paper presented at the Technologies for Learning Conference 2001. Retrieved June 15, 2010, from http://www.frontline-group.com/news/news_081301.html

Ganger, A. C., & Jackson, M. (2003). Wireless handheld computers in the preclinical undergraduate curriculum, *Medical Education Online, 8*(3). Retrieved July 2, 2010, from www.med-ed-online.org

Gao, X. (2007). Language learning experiences and learning strategy research: Voices of a mainland Chinese student in Hong Kong. *Innovations in Language Learning and Teaching, 1*(2), 193–207. doi:10.2167/illt011.0

Gardner, R. C. (1985). *Social psychology and second language learning: The role of attitudes and motivation. The Social Psychology of Language, 4*. London, UK: E. Arnold.

Garrett, B. M., & Jackson, C. (2006). A mobile clinical e-portfolio for nursing and medical students, using wireless personal digital assistants (PDAs). *Nurse Education in Practice, 6*(6), 339–346. doi:10.1016/j.nepr.2006.07.015

Garrison, D. R., & Kanuka, H. (2004). Blended learning: Uncovering its transformative potential in higher education. *The Internet and Higher Education, 7*(2), 95–105. doi:10.1016/j.iheduc.2004.02.001

Garrison, D. R., & Vaughan, N. D. (2008). *Blended learning in higher education: Framework, principles, and guidelines*. San Francisco, CA: Jossey-Bass.

Gay, G., & Hembrooke, H. (2004). *Activity-centered design: An ecological approach to designing smart tools and usable systems*. Cambridge, MA/ London, UK: MIT Press.

GCSU. (2006). *The GCSU iPod story*. Georgia College and State University.

Geddes, S. (2004). *Mobile learning in the 21st century: Benefit for learners*. The knowledge tree. Retrieved from http://knowledgetree.flexiblelearning.net .au/edition06/download/geddes.pdf

Gessler, S., & Kotulla, A. (1994). *PDAs as mobile WWW browsers*. In The Second International WWW Conference '94: Mosaic and the Web, Chicago, USA, October 1994, retrieved from http://www.ncsa.uiuc.edu/SDG/IT94/Proceedings/DDay/gessler/www pda.html

Giles, J. (2005). Internet encyclopaedias go head to head. *Nature, 438*, 900–901. doi:10.1038/438900a

Ginns, P., & Ellis, R. (2007). Quality in blended learning: Exploring the relationships between online and face-to-face teaching and learning. *The Internet and Higher Education, 10*(1), 53–64. doi:10.1016/j.iheduc.2006.10.003

Global, E. (n.d.). *Achievements of mobile learning today*. Retrieved July 10, 2009, from http://www.ericsson.com/ericsson/corpinfo/ programs/the_role_of_mobile_learning_in_ european_education/products/ workpackage2.shtml

Goodyear, P., Asensio, M., Jones, C., Hodgson, V., & Steeples, C. (2003). Relationships between conceptions of learning approaches to study and students' judgments about the value of their experiences of networked learning. *ALT-J. Research in Learning Technology*, *11*(1), 17–27.

Google. (2010). *Favorite places*. Retrieved April 10, 2010, from http://www.google.com/help/maps/favoriteplaces/gallery/#los-angeles-ca

Grasso, B. C., Genest, R., Yung, R., & Arnold, C. (2002). Reducing errors in discharge medication lists by using personal digital assistants. *Psychiatric Services (Washington, D.C.)*, *53*, 1325–1326. doi:10.1176/appi.ps.53.10.1325

Graziano, A. M., & Raulin, M. L. (2006). *Research methods: A process of inquiry* (6th ed.). Boston, MA: Allyn & Bacon.

Green, T., Alejandro, J., & Brown, A. (2009, June 10). The retention of experienced faculty in online distance education programs: Understanding factors that impact their involvement. *The International Review of Research in Open and Distance Learning North America*. Retrieved June 4, 2010, from http://www.irrodl.org/index.php /irrodl/article/view/683/1279

Group, T. L. T. (Teaching Learning and Technology Group). (2010). *Questionnaire*. Retrieved July 2, 2010, from www.tltgroup.org

Hafeez-Baig, A., & Danaher, P. A. (accepted for publication). Academics' attitudes towards mobile technologies in an Australian university: Implications for enhancing flexibility in cooperative and symbolic communities of Web based educational practice. In Kennedy, D. M., & Kommers, P. A. M. (Eds.), *Mobility: Adding flexibility to Web-based communities. Theme issue of the International Journal of Continuing Engineering Education and Life-long Learning*.

Hafeez-Baig, A., Gururajan, R., Danaher, P. A., & De George-Walker, L. R. (in press). Principles and pressures in managing student attitudes to innovative mobile learning: A view from an Australian distance education intensive university. In Wei, J., Ozok, A. A., & Yang, J. (Eds.), *Managing innovative distance learning for higher education. Theme issue of the*. International Journal of Management in Education.

Hafeez-Baig, A., & Danaher, P. A. (2007a). Future possibilities for mobile learning technologies and applications at the University of Southern Queensland, Australia: Lessons from an academic focus group. In D. P. Parsons & H. Ryu (Eds.), *Mobile learning technologies and applications (MoLTA) 2007: Proceedings of the conference on mobile learning technologies and applications (MoLTA) Massey University, Auckland, New Zealand February 19th 2007* (pp. 16-22). Auckland, New Zealand: Institute of Information and Mathematical Sciences, Massey University.

Hafeez-Baig, A., & Danaher, P. A. (2007b). Challenges and opportunities in facilitating student engagement and empowerment: Perspectives from Information Systems and education courses at the University of Southern Queensland, Australia. In C. Montgomerie & J. Seale (Eds.), *Proceedings of World Conference on Education multimedia, Hypermedia and Telecommunications 2007* (pp. 459-468). Chesapeake, VA: Association for the Advancement of Computing in Education.

Hafeez-Baig, A., Gururajan, R., Nazemi, S., De George-Walker, L. R., & Danaher, P. A. (2010). *Selected influences on students' uses of wireless handheld devices: An Iranian exploratory case study*. Paper presented at the Global Learn Asia Pacific conference, Shangri La's Rasa Sayang and Golden Sands Resorts, Penang, Malaysia.

Hair, J. F., Black, W. C., Babin, B. J., & Anderson, R. E. (2009). *Multivariate data analysis* (7th ed.). Upper Saddle River, NJ: Prentice Hall.

Hansson, T. (Ed.). (2008). *Handbook of research on digital Information Technologies: Innovations, methods, and ethical issues*. Hershey, PA: Information Science Reference/IGI Global.

Hargis, J., & Wilson, D. (2005). *Fishing for learning with a podcast net*. University of North Florida. Retrieved on November 1, 2006, from http://www.unf.edu/dept/cirt/tech/podcast /HargisPodcastArticle.pdf

Harley, D., Winn, S., Pemberton, S., & Wilcox, P. (2007). Using texting to support students' transition to university. *Innovations in Education and Teaching International, 44*(3), 229–241. doi:10.1080/14703290701486506

Harris, H., & Park, S. (2008). Educational usages of podcasting. *British Journal of Educational Technology, 39*(3), 548–551. doi:10.1111/j.1467-8535.2007.00788.x

Hayes, P., Joyce, D., & Pathak, P. (2004). Ubiquitous learning: An application of mobile technology in education. In L. Cantoni & C. McLoughlin (Eds.), *Proceedings of World Conference on Educational Multimedia, Hypermedia and Telecommunications 2004* (pp. 1811-1816). Chesapeake, VA: AACE. Retrieved July 10, 2009, from http://www.editlib.org/p/12724

Health Sciences Library. (2010). *Health related podcasts*. University of Colorado Denver. Retrieved March 21, 2010, from http://hsclibrary.uchsc.edu/podcasts

Herrington, A. (2009). Using a smartphone to create digital teaching episodes as resources in adult education. In Herrington, J., Herrington, A., Mantei, J., Olney, I., & Ferry, B. (Eds.), *New technologies, new pedagogies: Mobile learning in higher education* (pp. 28–35). Wollongong, Australia: University of Wollongong.

Herrington, A., & Herrington, J. (2007). *Authentic mobile learning in higher education.* Paper presented at the annual conference of the Australian Association for Research in Education. Retrieved March 6, 2010, from https://www.aare.edu.au/07pap/her07131.pdf

Herrington, J., Herrington, A., Mantei, J., Olney, I., & Ferry, B. (2009). Using mobile technologies to develop new ways of teaching and learning. In J. Herrington, A. Herrington, J. Mantei, I. Olney, & B. Ferry (Eds.), *New technologies, new pedagogies: Mobile learning in higher education.* (pp.1-14). Faculty of Education, University of Wollongong. Retrieved from http://nla.gov.au/nla.arc-101041

Higgs, J., & Jones, M. A. (2008). Clinical decision making and multiple problem spaces. In Higgs, J., Jones, M. A., Loftus, S., & Christensen, N. (Eds.), *Clinical reasoning in the health professions* (3rd ed., pp. 3–17). Oxford, UK: Elsevier Butterworth-Heinemann.

Higgs, J., & Titchen, A. (2000). Knowledge and reasoning. In Higgs, J., & Jones, M. (Eds.), *Clinical reasoning in the health professions* (2nd ed., pp. 23–32). Oxford, UK: Butterworth-Heinemann.

Hill, A. E., Davidson, B. J., & Theodoros, D. G. (2010). A review of standardised patients in clinical education: Implications for speech-language pathology programs. *International Journal of Speech-Language Pathology, Early Online,* 1-12.

Hjeltnes, T. A., & Hansson, B. (2005). *Cost-effectiveness and cost-efficiency in e-learning*. Norway: TISIP Research Foundation.

Horrigan, J. (2009). *The mobile difference*. Pew Research Center report.

Hramiak, A. (2005). A method for the analysis of data from online educational research. *Journal of Interactive Online Learning*, *4*(2), 82–93.

Huann, T. Y., & Thong, M. K. (2006). *Audioblogging and podcasting in education.* Edublog.net. Retrieved on November 1, 2006, from http://edublog.net/astinus/mt/files/docs/ Literature%20Review%20on%20audioblogging%20and%20podcasting.pdf

Huffstutler, S., Wyatt, T., & Wright, C. (2002). The use of handheld technology in nursing education. *Nurse Educator*, *27*(6), 271–275. doi:10.1097/00006223-200211000-00008

Hughes, G. (2007). Using blended learning to increase learner support and improve retention. *Teaching in Higher Education*, *12*(3), 349–363. doi:10.1080/13562510701278690

Hughes, M., & Daykin, N. (2002). Towards constructivism: Investigating students' perceptions and learning as a result of using an online environment. *Innovations in Education and Teaching International*, *39*(3), 217–224. doi:10.1080/13558000210150036

Hull, D. (1993). *Opening minds, opening doors: The rebirth of American education.* Waco, TX: Center for Occupational Research and Development.

Hundebol, J., & Helms, N. H. (2006). *Pervasive e-learning-in situ learning in changing contexts.*

Hung, D. (2001). Theories of learning and computer-mediated instructional technologies. *Educational Media International, 38*(4), 281–287. doi:10.1080/09523980110105114

Hung, M. C. M., Mui, W. S. W., Wah, W. Y. L., & Ching, L. M. Y. (2003). Moving from a personal to a social constructivist view of learning: Implications for support in the student teaching experience. *Asia Pacific Journal of Education, 23*(1), 85–98. doi:10.1080/0218879030230106

Hunter, J. L. (2008). Applying constructivism to nursing education in cultural competence: A course that bears repeating. *Journal of Transcultural Nursing, 19*(4), 354–362. doi:10.1177/1043659608322421

Huntsberger, M., & Stavitsky, A. (2007). The new podogogy: Incorporating podcasting into journalism education. *Journalism and Mass Communication Educator, 61*(4), 397–410.

Inman, E., Kerwin, M., & Mayes, L. (1999). Instructor and student attitudes toward distance learning. *Community College Journal of Research and Practice, 23*, 581–591. doi:10.1080/106689299264594

International Telecommunication Union. (2010). *ITU Statshot, issue 1*. Retrieved February 17, 2010, from http://www.itu.int/net/pressoffice /stats/2010/02/index.aspx

ISTE. (2007). *National Education Technology Standards (NET-S) and performance indicators for students*. Retrieved June 9, 2010, from http://www.iste.org/content/navigationmenu /nets/forstudents/2007standards/nets_for_ students_2007_standards.pdf

Ito, M. (2005). Technologies of the childhood imagination: Yu-gi-oh, media mixes, and everyday cultural production. In J. Karaganis & N. Jeremijenko (Eds.), *Structures of participation in digital culture.* Durham, NC: Duke University Press. Retrieved January 16, 2010, from http://www.itofisher.com/mito/archives/ technoimagination.pdf

Ito, M., Horst, H., Bittanti, M., Boyd, D., Herr-Stephenson, B., & Lange, P. … Robinson, L. (2008). *Living and learning with new media: Summary of findings from the digital youth project.* The John D. and Catherine T. MacArthur Foundation Reports on Digital Media and Learning. Retrieved from http://digitalyouth.ischool.berkeley.edu/report

Jackson, M., Ganger, A. C., Bridge, P. D., & Ginsburg, K. (2005). Wireless handheld computers in the undergraduate medical curriculum. *Medical Education Online, 10*, 5. Retrieved from http://www.med-ed-online.org

Jacob, E. K. (2004). Classification and categorization: A difference that makes a difference. *Library Trends, 52*(3), 515–541.

Januszewski, A., & Molenda, M. (2008). *Educational technology: A definition with commentary*. New York, NY: Lawrence Erlbaum Associate & Taylor and Francis Group.

Jarvelä, S., Lehtinen, E., & Salonen, P. (2000). Socio-emotional orientation as a mediating variable in the teaching-learning interaction: Implications for instructional design. *Scandinavian Journal of Educational Research, 44*(3), 293–306. doi:10.1080/713696677

Jeroski, S. (2005). *Research report: The Wireless Writing Program 2004-2005*. Vancouver, BC: Horizon Research & Development Inc.

Jeroski, S. (2006). *Research report: The Wireless Writing Program 2004-2006*. Vancouver, BC: Horizon Research & Development Inc.

JISC. (2005). Innovative practice with e-learning, a good practice guide to embedding mobile and wireless technologies into everyday practice. *JISC*. Retrieved from http://www.jisc.ac.uk/whatwedo/programmes / elearning_pedagogy/elp_practice.aspx

Johnson, P. A., & Yang, S. (2009). *Uses and gratifications of twitter: An examination of user motives and satisfaction of twitter use*. Association for Education in Journalism and Mass Communication.

Johnson, G. M. (2007). Learning style under two Web-based study conditions. *Educational Psychology, 27*(5), 617–634. doi:10.1080/01443410701309159

Johnston, B. (1999). Theory and research: Audience, language use and language learning. In Egbert, J., & Hanson-Smith, E. (Eds.), *CALL environments: Research, practice, and critical issues* (pp. 55–64). Alexandria, VA: Teachers of English to Speakers of Other Languages, Inc.

Jonassen, D. H., & Carr, C. S. (2000). Mindtools: Affording multiple knowledge representations for learning. In Lajoie, S. (Ed.), *Computers as cognitive tools* (2nd ed., pp. 165–196). Mahwah, NJ/ London, UK: L. Erlbaum.

Jones, A., & Issroff, K. (2007). Motivation and mobile devices: Exploring the role of appropriation and coping strategies. *ALT-J, 15*(3), 247–258. doi:10.1080/09687760701673675

Jones, D. (2006). *Enhancing the learning journey for distance education students in an introductory programming course.* Central Queensland University. Retrieved on November 1, 2006, from: http://cq-pan.cqu.edu.au/david-jones/ Publications/Papers_and_Books/react1.pdf

Jones, V., & Jo, J. H. (2004). *Ubiquitous learning environment: An adaptive teaching system using ubiquitous technology*. Paper presented at the ASCILITE, Perth, Australia.

Jordan, A., Carlile, O., & Stack, A. (2008). *Approaches to learning: A guide for teachers*. Maidenhead, UK: McGraw Hill/Open University Press.

Joscelyne, A. (1994). VR means virtual reconstruction: Cluny Abbey has been rebuilt. *Wired Magazine*, *2*(1), 114–115.

Judson, E. (2010). Improving technology literacy: Does it open doors to traditional content? *Educational Technology Research and Development*, *58*(3), 271–284. doi:10.1007/s11423-009-9135-8

Kaptelinin, V. (1996). Computer-mediated activity: Functional organs in social and developmental contexts. In Nardi, B. A. (Ed.), *Context and consciousness: Activity theory and human computer interaction* (pp. 45–68).

Katz, E. (1959). Mass communication research and the study of popular culture. *Studies in Public Communication*, *2*(1), 1–6.

Katz, E., Gurevitch, M., & Haas, H. (1973). On the use of the mass media for important things. *American Sociological Review*, *38*, 164–181. doi:10.2307/2094393

Katz, J., & Aakhus, M. (2002). *Perpetual contact: Mobile communication, private talk, public performance*. Cambridge, UK: Cambridge University Press. doi:10.1017/CBO9780511489471

Kayuga, S. (2007). Enhancing instructional efficiency of interactive e-learning environments: A cognitive load perspective. *Educational Psychology Review*, *19*, 387–399. doi:10.1007/s10648-007-9051-6

Keegan, D. (2002). *The future of learning: From e-learning to m-learning*. Forschungsbericht. FernUniversitat, Hagen. Retrieved from http://deposit.fernuni-hagen.de/1920/

Keegan, D., Kismihok, G., Mileva, N., & Rekkedal, T. (2009). *The role of mobile learning in European education*. Retrieved May 7, 2009, from http://www.ericsson.com/ericsson/corpinfo/ programs/the_role_of_mobile_learning_in_ european_education/products/ workpackage4.shtml

Kelly, F. S., McCain, T., & Jukes, I. (2009). *Teaching the digital generation: No more cookie-cutter high schools*. Thousand Oaks, CA: Corwin Press.

Kim, P., Miranda, T., & Olaciregui, C. (2008). Pocket school: Exploring mobile technology as a sustainable literacy education option for underserved indigenous children in Latin America. *International Journal of Educational Development*, *28*(4), 435–445. doi:10.1016/j.ijedudev.2007.11.002

King, R. (2009, November 3). Augmented reality helps sell the product. *BusinessWeek*. Retrieved April 10, 2010, from http://www.businessweek.com/technology/content/nov2009/tc2009112_198019.htm

Kirkwood, A. (1998). New media mania: Can information and communication technologies enhance the quality of open and distance learning? *Distance Education*, *19*(2), 228–241. doi:10.1080/0158791980190204

Kitchenham, A. D. (2009b). Diaries and journals. In Mills, A., Durepos, G., & Wiebe, E. (Eds.), *Encyclopedia of case study research* (pp. 299–301). Thousand Oaks, CA: Sage.

Kitchenham, A. D. (2009a). Laptops and teacher transformation. In Rogers, P., Berg, G., Boettcher, J., Howard, C., Justice, L., & Schenk, K. (Eds.), *Encyclopedia of distance learning* (pp. 1313–1318). Hershey, PA: IGI Global.

Kleijnen, M., Wetzels, M., & de Ruyter, K. (2004). Consumer acceptance of wireless finance. *Journal of Financial Services Marketing*, *8*(3), 206–217. doi:10.1057/palgrave.fsm.4770120

Knowles, M., Holton, E. F. III, & Swanson, R. A. (2005). *The adult learner: The definitive classic in adult education and human resource development* (6th ed.). Burlington, MA: Elsevier.

Koile, S., & Singer, D. (2006). Development of a tablet–PC based system to increase instructor-student classroom interactions and student learning. In Berque, D., Prey, J., & Reed, R. (Eds.), *The impact of pen-based technologies on education: Vingettes, evaluations and future directions*. Purdue University Press.

Kolb, D. A., & Fry, R. (1975). Toward an applied theory of experiential learning. In Cooper, C. (Ed.), *Theories of group process*. London, UK: John Wiley.

Krashen, S. D. (1981). *Second language acquisition and second language learning*. Oxford, UK: Pergamon Press.

Krashen, S. D. (1985). *The input hypothesis: Issues and implications*. London, UK: Longman.

Kress, G. R., & Van Leeuwen, T. (2001). *Multimodal discourse: The modes and media of contemporary communication*. London, UK/ New York, NY: Arnold; Oxford University Press.

Krippendorf, R. L., Simpson, D. E., & Schiedermayer, D. (1999). Promoting reflective teaching with personal digital assistants. *Academic Medicine*, *74*, 577. doi:10.1097/00001888-199905000-00042

Kristiansen, T. (2001). M-learning. Experiences from the use of WAP as a supplement in learning. In Szucs, A., Wagner, E., & Holmberg, C. (Eds.), *Learning without limits. Developing the next generation of education*. Oslo: Fornebu Knowation.

Kroeker, P. P. (2005). Technology to facilitate interaction in mobile learning. Unpublished master's thesis, Athabasca University.

Krug, S. (2006). *Don't make me think: A common sense approach to Web usability* (2nd ed.). Berkeley, CA: New Riders Publishing.

Kukulsha-Hulme, A., & Petit, J. (2008). Practitioners as educators: Emergent practice in personal mobile teaching, learning, work, and leisure. In Ally, M. (Ed.), *Mobile learning: Transforming the delivery of education & training* (pp. 135–155). Athabasca, Canada: Athabasca University.

Kukulska-Hulme, A., & Traxler, J. (Eds.). (2005). *Mobile learning: A handbook for educators and trainers open & flexible learning series*. London, UK/ New York, NY: Taylor & Francis Routledge.

Kukulska-Hulme, A., & Sharples, M. (2009). Mobile and contextual learning. *ALT-J: Research in Learning Technology*, *17*(3), 159–160.

Kukulska-Hulme, A., & Pettit, J. (2007a). *Self-service education: Smartphones as a catalyst for informal collective and individual learning*. In Mlearn '07 Mobile Learning Conference, 16-19th October 2007, Melbourne, Australia. Retrieved July 2, 2010, from http://oro.open.ac.uk/16621/

Kukulska-Hulme, A., & Pettit, J. (2007b). *Do smart devices make smart learners?* In MLearn '07 Mobile Learning Conference, 16-19th October 2007, Melbourne, Australia, pp.1-9. Retrieved July 2, 2010 from http://oro.open.ac.uk/16621/

Kumar, S., & Zahn, C. (2003). Mobile communications: Evolution and impact on business operations. *Technovation*, *23*, 515–520. doi:10.1016/S0166-4972(02)00120-7

Kurti, A., Spikol, D., & Milrad, M. (2008). Bridging outdoors and indoors educational activities in schools with the support of mobile and positioning technologies. *International Journal of Mobile Learning and Organisation*, *2*(2), 166–186. doi:10.1504/IJMLO.2008.019767

Lajoie, S. P., & Azevedo, R. (2006). Teaching and learning in technology-rich envrionments. In Alexander, P. A., & Winne, P. H. (Eds.), *Handbook of educational psychology* (2nd ed., pp. 803–821). Mahwah, NJ: Lawrence Erlbaum.

Lan, Y.-J., Sung, Y.-T., & Chang, K.-E. (2007). A mobile-device-supported peer-assisted learning system for collaborative early EFL reading. *Language Learning & Technology*, *11*(3), 130–151.

Lane, C., & Rollnick, S. (2007). The use of simulated patients and role-play in communication skills training: A review of the literature to August 2005. *Patient Education and Counseling*, *67*, 13–20. doi:10.1016/j.pec.2007.02.011

Lane, C. (2006). Podcasting at the UW: An evaluation of current use. *Catalyst*. Retrieved on November 1, 2006, from http://catalyst.washington.edu/projects/ podcasting_report.pdf

Laouris, Y., & Eteokleous, N. (2005). *We need an educationally relevant definition of mobile learning.* Paper presented at the 4th World Conference on Mobile Learning. Cape Town, South Africa. Retrieved on August 12, 2009, from http://www.mlearn.org.za/CD/papers /Laouris2.pdf

Laouris, Y., & Eteokleous, N. (2005). We need an educationally relevant definition of mobile learning. *Proceedings of the MLEARN 2005 Conference*. Melbourne, Australia. Retrieved from http://www.mlearn.org.za/CD/papers/ Laouris%20&%20Eteokleous.pdf

Laru, J., & Järvelä, S. (2008). Social patterns in mobile technology mediated collaboration among members of the professional distance education community. *Educational Media International*, *45*(1), 17–32. doi:10.1080/09523980701847131

Lau, F., & Bates, J. (2004). A review of e-learning practices for undergraduate medical education. *Journal of Medical Systems*, *28*(1), 71–87. doi:10.1023/B:JOMS.0000021522.30587.ff

Laurillard, D. (2002). *Rethinking university teaching: A conversational framework for the effective use of learning technologies* (2nd ed.). London, UK: Routledge. doi:10.4324/9780203304846

Laurillard, D. (2008). Technology enhanced learning as a tool for pedagogical innovation. *Journal of Philosophy of Education*, *42*(3-4), 521–533. doi:10.1111/j.1467-9752.2008.00658.x

Laurillard, D. (2006). E-learning in higher education. In Ashwin, P. (Ed.), *Changing higher education: The development of learning and teaching*. London, UK: Routledge.

Laurillard, D. (2005). Presentation given at the HEA 2005 conference: *Harnessing technology to improve the learning experience*. Higher Education Academy.

Lave, J. (1988). *Cognition in practice: Mind, mathematics, and culture in everyday life*. Cambridge, UK: Cambridge University Press. doi:10.1017/CBO9780511609268

Lazzari, M. (2008). Creative use of podcasting in higher education and its effect on competitive agency. *Computers & Education*, *52*(1), 27–34. doi:10.1016/j.compedu.2008.06.002

Learning Technology Standards Committee. (2002). *IEEE standard for learning object metadata. (IEEE Standard 1484.12.1)*. New York, NY: IEEE.

Lee, L. (2009). Promoting intercultural exchanges with blogs and podcasting: A study of Spanish-American telecollaboration. *Computer Assisted Language Learning*, *22*(5), 425–443. doi:10.1080/09588220903345184

Lee, M., & Chan, A. (2007). Pervasive, lifestyle integrated mobile learning for distance learners: An analysis and unexpected results from a podcasting study. *Open Learning*, *22*(3), 201–218.

Lee, M. J. W., McLoughlin, C., & Chan, A. (2008). Talk the talk: Learner-generated podcasts as catalysts for knowledge creation. *British Journal of Educational Technology*, *39*(3), 501–521. doi:10.1111/j.1467-8535.2007.00746.x

Lee, M. J. W., & McLoughlin, C. (2007). Teaching and learning in the Web 2.0 era: Empowering students through learner-generated content. *International Journal of Instructional Technology and Distance Learning, 4*(10). Retrieved on 24 January, 2008, from http://www.itdl.org/Journal/ Oct_07/article02.htm

Lefoe, G., Olney, I., Wright, R., & Herrington, A. (2009). Faculty development for new technologies: Putting mobile learning in the hands of the teachers. In J. Herrington, A. Herrington, J. Mantei, I. Olney, & B. Ferry (Eds.), *New technologies, new pedagogies: Mobile learning in higher education* (pp. 15-27). Wollongong: University of Wollongong. Retrieved from http://nla.gov.au/nla.arc-101041

Leung, L., & Wei, R. (1998). The gratifications of pager use: Sociability, information seeking, entertainment, utility, and fashion and status. *Telematics and Informatics*, *15*(4), 253–264. doi:10.1016/S0736-5853(98)00016-1

Library of Congress. (2006). *Library of Congress Authorities database*. Washington, D.C.: Library of Congress. Retrieved September 11, 2006, from http://authorities.loc.gov

Library of Congress. (2009) *LC authorities search of term: Landforms*. Retrieved July 9, 2009, from http://authorities.loc.gov

Light, D., McDermott, M., & Honey, M. (2002). *Project Hiller: The impact of ubiquitous portable technology on an urban school.* Retrieved December 13, 2006, from www2.edc.org/CCT/publications_report_ summary.asp?numPubld=129

Lim, K. (2005). *Now hear this–exploring podcasting as a tool in geography education.* Nanyang Technological University. Retrieved on November 1, 2006, from http://homepage.mac.com/voyager/ brisbane_kenlim.pdf

Lin, C. A. (2006). Predicting satellite radio adoption via listening motives, activity and format preference. *Journal of Broadcasting & Electronic Media, 50*(1), 140–159. doi:10.1207/s15506878jobem5001_8

Link, T. M., & Marz, R. (2006). Computer literacy and attitudes towards e-learning among first year medical students. *BMC Medical Education, 6*, 34. Retrieved from http://www.biomedcentral.com/ 1472-6920/6/34. doi:10.1186/1472-6920-6-34

Liu, T. C., Wang, H. Y., Llang, J. K., Chan, T. W., Ko, H. W., & Yang, J. C. (2003). Wireless and mobile technologies to enhance teaching and learning. *Journal of Computer Assisted Learning, 19*, 371–382. doi:10.1046/j.0266-4909.2003.00038.x

Livingstone, P. (2009). *1:1 learning: Laptop programs that work* (2nd ed.). Washington, DC: ISTE Publications.

Lohr, L. L., & Gall, J. E. (2008). Representation strategies. In Spector, J., Merrill, M., Van Merrienboer, J., & Driscoll, M. (Eds.), *Handbook of research on educational communications and technology* (3rd ed., pp. 85–96). New York, NY: Lawrence Erlbaum Associates.

Loke, S. (2006). *On representing situations for context-aware pervasive computing: Six ways to tell if you are in a meeting.* Paper presented at the Proceedings of PerCom. Workshops 2006.

Long, J. D. (in press). *Longitudinal analysis for the behavioral sciences using R.* Thousand Oaks, CA. *Sage (Atlanta, Ga.).*

Lonn, S., & Teasley, S. D. (2009). Podcasting in higher education: What are the implications for teaching and learning? *The Internet and Higher Education, 12*(2), 88–92. doi:10.1016/j.iheduc.2009.06.002

Lu, H., Lin, J., Lin, C., & Su, K. (2007). *A study of the construction of a mobile learning oriented mathematics learning activity.* 6th WSEAS International Conference on Applied Computer Science, Hangzhou, China. Retrieved March 27, 2008, from http://74.125.155.132/scholar?q=cache: HQ0M6UAVrVYJ:scholar.google.com/&hl=en&as_sdt=2000

Luo, J. (2004). Portable computing in psychiatry. *Canadian Journal of Psychiatry, 49*(1), 24–30.

Lysaght, R., & Hill, A. (in press). The use of standardized patients as an alternative to fieldwork. In McAllister, L., Paterson, M., Higgs, J., & Bithell, C. (Eds.), *Innovations in allied health fieldwork education: A critical appraisal.* The Netherlands: Sense Publishers.

Lyytinen, K., & Yoo, Y. (2002). Issues and challenges in ubiquitous computing. *Issues and Challenges in Ubiquitous Computing, 45*(12).

Ma, Q. (2009). *Second language vocabulary acquisition.* Bern, Germany: Peter Lang AG.

Maag, M. (2006). iPod, uPod? An emerging mobile learning tool in nursing education and students' satisfaction. *Proceedings of the 23rd Annual ASCILITE Conference: Who's Learning? Whose Technology?* University of Sydney. Retrieved November 20, 2009 from http://www.ascilite.org.au/conferences/ sydney06/proceeding/pdf_papers /p92.pdf

MacIntyre, B., & Feiner, S. (1996). Future multimedia user interfaces. *Multimedia Systems*, 4.

Maclellan, E. (2005). Conceptual learning: The priority for higher education. *British Journal of Educational Studies, 53*(2), 129–147. doi:10.1111/j.1467-8527.2005.00287.x

Maddrell, J. (2008). *The effect of backchannel interactions on cognitive load.* Retrieved July 2, 2009, from http://www.scribd.com/doc/3727549/BackchannelResearch-Paper-Jennifer-Maddrell

Magazine, P. C. (2010). Smartphone definition. Retrieved June 13, 2010, from http://pcmag.com/encyclopedia_term .php?gid=310

Mahmoud, S. S. (2008). A proposed model for distributed e-courses content through mobile technology architectures. *World Academy of Science. Engineering and Technology, 37*, 274–289.

Maine Education Policy Research Institute. (2003). *The Maine Learning Technology Initiative: Teacher, student, and school perspectives mid-year evaluation.* Retrieved January 2, 2007, from www.cepare.usm.maine.edu

Malaysian Communications and Multimedia Commission. (2010). *Facts and figures for cellular phone subscribers: Cellular phones in Malaysia.* Retrieved 23 February, 2010, from http://www.skmm.gov.my/facts_figures/ stats/ ViewStatistic.asp?cc=51298166&srid =9247989

Manhattan Research. (2008). *Health in the palm of your hand.* Retrieved August 18, 2009, from http://www.manhattanresearch.com/ newsroom/Press_Releases/ physician-mobile-trends- release.aspx

Maréchal, G. (2009). Autoethnography. In Mills, A., Durepos, G., & Wiebe, E. (Eds.), *Encyclopedia of case study research* (pp. 43–45). Thousand Oaks, CA: Sage.

Markett, C., Sanchez, I. A., Weber, S., & Tangney, B. (2006). Using short message service to encourage interactivity in the classroom. *Computers & Education, 46*(3), 280–293. doi:10.1016/j.compedu.2005.11.014

Martin, D. (2010). *Apple releases iPhone OS SDK 3.2 Beta 2 for iPad.* Retrieved July 2, 2010, from the http://reviews.cnet.com/8301-19512_7- 10450636-233.html

Mashable.com. (2010). *Detroit Red Wings make game programs interactive with QR codes.* Retrieved February 15, 2010, from http://mashable.com/2010/02/10/ red-wings-qr-codes/

Masters, K. (2008). M-learning: How much of what has been diffused? A systematic literature review. *Proceedings of World Conference on Educational Multimedia, Hypermedia and Telecommunications,* Chesapeake, VA. (pp. 5790-5795).

Matthews, D., & Schrum, L. (2003). High-speed Internet use and academic gratifications in the college residence. *The Internet and Higher Education, 6*(2), 125–144. doi:10.1016/S1096-7516(03)00020-4

Mauch, J. E., & Park, N. (2003). *Guide to the successful thesis and dissertation: A handbook for students and faculty* (5th ed.). New York, NY: Dekker.

Maxwell, J. A. (1996). *Qualitative research design: An interactive approach.* Thousand Oaks, CA: Sage Publications.

McGraw-Hill. (2010). *AccessMedicine weekly podcasts.* Retrieved May 21, 2010, from http://books.mcgraw-hill.com/podcast/acm/

McGreal, R. (Ed.). (2004). *Online education using learning objects.* London, UK: Routledge Falmer.

McGregor, D. (2007). *Developing thinking: Developing learning. A thinking skills guide for education.* Milton Keynes, UK: Open University Press.

McLoughlin, C., & Lee, M. J. W. (2008). The three p's of pedagogy for the networked society: Personalization, participation, and productivity. *International Journal of Teaching and Learning in Higher Education, 20*(1), 10–27.

McMahon, M. (1997). Social constructivism and the World Wide Web-a paradigm for learning. Ascilite Conference, December 7-10, Retrieved on 1 October, 2009, from http://www.ascilite.org.au/conferences/ perth97/papers/ Mcmahon/Mcmahon.html

Meng, P. (2005). *Podcasting & vodcasting: Definitions, discussions & implications.* A White paper by IAT Services at University of Missouri. Retrieved May 12, 2010, from http://edmarketing.apple.com/adcinstitute/wp- content/ Missouri_Podcasting_White_Paper.pdf

Merrill, M. D. (2002). First principles of instruction. *Educational Technology Research and Development, 50*(3), 43–59. doi:10.1007/BF02505024

Merrill, M. D. (2006). Levels of instructional strategy. *Educational Technology, 46*(4), 5–10.

Merrill, M. D. (2007). A task-centered instructional strategy. *Journal of Research on Technology in Education, 40*(1), 5–22.

Metcalf, D. S., & De Marco, J. (2006). *M-learning: Mobile learning and performance in the palm of your hand.* Amherst, MA: HRD Press.

Meyer, K. A. (2003). Face-to-face versus threaded discussions: The role of time and higher-order thinking. *Journal of Asynchronous Learning Networks, 7*(3), 55–65.

Michael-Spector, J., & Wang, X. (2002). Integrating technology into learning and working: Promising opportunities and problematic issues. *Journal of Educational Technology & Society, 5*(1).

Microsoft. (2010). *Smartphones*. Retrieved July 2, 2010, from http://www.microsoft.com/windosmobile/ engb/devices/smartphones.mspx?13=&14=&31.35=1&24=1&56=1&25=1

Middleton, A. (2009). Beyond podcasting: Creative approaches to designing educational audio. *ALT-J. Research in Learning Technology*, *17*(2), 143–155.

Milgram, P., & Kishino, F. (1994). A taxonomy of mixed reality visual displays. *IEICE Transactions on Information Systems E*, *77-D*(12), 1321–1329.

Milgram, P., Takemura, H., et al. (1994). Augmented reality: A class of displays on the reality-virtuality continuum. *SPIE Proceedings: Telemanipulator and Telepresence Technologies*. (pp. 282-292).

Miller, J., Shaw-Kokot, J. R., Arnold, M. S., Boggin, T., Crowell, K. E., & Allegri, F. (2005). A study of personal digital assistants to enhance undergraduate nursing education. *The Journal of Nursing Education*, *44*, 19–26.

Miller, G. A. (1956). The magical number seven, plus or minus two: Some limits on our capacity for processing information. *Psychological Review*, *63*, 81–97. doi:10.1037/h0043158

Miller, M., & Piller, M. (2005). Principal factors of an audio reading delivery mechanism – evaluating educational use of the iPod. In P. Kommers & G. Richards (Eds.) *Proceedings of World Conference on Educational Multimedia, Hypermedia and Telecommunications 2005*, Chesapeake, VA: AACE. (pp. 260-267).

Mills, A., Durepos, G., & Wiebe, E. (Eds.). (2009). *Encyclopedia of case study research*. Thousand Oaks, CA: Sage.

Mills, J. (2004). Faceted classification and logical division in information retrieval. *Library Trends*, *52*(3), 541–571.

MIT HARP. (2010). Retrieved February 10, 2010, from http://www.isites.harvard.edu/icb/icb.do?keyword=harp&pageid=icb.page69619

Mitchell. J. (2003). *E-business and online learning: Connections and opportunities for vocational education and training*. Australian National Training Authority, Melbourne. ISBN-1-74096-143-9

Mohammed, P., & Mohan, P. (2010). Combining digital games with culture: A novel approach towards boosting student interest and skill development in computer science programming. In A. Gadomski, B. Kramer, C. Lester, & M. Popescu (Eds.), *Proceedings of the Second International Conference on Mobile, Hybrid and On-line Learning, eL & mL, 2010,* St Maarten, Netherlands, Antilles.

Moller, L., Foshay, W., & Huett, J. (2008). The evolution of distance education: Implications for instructional design on the potential of the Web (Part 2: Higher education). *TechTrends*, *52*(4), 66–70. doi:10.1007/s11528-008-0179-0

Moore, M. G., & Kearsley, G. (1996). *Distance education*. Wadsworth Publishing Company.

Moore's law. (2010). In *Wikipedia, the free encyclopedia*. Retrieved June 14, 2010, from http://en.wikipedia.org/wiki/Moore's_law

Moran, S. (2008). After behaviourism, navigationism? *Irish Educational Studies*, *27*(3), 209–221. doi:10.1080/03323310802242138

Motiwalla, L. F. (2007). Mobile learning: A framework and evaluation. *Computers & Education*, *49*(3), 581–596. doi:10.1016/j.compedu.2005.10.011

Motlik, S. (2008). Technical evaluation report: Mobile learning in developing countries. *International Review of Research in Open and Distance Learning, 9*(2). Retrieved 10 July, 2009, from http://www.irrodl.org/index.php/irrodl/ article/view/564/1071

Motteram, G. (2006). Blended education and the transformation of teachers: A long-term case study in postgraduate UK higher education. *British Journal of Educational Technology*, *37*(1), 17–30. doi:10.1111/j.1467-8535.2005.00511.x

Moura, A., & Carvalho, A. (2008). Mobile learning: Teaching and learning with mobile phones and podcasts. *Proceedings of the 2008 Eighth IEEE International Conference on Advanced Learning Technologies*. Retrieved January 28, 2010, from http://adelinamouravitae.com.sapo.pt/ mobilespainfinal.pdf

Murphy-Berman, V., & Berman, J. J. (2003). *Cross-cultural differences in perspectives on the self.* Lincoln, NE/ London, UK: University of Nebraska Press.

Nah, K. C., White, P., & Sussex, R. (2008). The potential for using a mobile phone to access the Internet for learning EFL listening skills within a Korean context. *ReCALL, 20*(3), 331–347. doi:.doi:10.1017/S0958344008000633

Naismith, L., Naismith, P., Lonsdale, G., Vavoula, G., & Sharples, M. (2004). *Literature review in mobile technologies and learning*. Future Lab. Retrieved from http://www.futurelab.org.uk/resources/ publications_reports_articles/ literature_ reviews/ Literature_Review203/

Nationmaster.com. (2010). *Media statistics: Mobile phones by country*. Retrieved February 17, 2010, from http://www.nationmaster.com/graph /med_mob_pho-media- mobile-phones#source

Neo, M. (2003). Developing a collaborative learning environment using a Web-based design. *Journal of Computer Assisted Learning, 19*, 462–473. doi:10.1046/j.0266-4909.2003.00050.x

Neumeier, P. (2005). A closer look at blended learning–parameters for a blended learning environment for language teaching and learning. *ReCALL, 17*(2), 163–178. doi:.doi:10.1017/S0958344005000224

New England Journal of Medicine. (2010). *About podcasts*. Retrieved Mach 21, 2010, from http://content.nejm.org/misc/podcast.dtl

Nielsen/NetRatings. (2006). The 21st century UK digital consumer. *Nielsen/ NetRatings MegaPanel UK Digital Consumer Survey*.

Norman, D. A. (1998). *The design of everyday things* (1st MIT Press ed.). London, UK: MIT Press.

Noss, R. (2008). *Education 2.0? Designing the Web for teaching and learning.* Teaching and Learning Research Programme, www.tlrp.org/pub/documents/TELcomm.pdf

Nulden, U. (2001). Education: Research and practice. *Journal of Computer Assisted Learning, 17*, 363–375. doi:10.1046/j.0266-4909.2001.00193.x

Nusca, A. (2009). Smartphone vs. feature phone arms race heats up: Which did you buy? Retrieved August 20, 2009 from http://www.zdnet.com/blog/gadgetreviews /smartphone-vs-feature-phone-arms-race- heats-up-which-did-you-buy/6836

Nyiri, K. (2002). Towards a philosophy of m-learning. In *Proceedings of the 1st IEEE International Workshop on Wireless and Mobile Technologies in Education*, Växjö, Sweden, (pp. 121-124).

O'Malley, C., Vavoula, G., Glew, J. P., Taylor, J., Sharples, M., & Lefrere, P. (2003). *Guidelines for learning/teaching/tutoring in a mobile environment.* Retrieved July 2, 2010, from www.mobilearn.org/results/results.htm

Oblinger, D., & Oblinger, J. (2005). Is it age or IT: First steps toward understanding the Net Generation. In D. Oblinger & J. Obligner (Eds.), *Educating the Net Generation*. Educause 2005. Retrieved on 24 January, 2005, from http://www.educause.edu/ educatingthenetgen/

Oliver, M., & Trigwell, K. (2005). Can blended learning be redeemed? *E-Learning and Digital Media, 2*(1), 17–26.

Oliver, M. (2005). The problem with affordance. *e-learning, 2*(4), 402-413.

O'Malley, C., Vavoula, G., Glew, J. P., Taylor, J., Sharples, M., & Lefrere, P. (2003). *MOBIlearn WP4-guidelines for learning/teaching/tutoring in a mobile environment.* Retrieved March 21, 2010, from http://www.mobilearn.org/download /results/guidelines.pdf

Oppermann, R., & Specht, M. (2006). Situated learning in the process of work. In S. Netherlands (Ed.), *Engaged learning with emerging technologies* (pp. 69-89).

Osborne, H. (2008). *In other words... using text messages to improve medication adherence.* Retrieved January 8, 2010, from http://www.healthliteracy.com/ article.asp?PageID=7987

O'Shea, P., Mitchell, R., Johnston, C., & Dede, C. (2009). Lessons learned about designing augmented realities. *International Journal of Gaming and Computer-Mediated Simulations, 1*(1), 342–349.

Oxford, R., & Shearin, J. (1994). Language learning motivation: Expanding the theoretical framework. *Modern Language Journal, 78*(1), 12–28. doi:10.2307/329249

Paas, F., Ayres, P., & Pachman, M. (2008). Assessment of cognitive load in multimedia learning: Theory, methods, and applications. In Robinson, D. H., & Schraw, G. (Eds.), *Recent innovations in educational technology that facilitates student learning* (pp. 11–35). Charlotte, NC: Information Age Publishing.

Paavola, S., & Hakkarainen, K. (2005). The knowledge creation metaphor—an emergent epistemological approach to learning. *Science and Education, 14*, 535–557. doi:10.1007/s11191-004-5157-0

Pachler, N., Bachmair, B., & Cook, J. (2010). *Mobile learning: Structures, agency, practices*. London, UK: Springer.

Paczkowski. J. (2010). *Initial iPad demand greater than initial iPhone demand.* All things digital. Retrieved July 2, 2010, from http://digitaldaily.allthingsd.com/ 20100223/ initial-ipad-demand-greater-than- iphone-demand/

Palfrey, J., & Gasser, U. (2008). *Born digital: Understanding the first generation of digital natives*. New York, NY: Basic Books.

Palmer, D. (2005). A motivational view of constructivist- informed teaching. *International Journal of Science Education, 27*(15), 1853–1881. doi:10.1080/09500690500339654

Park, N. (2005). *User acceptance of e-learning in higher education: An application of technology acceptance model*. International Communication Association.

Park, N., Kwan, M. L., & Cheong, P. H. (2007). University instructors' acceptance of electronic courseware: An application of the technology acceptance model. *Journal of Computer-Mediated Communication, 13*(1), 163–186. .doi:10.1111/j.1083-6101.2007.00391.x

Park, N., Lee, S., Jae, E. C., & Roman, R. (2007). *User acceptance of a digital library system in developing countries: An application of the technology acceptance model.* Conference Papers-National Communication Association.

Parsons, D., & Ryu, H. (2006). *A framework for assessing the quality of mobile learning*. In Education (INSPIRE), Southampton Solent University, UK, 13 April 2006. Retrieved from http://citeseerx.ist.psu.edu/viewdoc/ downl oad?doi=10.1.1.108.2612&rep=rep1 &type=pdf

Patten, B., Armedillo Sánchez, I., & Tangney, B. (2006). Designing collaborative, constructionist and contextual applications for handheld devices. *Computers & Education, 46*(3), 292–308. doi:10.1016/j.compedu.2005.11.011

Perez, S. (2009). Matell's new Web enabled toys will offer augmented reality. Retrieved April 10, 2010, from http:// www.readwriteweb.com/archives/mattels_new_web-enabled_avatar_toys_will_offer_augmented_reality.php

Perse, E. M., & Dunn, D. G. (1998). The utility of home computers and media use: Implications of multimedia and connectivity. *Journal of Broadcasting & Electronic Media, 42*(4), 435–457.

Peters, K. (2007). M-learning: Positioning educators for a mobile connected future. *International Review of Research in Open and Distance Learning, 8*(2).

Peters, M., Weinberg, A., & Sarma, N. (2009). To like or not to like! Student perceptions of technological activities for learning French as a second language at five Canadian universities. *Canadian Modern Language Review, 65*(5), 869–896. doi:10.3138/cmlr.65.5.869

Peters, O. (1988). Distance teaching and industrial production: A comparative interpretation in outline. In Sewart, D., Keegan, D., & Holmberg, B. (Eds.), *Distance education: International perspectives* (pp. 95–111). London, UK/New York, NY: Croom Helm/St. Martin's Press.

Peterson, L., & Peterson, M. (1959). Short-term retention of individual verbal items. *Journal of Experimental Psychology, 58*, 193–198. doi:10.1037/h0049234

Pettit, J., & Kukulska-Hulme, A. (2008). Do smart devices make smart learners? In *Proceedings of the MLearn 2008 Conference*, 7-10 Oct 2008, Ironbridge Gorge, Shropshire, UK.

Pham-Nguyen, C., Garlatti, S., Lau, S., Barbry, B., & Vantroys, T. (2009a). An adaptive and context-aware scenario model based on a Web service architecture for pervasive learning systems. *International Journal of Mobile and Blended Learning, 1*(3), 41–69. doi:10.4018/ jmbl.2009092203

Pham-Nguyen, C., Garlatti, S., Lau, S., Barbry, B., & Vantroys, T. (2009b). Pervasive learning system based on a scenario model integrating Web service retrieval and orchestration. *International Journal of Interactive Mobile Technologies, 3*(2), 25–32.

Pham-Nguyen, C., & Garlatti, S. (2008a). *Context-aware scenarios for pervasive life-long learning*. Paper presented at the IEEE/WIC/ACM International Conference on Web Intelligence and Intelligent Agent Technology. WI-AT'08, Sydney, Australia.

Pham-Nguyen, C., & Garlatti, S. (2008b). *Pervasive technology-enhanced learning system integrating working and learning situations*. Paper presented at the 3rd International Conference on Information and Communication Technologies: From Theory to Applications. ICTTA 2008, Syria.

Pham-Nguyen, C., Lau, B.-Y.-S., Barbry, B., Vantroys, T., & Garlatti, S. (2008). *Pervasive learning system based on a scenario model integrating Web service retrieval and orchestration*. Paper presented at the mLearn 2008, the International Conference on Mobile Learning, University of Wolverhampton, School of Computing and IT, UK.

Pimmer, C., & Gröhbiel, U. (2008). *Mobile learning in corporate settings: Results from an expert survey*. Paper presented at the International conference on mobile learning. mLearn 2008.

Pinkwart, N., Hoppe, H. U., Milrad, M., & Perez, J. (2003). Educational scenarios for the cooperative use of personal digital assistants. *Journal of Computer Assisted Learning, 19*(3), 383–391. doi:10.1046/j.0266-4909.2003.00039.x

Pintrich, P. R. (1999). The role of motivation in promoting and sustaining self-regulated learning. *International Journal of Educational Research, 31*, 459–470. doi:10.1016/S0883-0355(99)00015-4

p-LearNet. (2006). *p-LearNet project*. Retrieved July 2, 2010, from http://p-learnet.univ- lille1.fr/web/guest/home

Powell, V. J. (1977). *Resource handbook*. Commerce, TX: Federation Bilingual Training Resources Center, East Texas State University.

Pozzi, F. (2007). The impact of m-learning in school contexts: An inclusive perspective. In Stephanidis, C. (Ed.), *Universal access in Human-Computer Interaction: Applications and services* (pp. 748–755). Berlin/Heidelberg, Germany: Springer-Verlag. doi:10.1007/978-3-540-73283-9_81

Premkumar, K., & Coupal, C. (2008). Rules of engagement-12 tips for successful use of clickers in the classroom. *Medical Teacher, 30*(2), 146–149. doi:10.1080/01421590801965111

Prensky, M. (2005). Listen to the natives. *Educational Leadership, 63*(4), 8–13.

Prensky, M. (2010). *Teaching digital natives: Partnering for real learning*. Thousand Oaks, CA: Corwin Press.

Prensky, M. (2001). Digital natives, digital immigrants. *Horizon, 9*(5), 1–6. doi:10.1108/10748120110424816

Prensky, M. (2005). What can you learn from a cell phone? Almost anything! *Journal of Online Education, 1*(5). Fischler School of Education and Human Services at Nova Southeastern University. Retrieved on March 6, 2008, from http://www.innovateonline.info/index.php?view=article&id=83

Prentice, R. (2000). Creativity: A reaffirmation of its place in early childhood education. *Curriculum Journal, 11*(2), 145–158. doi:10.1080/09585170050045173

Price, S. (2007). Ubiquitous computing. In Pachler, N. (Ed.), *Mobile learning: Towards a research agenda* (pp. 33–55). London, UK: WLE Centre, IoE.

Quality Matters. (2008-2010). *Quality Matters rubric standards 2008-2010 with assigned point values*. Retrieved June 5, 2010, from http://qminstitute.org/home/Public%20Library/About%20QM/RubricStandards 2008-2010.pdf

Quinn, C. (2000). M-learning: Mobile, wireless, in-your-pocket learning. *LINE Zine*. Retrieved August 18, 2009, from http://www.linezine.com/2.1/ features/cqmmwiyp.htm

Raacke, J., & Bonds-Raacke, J. M. (2008). MySpace and Facebook: Applying the uses and gratifications theory to exploring friend-networking sites. *Cyberpsychology & Behavior, 11*(2), 169–174. doi:10.1089/cpb.2007.0056

Ractham, P., & Zhang, X. (2006). Podcasting in academia-a new knowledge management paradigm within academic settings. *SIGMIS-CPR, 06*(April), 13–15.

Raelin, A. (Ed.). (2008). *Work-based learning: Bridging knowledge and action in the workplace*.

Ragusa, A. T., Chan, A., & Crampton, A. (2009). iPods aren't just for tunes: Exploring podcasting as a socio-cultural technology facilitating student experiences of higher education in rural Australia. *Information Communication and Society*, *12*(5), 678–690. doi:10.1080/13691180802471471

Ramos, A. J. O. (2005). *The viability of mobile SMS technologies for non-formal distance learning in Asia.* Retrieved July 12, 2009, from http://[REMOVED HYPERLINK FIELD]www.pandora-asia.org/downloads/05-AAOU_Ramos.pdf

Ravenscroft, A. (2003). From conditioning to learning communities: Implications of fifty years of research in e-learning interaction design. *ALT-J. Research in Learning Technology*, *11*(3), 4–18.

Ray, M. B. (2007). *Needs, motives, and behaviors in computer-mediated communication: An inductive exploration of social networking websites.* International Communication Association.

Remenyi, D. S. J., Williams, B., Money, A., & Swartz, E. (1998). *Doing research in business and management: An introduction to process and method.* London, UK: Sage Publications.

Research, T. D. G. (2005). *Podcasting users to approach 60 million US consumers by 2010.* Retrieved November 1, 2006, from http://www.tdgresearch.com/press044.htm

Rey, G. (2006). *Contexte en interaction homme-machine: Le contexteur.* Grenoble, France: L'université Joseph-Fourier-Grenoble I.

Rheingold, H. (2002). *Smart mobs, the next social revolution: Transforming cultures and communities in the age of instant access.* Cambridge, MA: Basic Books.

Robert Morris University. (2007). *Subject heading query log – internal memo.*

Robert Morris University. (2007). *Information security term and Library of Congress/Dewey Decimal System equivalencies document – internal memo.*

Robert Morris University. (2009). *ROBCAT catalog search.* Retrieved July 9, 2009, from http://robcat.rmu.edu/cgi-bin/Pwebrecon.cgi?DB=local&PAGE=First

Robert Morris University. (2009). *Database listing.* Retrieved July 9, 2009, from http://sentry.rmu.edu/web/cms/academics/library/research-assistance/Pages/databases.aspx

Robinson, J. M., Frey, J. G., Stanford-Clark, A. J., Reynolds, A. D., & Bedi, B. V. (2010). Chemistry by mobile phone (or how to justify more time at the bar). Retrieved July 2, 2010, from http://eprints.soton.ac.uk/28632/01/481.pdf

Rodger, S., Webb, G., Gilbert, J., Wrightson, P., McKeeken, J., & Devitt, L. (2008). Clinical education & practice placements in the allied health professions: An international perspective. *Journal of Allied Health*, *37*, 53–62.

Rogers, E. M. (2003). *Diffusion of innovations* (5th ed.). New York, NY: Free Press.

Romero, M., & Wareham, J. (2009). Just in time mobile learning model based on context awareness information. *Learning Technology Newsletter*, *11*(1 & 2), 4–6.

Romiszowski, A. J. (1992). *The selection and use of instructional media.* London, UK: Kogan Page.

Rothschild, J. M., Fang, E., Liu, V., Litvak, I., Yoon, C., & Bates, D. W. (2006). Use and perceived benefits of handheld computer-based clinical references. *Journal of the American Medical Informatics Association*, *13*(6), 619–626. doi:10.1197/jamia.M2103

Rourke, J. (2006). Social accountability in theory and practice. *Annals of Family Medicine*, *1*(4), S45–S48. doi:10.1370/afm.559

Rowell, L. (2010). An interview with Howard Rheingold. Retrieved March 28, 2010, from http://www.elearnmag.org/subpage.cfm?article=111-1&section=articles

Rubin, A. M., & Perse, E. M. (1987). Audience activity and television news gratifications. *Communication Research*, *14*(1), 58–84. doi:10.1177/009365087014001004

Ruggiero, T. E. (2000). Uses and gratifications theory in the 21st century. *Mass Communication & Society*, *3*(1), 3–37. doi:10.1207/S15327825MCS0301_02

Rumble, G. (1997). *The costs and economics of open and distance learning.* London, UK: Kogan Page.

Runyon, J. (2007). Quality in course design and impact on student achievement. Paper presented at the 104th Sloan-C International Conference on Online Learning. Retrieved May 13, 2010, from http://www.sloan-c.org/effective_practices/ quality-course-design-amp-impact-student- achievement

Russell, M., Bebell, D., & Higgins, J. (2004). *Laptop learning: A comparison of teaching and learning in upper elementary classrooms equipped with shared carts of laptops and permanent laptops*. Boston, MA: Boston College, Technology and Assessment Study Collaborative.

Ryu, H., & Parsons, D. (2009). *Innovative mobile learning: Techniques and technologies*. Hershey, PA: Information Science Reference. Retrieved April 7, 2009, from http://www.massey.ac.nz/~ryu/ acmelearn.pdf

Salmon, G., & Edirisingha, P. (2008). *Podcasting for learning*. Maidenhead, UK: Open University Press.

Savill-Smith, C., & Kent, P. (2003). *The use of palmtop computers for learning: A review of the literature*. London, UK: Learning and Skills Development Agency. Retrieved March 21, 2010, from http://www.m-learning.org/docs/ the_use_of_palmtop_computers _for_ learning_sept03.pdf

Scherr, D., Zweiker, R., Kollmann, A., Kastner, P., Schreier, G., & Fruhwald, F. M. (2006). Mobile phone-based surveillance of cardiac patients at home. *Journal of Telemedicine and Telecare*, *12*, 255–261. doi:10.1258/135763306777889046

Schmidt, J. (2009). Podcasting as a learning tool: German language and culture every day. *Unterrichtspraxis*, *41*(2), 186–194. doi:10.1111/j.1756-1221.2008.00023.x

Schnadelbach, H., Koleva, B., Flintham, M., Fraser, M., Izadi, S., & Chandler, P. … Rodden, T. (2002). *The Augerscope: A mixed reality interface for outdoors*. CHI'02, ACM Press, Minneapolis, USA, (pp. 9–16).

Scholosser, L. A., & Simonson, M. (2006). *Distance education: Definition and glossary of terms* (3rd ed.). United States: Information Age Publishing.

Schon, D. (1983). *The reflective practitioner*. New York, NY: Basic Books.

Schraefel, M. C., Andre, P., & Van Kllek, M. (2010). This time it's personal: From PIM to the perfect digital assistant. Retrieved July 2, 2010, from http://eprints.lancs.ac.uk/22928/1/getPDF.pdf

Schunk, D. H., & Zimmerman, B. J. (Eds.). (1998). *Self-regulated learning: From teaching to self-reflective practice*. New York, NY: Guilford Press.

Schwarz, R. D., & Sykes, R. (2004). *Psychometric foundations of formative assessment*. Conference paper presented at the National Educational Computing Conference, New Orleans, 2004 June. Retrieved October 5, 2007, from http://www.iste.org/Content/NavigationMenu/Research/NECC_Research_Paper_Archives/NECC_2004/Schwarz-Richard-NECC04.pdf

Seale, J. K. (2008). Developing e-learning experiences and practices: The importance of context. *ALT-J. Research in Learning Technology*, *16*(1), 1–3.

Sener, J. (2007). In search of student-generated content in online education. *e-mentor, 21*, 1-8.

Sharples, M. (2006). *Big issues in mobile learning: Report of a workshop by the Kaleidoscope Network of Excellence Mobile Learning Initiative*. LSRI, University of Nottingham.

Sharples, M. (2000). The design of personal mobile technologies for lifelong learning. *Computers & Education*, *34*, 177–193. doi:10.1016/S0360-1315(99)00044-5

Sharples, M., Taylor, J., & Vavoula, G. (2007). A theory of learning for the mobile age. In Andrews, R., & Haythornthwaite, C. (Eds.), *The Sage handbook of e-learning research* (pp. 221–247). London, UK: Sage Publications.

Sharples, M. (2005). Learning as conversation: Transforming education in the mobile age. *Proceedings of Conference on Seeing, Understanding, Learning in the Mobile Age* (pp. 147-152). Budapest, Hungary.

Shea, P., & Bidjerano, T. (2009). Community of inquiry as a theoretical framework to foster epistemic engagement and cognitive presence in online education. *Computers & Education*, *52*(3), 543–553. doi:10.1016/j.compedu.2008.10.007

Sheldon, P. (2008). Student favorite: Facebook and motives for its use. *Southwestern Mass Communication Journal*, *23*(2), 39–53.

Shih, Y. (2007). *Dynamic language learning: Comparing mobile language learning with online language learning.* Unpublished doctoral dissertation, Capella University.

Siemens, G. (2005). Connectivism: A learning theory for the digital age. *International Journal of Instructional Technology and Distance Learning, 2*(1). Retrieved on 24 January, 2008, from http://www.itdl.org/Journal/Jan_05/article01.htm

Simonson, M., Schlosser, L. A., & Hanson, D. (1999). Theory and distance education: A new discussion. *American Journal of Distance Education, 13*(1), 60–75. doi:10.1080/08923649909527014

Sims, D. E., Burke, C. S., Metcalf, D. S., & Salas, E. (2008). Research-based guidelines for designing blended learning. *Ergonomics in Design: The Quarterly of Human Factors Applications, 16*(1), 23–29. doi:10.1518/106480408X282764

Siobhan, T. (2007). *Pervasive scale: A model of pervasive, ubiquitous, and ambient learning.* Paper presented at the An International Workshop on Pervasive Learning, in conjunction with Pervasive 2007, Toronto, Ontario, Canada.

Smith, S., Salaway, G., & Borreson Caruso, J. (2009). *Undergraduate students and information technology, 2009. No. Research Study (Vol. 6).* Boulder, CO: Educause Center for Applied Research.

Smith, S. R., Cookson, J., McKendree, J., & Harden, R. M. (2007). Patient-centred learning–back to the future. *Medical Teacher, 29*(1), 33–37. doi:10.1080/01421590701213406

Smordal, O., & Gregory, J. (2002). *Mobile learning book – the KNOWMOBILE case study.* Retrieved March 21, 2010, from http://academic.research.microsoft.com/Paper/4947211.aspx

So, H.-J., & Brush, T. A. (2008, August). Student perceptions of collaborative learning, social presence and satisfaction in a blended learning environment: Relationships and critical factors. *Computers & Education, 51*(1), 318–336. doi:10.1016/j.compedu.2007.05.009

Soo, K. (1999). Theory and research: Learning styles, motivation and the CALL classroom. In Egbert, J., & Hanson-Smith, E. (Eds.), *CALL environments: Research, practice, and critical issues* (pp. 55–64). Alexandria, VA: Teachers of English to Speakers of Other Languages, Inc.

Steeples, C., & Jones, C. (Eds.). (2002). *Networked learning: Perspectives and issues.* London, UK: Springer.

Stevenson, K. R. (1999). *Evaluation report – year 3, Middle School Laptop Program.* Retrieved December 14, 2006, from www.beaufort.k12.sc.us/ district/eval-report3.htm

Stockwell, G. (2008). Investigating learner preparedness for and usage patterns of mobile learning. *ReCALL, 20*(3), 253–270. doi:.doi:10.1017/S0958344008000232

Stockwell, G. (2010). Using mobile phones for vocabulary activities: Examining the effect of the platform. *Language Learning & Technology, 14*(2), 95–110.

Suri, H. (1998). *A critique of contemporary methods of research synthesis.* Paper presented at the Conference of the Association for Active Educational Researchers, Adelaide, Australia. Retrieved on February 4, 2009, from www.aare.edu.au/98pap/sur98250.htm

Sutton-Brady, C., Scott, K. M., Taylor, L., Carabetta, G., & Clark, S. (2009). The value of using short-format podcasts to enhance learning and teaching. *ALT-J. Research in Learning Technology, 17*(3), 219–232.

Swan, K., & Ice, P. (2010). The community of inquiry framework ten years later: Introduction to the special issue. *The Internet and Higher Education, 13*(1-2), 1–4. doi:10.1016/j.iheduc.2009.11.003

Sweller, J. (1988). Cognitive load during problem solving: Effects on learning. *Cognitive Science, 12*(2), 257–285. doi:10.1207/s15516709cog1202_4

Sweller, J. (2004). Instructional design consequences of an analogy between evolution by natural selection and human cognitive architecture. *Instructional Science, 32,* 9–31. doi:10.1023/B:TRUC.0000021808.72598.4d

Sweller, J., Van Merriënboer, J., & Paas, F. (1998). Cognitive architecture and instructional design. *Educational Psychology Review, 10,* 251–296. doi:10.1023/A:1022193728205

Sweller, J. (2003). Evolution of human cognitive architecture. In Ross, B. (Ed.), *The psychology of learning and motivation* (pp. 215–266). San Diego, CA: Academic Press.

Sydorenko, T. (2010). Modality of input and vocabulary acquisition. *Language Learning & Technology*, *14*(2), 50–73.

Tallent-Runnels, M. K., Thomas, J. A., Lan, W. Y., Cooper, S., Ahern, T. C., Shaw, S. M., & Liu, X. (2006). Teaching courses online: A review of the research. *Review of Educational Research*, *76*(1), 93–135. doi:10.3102/00346543076001093

Tapscott, D. (2009). *Grown up digital: How the net generation is changing your world.* New York, NY: McGraw-Hill.

Technology Review. (2010). Microsoft adds augmented reality to Bing maps. Retrieved February 15, 2010, from http://www.technologyreview.com/blog/editors/24800/?ref=rss%20target=&a=f

Thorne, S., & Payne, S. (2006). Evolutionary trajectories, Internet-mediated expression, and language education. *CALICO Journal*, *22*(3), 371–397.

Thornton, P., & Houser, C. (2005). Using mobile phones in English education in Japan. *Journal of Computer Assisted Learning*, *21*, 217–228. doi:10.1111/j.1365-2729.2005.00129.x

Tooey, M. J., & Mayo, A. (2003). Handheld technologies in a clinical setting: State of the technology and resources. *AACN Clinical Issues*, *14*, 342–349. doi:10.1097/00044067-200308000-00009

Topolsky, J. (2010). *Apple iPad: The definitive guide (so far).* Retrieved July 2, 2010, from http://engadget.com/2010/01/29/apple- ipad-the-definitive-guide-so-far/

Torre, D., Simpson, D., Sebastian, J., & Elnicki, D. (2005). Learning/feedback activities and high-quality teaching: Perceptions of third year medical students during and inpatient rotation. *Academic Medicine*, *80*(10), 950–954. doi:10.1097/00001888-200510000-00016

Traxler, J. (2009a). Learning in a mobile age. *International Journal of Mobile and Blended Learning*, *1*(1), 1–12. doi:10.4018/jmbl.2009010101

Traxler, J. (2005). Defining mobile learning. *Proceedings IADIS International Conference Mobile Learning 2005*, Malta, (pp. 261-266).

Traxler, J. (2008). *Mobility, modernity and development.* Paper presented at first m4d Karlstadt, Sweden.

Traxler, J. (2009). The current state of mobile learning. In M. Ally, M. (Ed.), *Mobile learning: Transforming the delivery of education* (pp. 9-25). Athabasca, Canada: Athabasca University Press.

Traxler, J. (2009b). Students and mobile devices: Choosing which dream. In *Proceedings: ALT-C 2009 "In dreams begins responsibility" - choice, evidence and change*, 8-10 September 2009, Manchester.

Trifonova, A. (2003). *Mobile learning: Review of the literature.* (Technical report DIT-03-009). Trento, Italy*:* University of Trento, Department of Information and Telecommunications. Retrieved July 6, 2008, from http://eprints.biblio.unitn.it/ archive/00000359/

Trouche, L. (2003). *Managing the complexity of human/machine interaction in a computer based learning environment (CBLE): Guiding students' process command through instrumental orchestrations.* Paper presented at the Third CAME Symposium: Learning in a CAS Environment: Mind-Machine Interaction, Curriculum & Assessment, Reims, France.

Uday Bhaskar, N., & Govindarajulu, P. (2008). Implications of mobile technology usage on learners in a learning process. *International Journal of Computer Science and Network Security*, *8*(5), 251–259.

Uden, L. (2007). Activity theory for designing mobile learning. *International Journal of Mobile Learning and Organisation*, *1*(1), 81–102. doi:10.1504/IJMLO.2007.011190

Underkoffler, J., & Ishii, H. (1998). Illuminating light: An optical design tool with a luminous-tangible interface. In *CHI'98 Proceedings* (pp. 542–549). New York, NY: ACM Press/Addison-Wesley Publishing Co.

Uzunboylu, H., Cavus, N., & Ercag, E. (2009). Using mobile learning to increase environmental awareness. *Computers & Education*, *52*(2), 381–389. doi:10.1016/j.compedu.2008.09.008

van Barneveld, A., & Shaw, S. (2006). Mobile learning: A review of implementations. *Proceedings of World Conference on E-Learning in Corporate, Government, Healthcare, and Higher Education 2006,* Honolulu, Hawaii. (pp. 980-985).

van Lier, L. (1996). *Interaction in the language curriculum: Awareness, autonomy, and authenticity*. London, UK: Longman.

Vaughan, N. D., & Garrison, D. R. (2005). Creating cognitive presence in a blended faculty development community. *The Internet and Higher Education*, *8*(1), 1–12. doi:10.1016/j.iheduc.2004.11.001

Vavoula, G., Pachler, N., & Kukulska-Hulme, A. (Eds.). (2009). *Researching mobile learning: Frameworks, tools, and research designs*. New York, NY: Peter Lang.

Vavoula, G., & Sharples, M. (2009). Meeting the challenges in evaluating mobile learning: A 3-level evaluation framework. *International Journal of Mobile and Blended Learning*, *1*(2), 54–75. doi:10.4018/jmbl.2009040104

Vavoula, G.-N., & Sharples, M. (2008). *Challenges in evaluating mobile learning*. Paper presented at the mLearn 2008, the international conference on mobile learning, University of Wolverhampton, School of Computing and IT, UK.

Venkatesh, V., & Davis, F. D. (1996). A model of the antecedents of perceived ease of use: Development and test. *Decision Sciences*, *27*(3), 451–481. doi:10.1111/j.1540-5915.1996.tb01822.x

Venkatesh, V., Morris, M. G., Davis, G. B., & Davis, F. D. (2003). User acceptance of information technology: Toward a unified view. *Management Information Systems Quarterly*, *27*(3), 425–478.

Vlahakis, V., Ioannidis, N., & Karigiannis, J. (2002). Archeoguide: Challenges and solutions of a personalised augmented reality guide for archaeological sites. *Computer Graphics in Art, History and Archaeology. Special Issue of the IEEE Computer Graphics and Applications Magazine*, *22*(5), 52–60. doi:10.1109/MCG.2002.1028726

von Jan, U., Ammann, A., & Matthies, H. K. (2008). Generating and presenting dynamic knowledge in medicine and dentistry. In C. Bonk, et al. (Eds.), *Proceedings of World conference on E-Learning in Corporate, Government, Healthcare, and Higher Education 2008* (pp. 209-214). Chesapeake, VA.

Vygotsky, L. S. (1978). *Mind in society: The development of higher psychological process*. Cambridge, MA: Harvard University Press.

Wagner, E., & Wilson, P. (2005). Why learning professionals need to care about mobile learning. *Training & Development*, 40–43.

Wali, E., Winters, N., & Oliver, M. (2008). Maintaining, changing and crossing contexts: An activity theoretic reinterpretation of mobile learning. *ALT-J. Research in Learning Technology*, *16*(1), 41–57.

Wall Street Journal. (2009). *Esquire flirts with digital reality*. Retrieved January 6, 2010, from http://online.wsj.com/article/SB10001424052748704222704574501122991439500.html

Walsh, M. (2010). Summer of smartphones II. *MobileMarketingDaily*.

Walton, G., Childs, S., & Blenkinsopp, E. (2005). Using mobile technologies to give health students access to learning resources in the UK community setting. *Health Information and Libraries Journal*, *22*, 51–65. doi:10.1111/j.1470-3327.2005.00615.x

Wang, M., Shen, R., Novak, D., & Pan, X. (2008). The impact of mobile learning on students' learning behaviours and performance: Report from a large blended classroom. *British Journal of Educational Technology*, *40*(4), 673–695. doi:10.1111/j.1467-8535.2008.00846.x

WARC. (2009). *Augmented reality attracts big customers*.Retrieved December 13, 2009, from http://www.warc.com/News/TopNews.asp?ID=25974&Origin=WARCNewsEmail

Waycott, J. (2002). Reading with new tools: An evaluation of personal digital assistants as tools for reading course materials. *ALT-J*, *10*(2), 38–50. doi:. doi:10.1080/0968776020100205

Waycott, J., & Kukulsha-Hulme, A. (2003). Students' experiences with PDAs for reading course materials. *Personal and Ubiquitous Computing*, *7*(1), 30–43. doi:. doi:10.1007/s00779-002-0211-x

Waycott, J. (2005). Appropriating tools and shaping activities: The use of PDAs in the workplace. In Hamill, L., & Amparo, L. (Eds.), *Mobile world: Past, present and future* (pp. 119–139). London, UK/ New York, NY: Springer.

Wei, R., & Lo, V. (2006). Staying connected while on the move: Cell phone use and social connectedness. *New Media & Society*, *8*(1), 53–72. doi:10.1177/1461444806059870

Weisner, M. (1991). The computer for the 21st century. *Scientific American, 265*(3), 94–104. doi:10.1038/scientificamerican0991-94

Welch, R. B. (1978). *Perceptual modification.* New York, NY: Academic Press.

Wellner, P., Mackay, W., & Gold, R. (Eds.). (1993). Special issue on computer augmented environments: back to the real world. *Communications of the ACM, 36*(7), 212–231.

Wertsch, J. V. (2002). Computer mediation, PBL, and dialogicality. *Distance Education, 23*(1), 105–108. doi:10.1080/01587910220124008

White, A., Allen, P., Goodwin, L., Breckinridge, D., Dowell, J., & Garvy, R. (2005). Infusing PDA technology into nursing education. *Nurse Educator, 30*, 150–154. doi:. doi:10.1097/00006223-200507000-00006

Wiggins, G., & McTighe, J. (2005). *Understanding by design.* Alexandria, VA: Association for Supervision & Curriculum Development.

Wikipedia. (2010a). *iPhone.* Retrieved July 2, 2010, from http://en.wikipedia.org/wiki/IPhone

Wikipedia. (2010b). *PDA.* Retrieved July 2, 2010, from http://wikipedia.org/wiki/PDA

Wiley, D. A. (2000). Connecting learning objects to instructional design theory: A definition, a metaphor, and a taxonomy. In D. A. Wiley (Ed.), *The instructional use of learning objects.* Retrieved from http://reusability.org/read/

Wiley, D. A. (Ed.). (2001). Connecting learning objects to instructional design theory: A definition, a metaphor, and a taxonomy. *The Instructional Use of Learning Objects: Online version.* Retrieved October 5, 2007, from http://reusability.org/read/chapters/wiley.doc

Wilkinson, S., Dennick, R., & Rae, H. (2006). A Touchstone for online summative assessment. *Issues and News on Learning and Teaching in Medicine, Dentistry and Veterinary Medicine*, 32-33.

Willett, R. (2007). Technology, pedagogy and digital production: A case study of children learning new media skills. *Learning, Media and Technology, 32*(2), 167–181. doi:10.1080/17439880701343352

Windowshop from your mobile. (2009). Retrieved January 6, 2010, from http://googleblog.blogspot.com/2009/12/explore-whole-new-way-to-window-shop.html

Wise, L., & Quealy, J. (2006). At the limits of social constructivism: Moving beyond LMS to re- integrate scholarship. Paper presented at the 23rd Annual ASCILITE Conference: Who's learning? Whose technology? Retrieved on 1st February, 2008, from http://www.ascilite.org.au/conferences/ sydney06/proceeding/pdf_papers/p158.pdf

Wishart, J. (2008). Challenges faced by modern foreign language teacher trainees in using handheld pocket PCs (PDAs) to support their teaching and learning. *ReCALL, 20*(3), 348–360. doi:.doi:10.1017/S0958344008000736

Wishart, J. (2006). *Personal digital assistants–teachers prefer the personal.* Paper given at Mlearn'06, 5th World Conference on Mobile Learning, October 22-26th 2006, Banff, Ab, Canada, 15.

Wu, T. (2009). Flexible learning: Designing blended language instruction for digital natives. In Gibson, I. (Eds.), *Proceedings of Society for Information Technology & Teacher Education international conference 2009* (pp. 3352–3355). Chesapeake, VA: Association for the Advancement of Computers in Education.

Young, J. R. (2004, August 13). Students say technology has little impact on teaching. *The Chronicle of Higher Education, 50*, A28.

Young, J. R. (2009, November 22). Teaching with twitter: Not for the faint of heart. *The Chronicle of Higher Education*, 56.

Zawacki-Richter, O., Brown, T., & Delport, R. (2008). *Mobile learning–a new paradigm shift in distance education?* Retrieved March 6, 2010, from http://auspace.athabascau.ca:8080/dspace/bitstream/2149/1258/1/Zawacki-Richter_Brown_Delport%20mlearn06.pdf

Zembylas, M. (2005). Three perspectives on linking the cognitive and the emotional in science learning: Conceptual change, socio-constructivism and poststructuralism. *Studies in Science Education, 41*, 91–116. doi:10.1080/03057260508560215

Zhou, Y. (2008). Voluntary adopters versus forced adopters: Integrating the diffusion of innovation theory and the technology acceptance model to study intra-organizational adoption. *New Media & Society*, *10*(3), 475. doi:10.1177/1461444807085382

Zikmund, W. G., Babin, B. J., Carr, J. C., & Griffin, M. (2009). *Business research methods* (8th ed.). Florence, KY: South-Western College Publishing.

Zuber-Skerritt, O. (1993). Improving learning and teaching through action learning and action research. *Higher Education and Development*, *12*(1), 45–58. doi:10.1080/0729436930120105

Zurita, G., & Nussbaum, M. (2007). A conceptual framework based on activity theory for mobile CSCL. *British Journal of Educational Technology*, *38*(2), 211–235. doi:10.1111/j.1467-8535.2006.00580.x

About the Contributors

Andrew Kitchenham is an Associate Professor in Educational Technology in the School of Education at the University of Northern British Columbia. Dr. Kitchenham's research examines teacher transformation through the use, integration, and teaching of technology. He has published numerous articles, conference papers, technical reports, and book chapters on diverse topics from technology to special education to rural professionals to gender differences. This book is one of two dealing with e-learning.

* * *

Zoraini Wati Abas is the Director of the Centre for Tutor Management and Development, Open University Malaysia. She graduated from Northern Illinois University where she received her Bachelor of Science in Biological Sciences and specialized in Instructional Technology at the Master and doctoral levels. She has been involved at all levels of education, from pre-school to higher education. She has published in journals, is an author of several books and has compiled several computer dictionaries. As a weekly columnist, she has written more than 800 articles in the national newspapers on computer education, Internet and e-learning. Her research interests include utilization and effectiveness of learning technologies in general, and e-learning, as well as mobile learning, in particular. She is the founding President of the South East Asian Association for Institutional Research (SEAAIR) and is an Executive Committee Member & Founder of the Global Learn Asia Pacific Conference.

Tom Adamich has been a certified teacher-librarian since 2000 and a librarian since 1991. A graduate of the Kent State University (Ohio) School of Library/Information Science and KSU College of Education School Library Media Certification, Tom was a teacher-librarian and consultant for the Indian Valley Local Schools (Ohio) from 1999-2009 and is president of the Visiting Librarian Service, a contract librarian firm he has operated on a full or part-time basis since 1993. Tom also served as the Cataloging Librarian/ Government Documents Librarian and Assistant Professor of Learning Resources at Robert Morris University in Pittsburgh, Pennsylvania from 2005-2008. Tom is currently Head of Metadata Services/Government Documents Librarian, Muskingum University, where he administers both the library's technical services operations and manages its participation in the Federal Depository Library Program. His primary research focus, since 1998, has been K-20 cataloging/metadata creation and its connection to critical thinking skill development, and standards-based education.

Trish Andrews has worked in the field of higher education since the early 1990s and has considerable experience in the areas of open, distance, flexible, blended, and mobile learning. She has extensive experience in supporting innovative curriculum development with a particular focus on integrating technologies into higher education programs. Trish has had several educational development and research grants including current ALTC grants in the area of rich media and learning spaces and has published widely in the area of educational innovation. Her current activities and research interests include investigating the student voice in relation to ICTs for teaching and learning, the integration of ICTs including rich media into teaching and learning, mobile learning, learning spaces, and academic development.

Jennifer Banas is an assistant professor in the College of Education at Northeastern Illinois University (NEIU). At NEIU, Dr. Banas designs curriculum and assessments for undergraduate health education courses and serves on the online teaching quality task force. Prior to NEIU, Dr. Banas served as dean and an instructional designer for a fully mobile graduate college of education, the American College of Education. She holds an Ed.D. in Instructional Design and Technology and a Masters in Curriculum and Instruction. Her previous positions include: high school health teacher, hospital community wellness instructor, county health educator, and research project coordinator. Her research interests lie in the efficient and effective design of electronic learning environments, tailored communications, technological pedagogical content knowledge, and the motivational design of instruction.

Carly J. Born, Academic Technologist, offers technical and pedagogical support to the language programs at Carleton College. Born acquired her proficiency in Japanese while living and working in Japan for three years. After returning to the United States, she attended the Monterey Institute of International Studies, where in 2001 she earned a Master's in Teaching Japanese as a Foreign Language, a Certificate in Language Program Administration and a specialty in Computer-Assisted Language Learning. Her language skills complement a longstanding interest in computing technology. At Carleton, her research interests and development projects include the use of mobile devices and pen-based input technologies in foreign language education. Born has been active in professional associations relating to foreign language education and the use of technology in the field. In addition, she is an active practitioner and instructor of several Japanese martial arts.

Amel Bouzeghoub is an associate professor in the computer science department of Telecom Sud-Paris, Telecom Institute, France. She has a post-doctoral degree in computer science. She is member of the teams SIMBAD (Semantic Interoperability for Mobile, collaBorative and ADaptive applications) and SAMOVAR, CNRS UMR 5157. She is mainly interested in semantic and context-aware Information Systems based on a Semantic Web approach, adaptation, user modelling, semantic metadata, and semantic information retrieval. One important target domain is technology-enhanced learning systems based on semantic and adaptive composition of learning objects and pervasive learning systems.

Tim Brown is an assistant professor of Radio/Television at the Nicholson School of Communication at the University of Central Florida. Dr. Brown's primary teaching areas are broadcast journalism, and the effects of new media on communication in society. His research interests include new media effects in journalism and education. Prior to entering academia, Dr. Brown was a television news anchor and reporter for 12 years in the southeastern United States. He taught broadcast journalism at the College of Mass Communication and Information Studies at the University of South Carolina for three years before moving to the University of Central Florida in 2004.

Davina Calbraith has been a Registered Nurse for 15 years during which time she has led many clinical and educational research projects, and a cardiothoracic directorate teaching team. Davina is a Senior Lecturer, has a Masters Degree in Clinical Education, and was awarded an Honorary Teaching Fellowship. She has been a reviewer and subject panel advisor for several journals over the past 15 years (including - Nursing Standard, Nurse Researcher, Journal of Advanced Nursing, International Journal of Knowledge and Learning Objects). She has chaired IATED sponsored conferences and reviewed international conference papers for TCEA (Texas Computer Education Association) and iPed (International Inquiring Pedagogies Conference). Her special interest in learning objects and the practical application of skills formed the basis of her PhD thesis in Medicine. Having developed many pathophysiology learning objects for medics at Nottingham University, Davina is currently developing online critical care nursing curricula for Anglia Ruskin University in the UK.

Crispin Dale is a Senior Lecturer in the School of Sport, Performing Arts and Leisure at the University of Wolverhampton and has taught tourism and hospitality at undergraduate and postgraduate level for a number of years. Crispin has published widely in peer-reviewed journals. His research has focused upon strategic management and business development in the tourism, hospitality, sport, and events industries. Crispin is also involved in research concerning the use of technology in learning and teaching with a specific focus on podcasting and virtual learning environments to support student learning.

P. A. Danaher is an associate professor in education research in the Faculty of Education at the Toowoomba campus of the University of Southern Queensland, Australia. He is the sole and co-author and editor of two research books, six edited books, four conference refereed proceedings and 30 journal theme issues. His research and editing interests include: the education of mobile learners, educational research, university learning and teaching, lifelong learning, teacher education, educators' work and identities, professional learning, rural education, open and distance education, and vocational education and training. He teaches undergraduate and postgraduate courses about educational research ethics and methods.

Bronwyn Davidson is a Senior Lecturer in Speech Pathology in the School of Health and Rehabilitation Sciences (SHRS) at the University of Queensland and Chair of the SHRS Clinical Education Committee, a multidisciplinary group of professional and academic staff from disciplines of Audiology, Physiotherapy, Occupational Therapy and Speech Pathology. Dr. Davidson is an active researcher in the Communication Disability Centre at the University of Queensland and also has strong research interests in the field of practice education. She has been the recipient of both University of Queensland and National Teaching and Learning Grants. In 2010, Dr Davidson became an inaugural University of Queensland Teaching Fellow and is currently undertaking a project on engaging students in innovative interprofessional clinical learning.

Linda De George-Walker is a lecturer in the Faculty of Education at the Toowoomba campus of the University of Southern Queensland, Australia. She teaches undergraduate and postgraduate courses in educational psychology, special education, and guidance and counselling. In addition to research interests in teacher self-efficacy and school-based mental health, she has interests in teaching and learning in higher education, particularly blended learning, and she has several refereed publications in this area.

Reg Dennick is Professor of Medical Education at the University of Nottingham. He is a biochemist turned educator and has spent many years in teaching, research and publishing in the areas of biochemistry, science education, computer assisted learning, multimedia CD-ROMs and medical education. He is Assistant Director of Medical Education and Course Director for the Master of Medical Sciences (Medical Education) programme where he supervises Masters and doctoral students. He is the holder of teaching and learning awards and is an educational consultant to a number of professional and higher educational bodies.

Kien Ngoc Do is a PhD student in computer science at the Computer Science department of Telecom SudParis. His research areas are technology-enhanced learning systems, context-awareness, adaptation and situation-based learning.

Serge Garlatti is a Full Professor in the computer science department of Telecom Bretagne, Telecom Institute, France. He heads a research team, named SCRIPtureS for SemantiC Retrieval and Composition of Information, Pictures and Services and belongs to the CNRS laboratory LABSTICC, LabSTICC UMR 3192. This team is mainly interested in semantic and context-aware Information Systems based on a Semantic Web approach, adaptation, user modelling, semantic metadata, semantic information and image retrieval and Web service composition. One important target domain is technology-enhanced learning systems based on scenarios and pervasive learning systems.

Amanda Groff is an instructor of anthropology who specializes in archaeology. She received her MA in Anthropology from the University of Central Florida in Fall 2007, as well as a certificate in Mayan Studies in 2005. Since 2002, Ms. Groff has participated in many archaeological investigations, including work in Belize, Egypt and in the U.S. She currently teaches several online courses, including "Sex, Gender, and Culture and the Human Species."

Raj Gururajan is a professor in Information Systems in the Faculty of Business at the Toowoomba campus of the University of Southern Queensland (USQ), Australia. His main areas of research are user behaviour issues in using wireless technologies, the management of technologies and the diffusion of technologies at enterprise level. Currently he is applying these interests to health settings. He has published over 100 refereed articles, won over $1 million in grant income, managed six funded projects and supervised over 10 research students. He is also the leader of the technology management cluster in the Faculty of Business at USQ.

Abdul Hafeez-Baig holds a PhD in the domain of healthcare and Information Systems from the University of Southern Queensland (USQ), Australia and Masters of Information Systems and Business Administration degrees from Griffith University, Australia. Since joining USQ in 2004, he has published many scholarly refereed publications in the domains of health and education regarding technology management. Abdul is very conversant with wireless technology as well as emerging technologies. He teaches Information Systems concepts to both undergraduate and postgraduate students, including MBA students. He is particularly interested in wireless healthcare applications, adoption, the infusion and diffusion of m-learning, health care and InformationTechnology.

Anne Hill is an associate lecturer in a teaching-focussed position and clinic manager in the Division of Speech Pathology at The University of Queensland (UQ). She has many years of experience in both clinical practice and in clinical education. Her current role involves classroom and clinical teaching, management of students and clinical educators within the UQ clinics, and liaising with workplace educators to ensure clinical placement quality and training of educators in the principles and practices of supporting students in the clinical setting. She collaborates with staff in occupational therapy and physiotherapy to provide this training in an interprofessional model, delivered in face-to-face, online, and videoconference formats. Anne has established a program of innovative scholarship of teaching research which has a focus in standardised patient use and is also engaged in research with occupational therapy colleagues to investigate student learning within interprofessional (IP) clinics.

Tina Lim is currently the Acting Director of the Institute of Quality, Research & Innovation at Open University Malaysia. She obtained her PhD in Web-based learning & Instructional Design from Universiti Sains Malaysia and a Master in Education (Curriculum & Instruction) from the University of Houston, USA. Her Bachelor's degree is in Science with Education (Honours) and she also holds a Postgraduate certificate in TESL. Dr Tina Lim has presented numerous papers and contributed journal articles both at national and international levels. Her research interests include open and distance learning, flexible learning, online learning, continuous professional development, program evaluation, classroom interaction, and science & mathematics education. She has worked as a secondary school teacher, a teacher educator, a facilitator to open and distance learners, and a research manager.

Andrea Lisa Nixon, Ph.D., serves as the Director of Curricular and Research Support at Carleton College. In this capacity, she conducts educational research projects and coordinates institutional efforts to more effectively provide curricular and research support. Her most recent research project is longitudinal study of the ways in which students engage resources and expertise on campus as they work on course assignments. Prior to this, Nixon worked in the academic computing field for 20 years, chiefly at Carleton. During that time she worked with faculty, students, librarians and instructional technologists on the effective integration of Information Technologies into the curriculum of residential liberal arts colleges. In 2008, Nixon earned a PhD from the Department of Educational Policy and Administration at the University of Minnesota.

Cuong Pham-Nguyen is a researcher in computer science at the Computer Science department of Telecom Bretagne. He obtained his PhD in 2010 from Telecom Bretagne in the field of pervasive workplace learning systems design based on a scenario model. His research areas are technology-enhanced learning systems, context-awareness, and adaptation. At present, his research focuses mainly on an integration of the social Semantic Web in mobile and pervasive learning systems.

Grant Potter is the e-learning Coordinator at the UNBC Centre for Teaching, Learning, & Technology. Grant has been pursing research and development of educational technologies for the last 12 years through his work in the Canadian Arctic, China, and Northern British Columbia. Grant leverages technologies to extend the teaching and learning potential of UNBC programs and facilitates faculty development through regular workshops, seminars, and consultations. In addition to academic pursuits, Grant also holds a number of IT certifications and is an active contributor to various open-source software projects. Grant has presented keynote addresses, presentations, seminars, and workshops at national and international conferences on a range of educational technologies.

Ghislaine Povey is a part-time Senior Lecturer within the School of Sport, Performing Arts and Leisure, having worked full time for 14 years before having her daughter. She holds a MSc in Tourism and Hospitality Education from the University of Surry (1996). Ghislaine joined University of Wolverhampton in 1992, and has been a member of SSPAL since its inception. Prior to that she was a lecturer in tourism at Bournville College of Further Education. She has extensive international industrial experience. She has made numerous appearances on TV and radio discussing regional food and particularly, fagots. She is currently working on her PhD looking at the interpretation of kitchens in heritage sites.

Kalyani Premkumar is an Associate Professor and Associate Director of the Educational Support & Development Unit, College of Medicine, University of Saskatchewan, Saskatoon, Canada. She earned her medical degree (MBBS) and a postgraduate degree specializing in Physiology (MD) from Christian Medical College, University of Madras, India. She has a Masters degree in Medical Education (MSc) and a doctoral degree in Educational Technology from the University of Calgary. Her areas of research interest include evaluation of multimedia resources for health professionals, use of technology in medical teaching/learning, and complementary and alternative medicine (CAM).

Ruzita Ramly is the Chief Librarian at the Tan Sri Dr Abdullah Sanusi Digital Library, Open University Malaysia. She obtained her Master of Science in Information Management Education from Syracuse University, USA. Her Bachelor's degree is in Landscape Architecture from Louisiana State University, USA and she also holds a Postgraduate Diploma in Library Science from Universiti Teknologi MARA, Malaysia. Ruzita Ramly has made several presentations at national levels in library and Information Science conferences on digital libraries and library Web 2.0. Apart from her interests in the librarianship and knowledge management, other interests include open and distance learning, flexible learning, and online learning. She has worked as a librarian, mostly focusing on user education programmes and customer services.

Danielle Sloane is a Speech Pathologist who graduated with honours from the University of Queensland in 1996. She has worked as a clinician and clinical educator in a variety of health settings, both in Australia and in the United Kingdom. Throughout her clinical career, she has followed an interest in the voice and its disorders. Her recent work in the Division of Speech Pathology at the University of Queensland has drawn on her knowledge and clinical experience in this area, to develop an innovative standardised patient program that supports students' clinical learning about the assessment and management of voice disorders.

Dawn Stevens has a Bachelor's degree in French Language and Literature and a college diploma in Writing and New Media Technologies. She recently completed a Master of Education degree, for which she explored leadership and mobile learning. Dawn's research interests include mobile application user interface design, the application of m-learning to second language acquisition, and m-learning for corporate training.

Christopher Tassava, PhD, works as the Associate Director of Corporate & Foundation Relations at Carleton College, in which role he assists faculty and staff in the development and execution of grant-funded projects ranging from scholarly research and pedagogical innovation to financial aid and infrastructure construction. He has worked closely with his coauthors on several grant-related projects. Prior to coming to Carleton in 2005, Tassava worked at a for-profit university, where he supported online learning. Tassava holds a PhD in history from Northwestern University; he also teaches online U.S. history courses for a university in St. Paul.

Lynn Woodhouse is a Clinical Educator and Researcher in the Division of Speech Pathology at The University of Queensland (UQ). Her PhD is in lipreading skills and in the development of phonology in children with hearing impairment and included the development of a test "Lipreading Assessment of Children with Hearing Impairment (LACHI)." Her clinical experience has centered on paediatric school based clinics. She was also responsible for the production of an innovative series of videos to assist parents in understanding speech and language development in children. Her current research interests include utilising mobile technology in effective delivery of speech pathology course content for students and the use of peer learning strategies in a clinical setting.

Song Yanjie is a researcher on Knowledge Building Teacher Network Project and University Strategic Research Theme Sciences of Learning Project at Faculty of Education, The University of Hong Kong. Her research interest focuses on theories and practices of mobile technology use in education, computer assisted language learning, knowledge building pedagogies, technology affordances, qualitative research methods, design research methods, and learning sciences. She has published a few journal papers journals such as *European Journal of Computer Assisted Language Learning (ReCALL)*, *International Journal of Mobile Learning and Organization,* and *TechTrends*. She also published a number of book chapters, and refereed international conference papers.

Index

Symbols

A

B

C

Breinigsville, PA USA
31 March 2011
258903BV00007B/7/P